Lecture Notes in Bioinformatics 8859

Subseries of Lecture Notes in Computer Science

T0212813

Pedro Mendes   Joseph O. Dada
Kieran Smallbone (Eds.)

Computational Methods in Systems Biology

12th International Conference, CMSB 2014
Manchester, UK, November 17-19, 2014
Proceedings

 Springer

Volume Editors

Pedro Mendes
Joseph O. Dada
Kieran Smallbone

The University of Manchester, Manchester Institute of Biotechnology
131 Princess Street, Manchester M1 7DN, UK
E-mail: {pedro.mendes, joseph.dada, kieran.smallbone}@manchester.ac.uk

ISSN 0302-9743   e-ISSN 1611-3349
ISBN 978-3-319-12981-5   e-ISBN 978-3-319-12982-2
DOI 10.1007/978-3-319-12982-2
Springer Cham Heidelberg New York Dordrecht London

Library of Congress Control Number: 2014952778

LNCS Sublibrary: SL 8 – Bioinformatics

Typesetting: Camera-ready by author, data conversion by Scientific Publishing Services, Chennai, India

Printed on acid-free paper

Springer is part of Springer Science+Business Media (www.springer.com)

Preface

This volume contains the papers presented at CMSB 2014. The 12th International Conference on Computational Methods in Systems Biology was held during November 17–19, 2014, at the Manchester Institute of Biotechnology of the University of Manchester.

The conference is an annual event that brings together computer scientists, biologists, mathematicians, engineers, and physicists from all over the world who share an interest in the computational modeling and analysis of biological systems, pathways, and networks. It covers computational models for all levels, from molecular and cellular, to organs and entire organisms.

There were 31 regular and 18 poster submissions. Each regular submission was reviewed by at least two, and on average 2.77, Program Committee members. Each poster submission was reviewed by an average of 1.38 Program Committee members. Selected poster flashes were all reviewed by three Program Committee members. The committee decided to accept 16 regular papers, and all the submitted posters. The program also included three invited talks, by Ruth Baker, Dagmar Iber, and Magnus Rattray.

We thank the Program Committee for their hard work in reviewing submissions. We especially thank François Fages, Monika Heiner, and Carolyn Talcott for their advice on matters relating to the organization of the conference. We acknowledge support by the EasyChair conference system during the reviewing process and the production of these proceedings, see http://www.easychair.org (managed by our Manchester colleague Andrei Voronkov and his team). We thank Tommaso Mazza and the IEEE Computer Society Technical Committee on Simulation for supporting the best student paper award. We thank the Manchester Institute of Biotechnology for providing the conference venue.

September 2014

Pedro Mendes
Joseph O. Dada
Kieran Smallbone

Organization

Steering Committee

Finn Drablos	NTNU, Trondheim, Norway
François Fages	Inria, Paris-Rocquencourt, France
Monika Heiner	Brandenburg University at Cottbus, Germany
Tommaso Mazza	IRCCS Casa Sollievo della Sofferenza - Mendel, Rome, Italy
Satoru Miyano	University of Tokyo, Japan
Gordon Plotkin	University of Edinburgh, UK
Corrado Priami	Microsoft Research and University of Trento, Italy
Carolyn Talcott	SRI International, Menlo Park, USA
Adelinde Uhrmacher	University of Rostock, Germany

Program Committee

Paolo Ballarini	Ecole Centrale Paris, France
Julio Banga	CSIC, Vigo, Spain
Claudine Chaouiya	IGC, Oeiras, Portugal
Joseph O. Dada	University of Manchester, UK
Alberto de la Fuente	Leibniz Institute for Farm Animal Biology, Germany
Diego di Bernardo	Telethon Institute of Genetics and Medicine, Naples, Italy
François Fages	Inria, Paris-Rocquencourt, France
Monika Heiner	Brandenburg University at Cottbus, Germany
Andrzej M. Kierzek	University of Surrey, UK
Ross King	University of Manchester, UK
Ursula Kummer	University of Heidelberg, Germany
Reinhard Laubenbacher	University of Connecticut Health Center, USA
Tommaso Mazza	Instituto C.S.S. Mendel, Italy
Pedro Mendes (chair)	University of Manchester, UK and University of Connecticut Health Center, USA
Kieran Smallbone	University of Manchester, UK
Jörg Stelling	ETH Zurich, Switzerland
Carolyn Talcott	SRI International, Stanford, USA
Adelinde Uhrmacher	University of Rostock, Germany
Paola Vera-Licona	University of Connecticut Health Center, USA
Verena Wolf	Saarland University, Germany

Organizing Committee

Joseph O. Dada	University of Manchester, UK
Pedro Mendes	University of Manchester, UK and University of Connecticut Health Center, USA
Kieran Smallbone	University of Manchester, UK

Additional Reviewers

A.T. Bittig	A. Naldi	D. Šafránek
T. Helms	D. Peng	T. Krüger
A. Horváth	M. Schwarick	C. Rohr
K. Krivine	S. Soliman	

Invited Talks
(Abstracts)

Experimental and Modelling Investigation of Monolayer Development with Clustering

Ruth Baker

Mathematical Institute, Radcliffe Observatory Quarter,
Woodstock Road, Oxford, OX2 6GG, UK

Abstract. Standard differential equation models of collective cell behaviour, such as the logistic growth model, invoke a mean-field assumption which is equivalent to assuming that individuals within the population interact with each other in proportion to the average population density. Implementing such assumptions implies that the dynamics of the system are unaffected by spatial structure, such as the formation of patches or clusters within the population. Recent theoretical developments have introduced a class of models, known as moment dynamics models, that aim to account for the dynamics of individuals, pairs of individuals, triplets of individuals, and so on. Such models enable us to describe the dynamics of populations with clustering, however, little progress has been made with regard to applying moment dynamics models to experimental data. Here, we report new experimental results describing the formation of a monolayer of cells using two different cell types: 3T3 fibroblast cells and MDA MB 231 breast cancer cells. Our analysis indicates that the 3T3 fibroblast cells are relatively motile and we observe that the 3T3 fibroblast monolayer forms without clustering. Alternatively, the MDA MB 231 cells are less motile and we observe that the MDA MB 231 monolayer formation is associated with significant clustering. We calibrate a moment dynamics model and a standard mean-field model to both data sets. Our results indicate that the mean-field and moment dynamics models provide similar descriptions of the 3T3 fibroblast monolayer formation whereas these two models give very different predictions for the MDA MD 231 monolayer formation. These outcomes indicate that standard mean-field models of collective cell behaviour are not always appropriate and that care ought to be exercised when implementing such a model.

From Networks to Function — Computational Models of Organogenesis

Dagmar Iber

Department of Biosystems Science and Engineering (D-BSSE)
ETH Zürich, Mattenstraße 26, 4058 Basel, Switzerland

Abstract. One of the major challenges in biology concerns the integration of data across length and time scales into a consistent framework: how do macroscopic properties and functionalities arise from the molecular regulatory networks and how do they evolve? Morphogenesis provides an excellent model system to study how simple molecular networks robustly control complex pattern forming processes on the macroscopic scale in spite of molecular noise, and how important functional variants can evolve from small genetic changes. Recent advancements in 3D imaging technologies, computer algorithms, and computer power now allow us to develop and analyse increasingly realistic models of biological control. To also incorporate cellular dynamics and cell-cell interactions in our simulations, we have now also developed a software tool that allows us to solve our regulatory network models on dynamic 2D and 3D tissue domains at cellular resolution. I will present our recent work where we use data-based modeling to arrive at predictive models to address the mechanism of branching in lungs and kidneys, the mechanism by which an asymmetry emerges in our hand (thumb to pinky), as well as a mechanism by which proportions are maintained in differently sized embryos.

Integrating mRNA and Polymerase Time Course Data to Model the Dynamics of Transcription

Magnus Rattray

Faculty of Life Sciences, The University of Manchester, Oxford St.,
Manchester, M13 9PL, UK

Abstract. We are developing methods to model transcription using
mRNA expression (RNA-Seq) and RNA polymerase (pol-II ChIP-Seq)
time course data. In our first application we model the motion of RNA
polymerase during pre-mRNA elongation. We model the pol-II dynamics by using a spatio-temporal Gaussian process to described changes in
pol-II density profiles across sites of the transcribed region [1]. We apply
our model to infer the elongation speed and promoter-proximal pol-II
activity for early targets of estrogen receptor in MCF7 breast cancer
cells. Bayesian methods are used to infer the model parameters and associate our parameter estimates with levels of confidence. By clustering
the inferred promoter-proximal pol-II activity profiles we can associate
early-activated target genes with specific transcription factor binding
patterns.

In our second application we link the pol-II dynamics with mRNA
production and degradation in the same system using a simple linear
differential equation. We again represent the pol-II dynamics as a Gaussian process and are able to exactly compute the data likelihood by exploiting the fact that a linear operation on a Gaussian process remains
a Gaussian process. We find that for a certain number of target genes it
is necessary to include an RNA-processing delay to get a reasonable fit
to the data. We use Bayesian inference to infer the delay parameter and
identify genes with strong evidence of a significant delay, about 11% of
the genes where the signal is strong enough to fit the model. This delay
appears to be related to splicing: we find that short genes tend to exhibit
longer splicing-associated delay and there is also a positive association
with genes that have a relatively long final intron.

References

1. wa Maina, C., Matarese, F., Grote, K., Stunnenberg, H.G., Reid, G., Honkela, A.,
   Lawrence, N., Rattray, M.: Inference of RNA Polymerase II Transcription Dynam-
   ics from Chromatin Immunoprecipitation Time Course Data. PLoS Computational
   Biology 10(5), e1003598

Table of Contents

Models and Their Biological Applications

Computational Approaches for Synthetic Biology

Flash Posters

On Defining and Computing "Good" Conservation Laws

François Lemaire and Alexandre Temperville

Université Lille 1, LIFL, UMR CNRS 8022, Lille, France
francois.lemaire@univ-lille1.fr, a.temperville@ed.univ-lille1.fr

Abstract. Conservation laws are a key-tool to study systems of chemical reactions in biology. We address the problem of defining and computing "good" sets of conservation laws. In this article, we chose to focus on sparsest sets of conservation laws. We present a greedy algorithm computing a sparsest set of conservation laws equivalent to a given set of conservation laws. Benchmarks over a subset of the curated models taken from the BioModels database are given.

Keywords: conservation analysis, sparse conservation laws, biological models, sparse null space, greedy algorithm.

1 Introduction

Many biological processes can be modelled by systems of chemical reactions. In order to study such systems, one usually computes its (linear) conservation laws (*i.e.* linear combinations of number of species) which have the property of being constant along the time. In this article, we only consider (linear) conservation laws. For a given chemical reaction system, a complete set of conservation laws is easily computed by computing a basis of the kernel of the transpose of the stoichiometry matrix of the system [10].

This paper tries to answer to the difficult question: "what is a good conservation law ?". Consider for example the well known enzymatic degradation given by the reactions $E + S \leftrightarrow C$ and $C \to E + P$. It admits for example $E + C$ and $S + C + P$ as conservation laws. One could as well consider their sum (*i.e.* $E + 2C + S + P$), their difference (*i.e.* $E - S - P$), ... On the example, the laws $E + C$ and $S + C + P$ seem less artificial and closer to the physics of the system, than the two laws $E + 2C + S + P$ and $E - S - P$. Indeed the law $E + C$ corresponds to the conservation of the enzyme E, and $S + C + P$ corresponds to the conservation of the substrate S (which is either in the form S, C or P).

In an attempt to define "good conservation laws", we think that a good conservation law should have many zero coefficients (*i.e.* sparse laws) and many positive coefficients. Concerning the sparse property, we think that a practitioner would understand better sparse laws than dense laws, since sparse laws are shorter and thus easier to read. Moreover, a sparse conservation law can also be useful when doing substitutions in differential equations to preserve the

P. Mendes et al. (Eds.): CMSB 2014, LNBI 8859, pp. 1–19, 2014.

sparsity of the differential equations. For example, if one has a sparse system $\dot{X} = F(X)$ (where X is a vector of species X_1, X_2, ...), one can use a conservation law involving X_1 (say $X_1 + X_5 - X_8 = c_0$) to discard the variable X_1 by substituting X_1 by an expression in the other species ($c_0 - X_5 + X_8$ on the example). Consequently, a sparse conservation law will more likely preserve the sparsity of the differential equations. Concerning positive coefficients, we think that conservation laws with positive coefficients are more likely to represent a conservation of matter.

Those two criteria (sparsity and positiveness) are sometimes impossible to satisfy at the same time. For example, if we have a basis of two conservation laws $X_1 + X_2 + X_3$ and $X_2 + X_3 + X_4$, then the difference $X_1 - X_4$ is sparser than any of the two laws but involves a negative coefficient. Moreover, in some particular examples, there are no conservation laws with positive coefficients only (like in $A + B \to \emptyset$ which only admits $A - B$ as a conservation law).

In this paper, we have chosen to compute a sparsest basis of conservation laws, leaving the positivity argument for a further work. As a consequence, our approach differs from computing minimal semi-positive P-invariants (*i.e.* conservation laws with non-negative coefficients with minimal support [13]).

Our approach corresponds to the well known Nullspace Sparsest Problem (NSP) which is proven to be NP-hard in [3]. NSP consists in finding a matrix with the fewest nonzeros, whose columns span the null space of a given matrix. Approximate algorithms to solve NSP are given in [3,4].

We chose to develop our method by testing it on the Biomodels database [1]. Our hope was that biological models might have special properties and might be solved easily even if the problem is NP-hard. Even if we could not exhibit special properties of the biological models, our method computes the sparsest basis of conservation laws for most curated models of the Biomodels database (see Section 5), thus validating our approach.

Some usual linear algebra algorithms can sometimes produce a sparser basis, with no guarantee it is a sparsest one. The Hermite normal form is such a technique, the (reduced or not) row echelon form is another. In the context of $\mathbb{Z}$-lattices, [5] introduces and computes "short" (pseudo-) bases using the LLL algorithm [7] and a variant of the Hermite normal form. In a numerical context (*i.e.* using floating point coefficients), there are methods to compute sparser basis (as in [2] where the turnback algorithm computes a sparse and banded null basis of a large and sparse matrix).

[14,10] present method based on numerical computations (QR decomposition, SVD,...) to compute exact conservation laws of large systems. Finally, we wonder if the techniques used in the extreme pathways [12,11] could be used to compute sparse conservation laws.

The paper is organized in the following way. Section 2 presents, on an example, the idea of our algorithm ComputeSparsestBasis(B), which computes a sparsest basis equivalent to B. Section 3 presents the algorithm ComputeSparsestBasis and its sub-algorithms, and Appendix A details their proofs. Section 4 details the implementation and improvements. Finally, Section 5 shows our benchmarks.

2   A Worked Out Example

We illustrate our method on the model number 361 (BIOMD0000000361) of the BioModels database [1,8]. For clarity reasons, one renames the species in the following way : $VIIa\_TF \to V$, $VIIa\_TF\_X \to VX$, $VIIa\_TF\_Xa \to VX_a$, $TFPI \to T$, $Xa\_TFPI \to X_aT$, $Xa\_TFPI\_VIIa\_TF \to X_aTV$, $Xa \to X_a$.
    The model contains the five following chemical reactions:

$$X + V \leftrightarrow VX, \qquad VX \to VX_a, \qquad VX_a \leftrightarrow X_a + V$$
$$X_a + T \leftrightarrow X_aT, \qquad V + X_aT \leftrightarrow X_aTV.$$

By choosing the vector of species $^t(X_aTV, X_a, X, VX_a, VX, T, V, X_aT)$, one can compute the stoichiometry matrix M and a basis of conservation laws (written row by row) $B = \begin{pmatrix} 1 & 1 & 1 & 1 & 1 & 0 & 0 & 1 \\ 1 & 0 & 0 & 1 & 1 & 0 & 1 & 0 \\ 0 & -1 & -1 & -1 & -1 & 1 & 0 & 0 \end{pmatrix}$ by computing a basis of the nullspace of the transpose of M (denoted ^tM). Thus, the matrix B represents the three conservation laws:

1. $X_aTV + X_a + X + VX_a + VX + X_aT$
2. $X_aTV + VX_a + VX + V$
3. $-X_a - X - VX_a - VX + T$

Our method for decreasing the number of nonzeros consists in finding a linear combination $w = {}^tvB$ of the rows of B such that w contains less nonzeros than one row B_i of B. If one can find such a combination w, it feels natural to replace the row B_i by w in order to decrease the total number of nonzeros in B. Repeating this process until no such linear combination can be found, one obtains a sparsest basis in terms of nonzeros. This approach is greedy and is justified in Section 3.
    However, replacing a row of B by w should only be done if one maintains a basis. This last requirement is fulfilled by replacing the row B_i of B by w only if $v_i \neq 0$ (which loosely speaking means that the information in the row B_i has been kept).
    Consider the linear combination $w = {}^tvB$ with $v = (\alpha, \beta, \gamma)$, so one has $w = (\alpha + \beta, \alpha - \gamma, \alpha - \gamma, \alpha + \beta - \gamma, \alpha + \beta - \gamma, \gamma, \beta, \alpha)$. The number of nonzeros of w clearly depends on the values of α, β and γ. In order to compute the number of nonzeros of w, one considers all possible cases, corresponding to the cancellation or the non cancellation of each element of w. In theory, if w has n components, one has 2^n cases to consider. For example, if we request w to have the form $(\neq 0, 0, 0, 0, 0, \neq 0, 0, \neq 0)$ one considers the following system of equations and inequations:

$$\begin{cases} \alpha + \beta \neq 0 & \text{(column 1)} \\ \alpha - \gamma = 0 & \text{(columns 2 and 3)} \\ \alpha + \beta - \gamma = 0 & \text{(columns 4 and 5)} \\ \gamma \neq 0 & \text{(column 6)} \\ \beta = 0 & \text{(column 7)} \\ \alpha \neq 0 & \text{(column 8)} \end{cases}$$

This last system admits for example the solution $\alpha = \gamma = 1$ and $\beta = 0$. The corresponding linear combination $w = {}^t vB$ is $w = B_1 + B_3 = (1,0,0,0,0,1,0,1)$ which contains 3 nonzeros, and is thus better than the rows B_1, B_2 and B_3 (which respectively involves 6, 4, and 5 nonzeros). Since w involves the rows B_1 and B_3 (*i.e.* $\alpha \neq 0$ and $\beta \neq 0$), one can replace B_1 or B_3 by w. Note that replacing B_2 by w would lead to a matrix of rank 2, meaning that our basis has been lost. For example, replacing B_1 by w, one obtains an equivalent basis:

$$B' = \begin{pmatrix} 1 & 0 & 0 & 0 & 0 & 1 & 0 & 1 \\ 1 & 0 & 0 & 1 & 1 & 0 & 1 & 0 \\ 0 & -1 & -1 & -1 & -1 & 1 & 0 & 0 \end{pmatrix}.$$

In practice, one does not enumerate all the possible patterns of zeros and nonzeros for the vector w. Instead, one considers the columns of B one by one from left to right, and one builds systems of equations (corresponding to the zeros in w) and inequations (corresponding to the nonzeros in w). Since each column of B yields two cases, one builds a binary tree of systems of equations and inequations. By doing this, many branches are hopefully cut before all the columns of B have been treated. For example, if one tries to cancel the first five columns of w, one gets the system of equations $\begin{cases} \alpha + \beta & = 0 \ (1^{st} \text{ column of } B) \\ \alpha & -\gamma = 0 \ (2^{nd}, 3^{rd} \text{ columns of } B) \\ \alpha + \beta -\gamma = 0 \ (4^{th}, 5^{th} \text{ columns of } B) \end{cases}$. which only admits the useless solution $\alpha = \beta = \gamma = 0$.

Let us continue the improvement of B'. Following the same ideas as above, one can find the following linear combinations of the rows of B':

$$w' = B_1' - B_2' - B_3' = (0,1,1,0,0,0,-1,1)$$

which has less nonzeros than B_3'. By replacing B_3' by w, one gets the basis

$$B'' = \begin{pmatrix} 1 & 0 & 0 & 0 & 0 & 1 & 0 & 1 \\ 1 & 0 & 0 & 1 & 1 & 0 & 1 & 0 \\ 0 & 1 & 1 & 0 & 0 & 0 & -1 & 1 \end{pmatrix}.$$

Finally, further computations would show that there does not exist a linear combination of the rows allowing to improve our basis B''. In that case, one knows that our basis is a sparsest one. If one replaces the third line of B'' by $(1\,1\,1\,1\,1\,0\,0\,1)$, then one gets three conservation laws with a total of 13 non-negative coefficients (instead of 11 nonzero coefficients for B''). One can show that computing the minimal semi-positive P-invariants (as done in [13]) would retrieve these three conservation laws with 13 nonzeros.

To summarize, our method adopts a greedy approach by successively improving the initial basis (each time by only changing one row), until it reaches a sparsest basis. Each improvement consists in a binary tree exploration where each node is a system of equations and inequations.

3   The Algorithm **ComputeSparsestBasis**

3.1   Sparsest Basis of Conservation Laws

As mentioned in the introduction, we have chosen to compute sparsest basis of conservation laws. We define that notion precisely in this part.

Let B be a matrix of dimensions $m \times n$ over $\mathbb{Q}$, with $m \leq n$. The matrix B is called a *basis matrix* if B is a full rank matrix *i.e.* $\mathrm{Rank}(B) = m$. In this paper, a basis matrix contains a basis of conservation laws written row by row. Let B and B' be basis matrices of same dimensions. B and B' are *equivalent* if and only if there exists an invertible matrix Q such that $B = QB'$. Let M (resp. v) be a matrix (resp. vector), we denote $\mathcal{N}(M)$ (resp. $\mathcal{N}(v)$) the number of nonzero coefficients of M (resp. v). Let v be a vector, we denote $\mathcal{N}_k(v)$ the number of nonzero coefficients in the first k coefficients of v. A basis matrix B' is a *sparsest basis matrix* if and only if for any basis matrix B equivalent to B, one has $\mathcal{N}(B') \leq \mathcal{N}(B)$.

For any basis matrix B, it is clear that there exists a sparsest matrix B' equivalent to B. Indeed, consider the set of all equivalent matrices to B, and pick one matrix B' in that set such that $\mathcal{N}(B')$ is minimal.

3.2   A Greedy Approach

Our method follows a greedy approach. Given a basis matrix B, one looks for a vector v and an index i such that $\mathcal{N}(^tvB) < \mathcal{N}(B_i)$ and $v_i \neq 0$. If such v and i exists, one can decrease the number of nonzeros of B by replacing the row B_i by ^tvB. Moreover, the rank of B does not drop since one has $v_i \neq 0$. When such suitable v and i do no exist, our method stops and claims that our basis has become a sparsest one. This last claim is not obvious, since one could have fallen in a local minimum. The following theorem justifies our greedy approach.

Theorem 1. *A basis matrix B is not a sparsest one if and only if there exist a vector v and an index j such that $\mathcal{N}(^tvB) < \mathcal{N}(B_j)$ and $v_j \neq 0$.*

Proof. $\Leftarrow$: Taking $B' = B$ and replacing the row B'_j by ^tvB, one gets a matrix B' equivalent to B such that $\mathcal{N}(B') < \mathcal{N}(B)$, which proves that B is not a sparsest one.

$\Rightarrow$: Assume B has dimensions $m \times n$. There exists B' equivalent to B such that $\mathcal{N}(B') < \mathcal{N}(B)$. By permuting the rows of B and the rows of B', one can assume $\mathcal{N}(B_1) \geq \mathcal{N}(B_2) \geq \cdots \geq \mathcal{N}(B_m)$ and $\mathcal{N}(B'_1) \geq \mathcal{N}(B'_2) \geq \cdots \geq \mathcal{N}(B'_m)$. As $\mathcal{N}(B) = \sum_{i=1}^{m} \mathcal{N}(B_i) > \sum_{i=1}^{m} \mathcal{N}(B'_i) = \mathcal{N}(B')$, there exists an index k such that $\mathcal{N}(B_k) > \mathcal{N}(B'_k)$. Since B and B' are equivalent, each row of B' is a linear combination of rows of B. If all the $m - k + 1$ rows $B'_k, B'_{k+1}, \cdots, B'_m$ were linear combinations of the $m - k$ rows $B_{k+1}, \ldots, B_m$, then B' would not be a full rank matrix. Thus, there exist a vector v and indices j, l with $j \leq k \leq l$ such that $B'_l = {}^tvB$ with $v_j \neq 0$. Since $\mathcal{N}(B'_l) \leq \mathcal{N}(B'_k) < \mathcal{N}(B_k) \leq \mathcal{N}(B_j)$, one has $\mathcal{N}(^tvB) = \mathcal{N}(B'_l) < \mathcal{N}(B_j)$ with $v_j \neq 0$. $\quad\square$

3.3   Description of a Task

As explained in Section 2, our method builds a (binary) tree of systems of equations and inequations. In practice, one only stores the leaves of the tree in construction. One introduces the notion of *task* which basically represents one leaf of the tree. In order to cut useless branches as soon as possible, one also requires a task to satisfy the extra properties **LCP** and **IZP** of Definition 1. Let v be a vector of dimension n. The notation $v \not\equiv 0$ means that $\forall i \in [\![1, n]\!]$, $v_i \neq 0$.

Definition 1. *Let B be a basis matrix. Let A, Λ be matrices with m columns. Let*
$$(\mathcal{S}) : \begin{cases} Ax = 0 \\ \Lambda x \not\equiv 0 \end{cases} \text{be a system in the variable } x. \text{ Let } c \text{ be the number of rows of } \Lambda.$$
*Let k be the sum of the number of rows of A and Λ. A **task** $t = \text{TASK}[A, \Lambda, c, k]$, stemming from B, is defined as follow :*

- *the union of the rows of A and Λ coincides with the first k columns of B (up to the order),*
- *(A, Λ) satisfies the so-called **LCP** property (Linear Combination Property) i.e. there exists a solution v of $(\mathcal{S})$ with at least two nonzero coefficients,*
- *(A, Λ, c, k) satisfies the so-called **IZP** property (Increase Zeros Property) i.e. there exist a solution v of $(\mathcal{S})$ and an index j such that $v_j \neq 0$ and $\mathcal{N}_k(^tvB) < \mathcal{N}(B_j)$.*

Proposition 1. *Consider a task $t = \text{TASK}[A, \Lambda, c, k]$ stemming from a basis matrix B, and the system $(\mathcal{S})$, as defined in Definition 1. Consider $\mathcal{U} = \{i \in [\![1, m]\!], c < \mathcal{N}(B_i)\}$. Then one has the following properties:*

1. *$A \in \mathbb{Q}^{(k-c) \times m}$ and $\Lambda \in \mathbb{Q}^{c \times m}$ with $0 \leq c \leq k \leq n$,*
2. *For each nonzero solution v of $(\mathcal{S})$, $\mathcal{N}_k(^tvB) = c$ i.e. c is the number of nonzeros in the first k coefficients of any solution of $(\mathcal{S})$,*
3. *There exist a solution v of $(\mathcal{S})$ and $j \in \mathcal{U}$ such that $v_j \neq 0$ and $c < \mathcal{N}(B_j)$.*

Proof. 1. Trivial.
2. Take a nonzero solution v of $(\mathcal{S})$ and consider $w = {}^tvB$. Consider a column j of B with $j \leq k$. Then the transpose of this column is either a row of A or a row of Λ. If it is a row A_i of A (resp. a row Λ_i of Λ), then the jth coefficient of w equals zero (resp. is nonzero) since $A_i v = 0$ (resp. $\Lambda_i v \neq 0$). Consequently, the number of nonzero elements among the first k coefficients of $w = {}^tvB$ equals c (*i.e.* the number of rows of Λ).
3. It is a consequence of the **IZP** property and $\mathcal{N}_k(^tvB) = c$. $\square$

The task $t_0 = \text{TASK}[\text{the } 0 \times m \text{ matrix, the } 0 \times m \text{ matrix}, 0, 0]$ is called the *initial task*. A task $t = \text{TASK}[A, \Lambda, c, k]$ stemming from a basis matrix B of dimensions $m \times n$ is called a *solved task* if $k = n$.

Using Proposition 1, a solved task ensures the existence of a vector v and an index i such that $\mathcal{N}(^tvB) < \mathcal{N}(B_i)$ and $v_i \neq 0$, allowing the improvement of B.

3.4   The Algorithms

In this section, one presents the pseudo codes of the algorithms, and gives some hints on the way they work. The rigorous proofs of the algorithms are given in Appendix A.

Algorithm ComputeSparsestBasis(B). It is the main algorithm. It takes a basis B as an input and returns a sparsest basis B' equivalent to B. It relies on the algorithm EnhanceBasis(B) which either returns an equivalent basis B' with $\mathcal{N}(B') < \mathcal{N}(B)$ or proves that B was a sparsest basis. Thus, ComputeSparsestBasis(B) iterates calls to EnhanceBasis(B) until the basis is a sparsest one.

Input: B a basis matrix of dimensions $m \times n$
Output: B', a sparsest basis matrix equivalent to B
1 **begin**
2 | $B' \leftarrow B$; $a \leftarrow$ **true** ;
3 | **while** a **do**
4 | | $a, B' \leftarrow$ EnhanceBasis(B') ;
5 | **return** B' ;

Algorithm 1. ComputeSparsestBasis(B)

Algorithm EnhanceBasis(B). It relies on the algorithms BasisToSolvedTask and EnhanceBasisUsingSolvedTask. The algorithm BasisToSolvedTask(B) builds a solved task stemming from B if it exists, or returns the empty set if no such solved task exists. If such a solved task can be computed, Algorithm EnhanceBasisUsingSolvedTask is used to improve the basis B.

Input: B a basis matrix of dimensions $m \times n$
Output: One of the two cases: **false** and B if B is a sparsest basis matrix ;
       true and a basis matrix B' equivalent to B such that $\mathcal{N}(B') < \mathcal{N}(B)$
       otherwise
1 **begin**
2 | $t \leftarrow$ BasisToSolvedTask(B) ;
3 | **if** $t \neq \emptyset$ **then**
4 | | $B' \leftarrow$ EnhanceBasisUsingSolvedTask(t, B) ;
5 | | **return true**, B' ;
6 | **else**
7 | | **return false**, B ;

Algorithm 2. EnhanceBasis(B)

Algorithm BasisToSolvedTask(B). It looks for a solved task by exploring a binary tree. It makes use of a stack, initially filled with the initial task. At each step of the loop, a task t (where k columns of B have been processed) is popped and two new candidate tasks t_1 and t_2 are built by processing the column $k+1$ of B. These candidates are pushed onto the stack if they are actually tasks (which is checked by Algorithm IsTask).

Input: a basis matrix B
Output: a solved task t, stemming from B, or $\emptyset$ if no such solved task exists
        (*i.e.* B is a sparsest basis)

1 **begin**
2 | Let S_t be an empty stack ;
3 | Push the initial task t_0 onto S_t ;
4 | **while** $S_t \neq \emptyset$ **do**
5 | | Pop a task $t = \text{TASK}[A, \Lambda, c, k]$ from S_t ;
6 | | **if** $k < n$ **then**
7 | | | // *The task t is not solved*
8 | | | Let w be the transpose of the $(k+1)^{th}$ column of B ;
9 | | | $A' \leftarrow \begin{pmatrix} A \\ w \end{pmatrix}$; $\Lambda' \leftarrow \begin{pmatrix} \Lambda \\ w \end{pmatrix}$;
10 | | | $t_1 \leftarrow [A', \Lambda, c, k+1]$;   // t_1 *may be a task*
11 | | | **if** IsTask(t_1, B) **then** Push t_1 onto S_t ;
12 | | | ;
13 | | | $t_2 \leftarrow [A, \Lambda', c+1, k+1]$;   // t_2 *may be a task*
14 | | | **if** IsTask(t_2, B) **then** Push t_2 onto S_t ;
15 | | | ;
16 | | **else**
17 | | | **return** t ;
18 | **return** $\emptyset$;

Algorithm 3. BasisToSolvedTask(B)

Algorithm EnhanceBasisUsingSolvedTask(t, B). It basically finds a vector v and an index i such that $\mathcal{N}(^tvB) < \mathcal{N}(B_i)$ and $v_i \neq 0$, which necessarily exist since t is a solved task. It then builds an improved basis B' by making a copy of B and replacing the row B'_i by ^tvB.

Algorithm IsTask(t). It checks where a candidate task t is indeed a task, by checking if t satisfies the **LCP** and **IZP** properties. The goal of this function is to detect as soon as possible useless tasks (*i.e.* tasks that will not help improving our basis).

Algorithm NextVector(u). The goal of NextVector is to iterate the p-tuples of $\mathbb{Z}^p$. This is needed in Algorithm EnhanceBasisUsingSolvedTask to obtain a solution v of the system $(\mathcal{S})$, which is composed of equations (*i.e.* $Ax = 0$) and inequations (*i.e.* $\Lambda x \neq 0$). Indeed, the only way we have found to obtain solutions of $Ax = 0$ also satisfying $\Lambda x \neq 0$ consists in iterating some solutions of $Ax = 0$ until $\Lambda x \neq 0$ is satisfied.

Input: a solved task $t = \text{TASK}[A, \Lambda, c, n]$, stemming from B of size $m \times n$
Output: B' a basis matrix equivalent to B such that B and B' only differ by one row and $\mathcal{N}(B') < \mathcal{N}(B)$

1 **begin**
2     $B' \leftarrow B$;
3   Compute a basis of $\text{Ker}(A)$ and store it columnwise in the matrix K of dimensions $m \times p$, where $p = m - \text{Rank}(A)$;
4   Compute $\mathcal{U} = \{i \in [\![1, m]\!], c < \mathcal{N}(B_i)\}$;
5   **if** $p = 1$ **then**
6        $v \leftarrow$ the unique column of K ;
7   Choose $i \in \mathcal{U}$ such that $v_i \neq 0$;
8   **else**
9        $i \leftarrow 0$; $u \leftarrow 0$;   // *u is the zero vector of dimension p*
10   **while** $i = 0$ **do**
11          $u \leftarrow \text{NextVector}(u)$;
12          $v \leftarrow Ku$;
13   **if** $\Lambda v \not\equiv 0$ **then**
14   // *v is a nonzero solution of* $(\mathcal{S})$
15   Choose $i \in \mathcal{U}$ such that $v_i \neq 0$ if it exists ;
16     $B'_i \leftarrow {}^t v B$;
17   Multiply B'_i by the LCM of the denominators of the elements of B'_i ;
18   Divide B'_i by the GCD of the elements of B'_i ;
19   **return** B' ;

Algorithm 4. EnhanceBasisUsingSolvedTask(t, B)

Input: $t = [A, \Lambda, c, k]$, satisfying all conditions of a task stemming from a basis matrix B of dimensions $m \times n$, except **LCP** and **IZP** properties
Output: true if t satisfies **LCP** and **IZP** (*i.e.* t is a task), **false** otherwise

1 **begin**
2   // **LCP** *(resp.* **IZP***) is true if the tests lines 6 (resp. 11) and 3 are false*
3   **if** $\left(\exists i \in [\![1, c]\!], \text{Rank}\begin{pmatrix} A \\ \Lambda_i \end{pmatrix} = \text{Rank}(A)\right)$ *or* $(\text{Rank}(A) = m)$ **then**
4   **return false** ;
5   // *One has* $\left(\forall i, \text{Rank}\begin{pmatrix} A \\ \Lambda_i \end{pmatrix} = \text{Rank}(A) + 1\right)$ *and* $(\text{Rank}(A) \leq m - 1)$
6   **if** $(\text{Rank}(A) = m - 1)$ *and* A *contains at least one zero column* **then**
7   **return false** ;
8   // *One has* $(\text{Rank}(A) \leq m - 2)$ *or* A *does not have any zero column*
9   Compute A_{RREF}, the RREF form of A ;
10   Compute $\mathcal{U} = \{i \in [\![1, m]\!], c < \mathcal{N}(B_i)\}$;
11   **if** $\forall j \in \mathcal{U}$, A_{RREF} *is row-unit of index j* **then**
12   **return false** ;
13   **return true** ;

Algorithm 5. IsTask(t, B)

Input: an integer vector u of dimension p
Output: the vector following u for some fixed ordering on $\mathbb{Z}^p$

Algorithm 6. NextVector(u)

4   Complexity and Improvements

4.1   Complexity

The bottleneck of ComputeSparsestBasis is located in BasisToSolvedTask. Indeed, the number of while loops performed in BasisToSolvedTask can be close to $2^{n+1} - 1$ in the worst case (*i.e.* when the binary tree is almost completely explored). It is the only place where an exponential complexity occurs, since all other operations rely on linear algebra operations (such as computing a nullspace, a RREF, ...).

However, many branches are cut thanks to the line 11 in the algorithm IsTask. The sparser the matrix B is, the more branches are cut. Indeed, let us denote $d = \max\{\mathcal{N}(B_i), i \in [\![1, m]\!]\}$. Suppose that, in our binary tree, the left (resp. right) child corresponds to adding an equation (resp. an inequation). If the number of inequations c of some task is greater than or equal to d, the set $\mathcal{U}$ at line 10 in IsTask is empty, thus IsTask returns false. This implies that only the branches starting from the root and going through the right children at most d times will be explored. The number of processed nodes at depth k is equal to $\sum_{i=0}^{d} \binom{k}{i}$. Thus the total number of processed nodes is bounded

by $\sum_{k=0}^{n} \sum_{i=0}^{d} \binom{k}{i} = \sum_{i=0}^{d} \sum_{k=i}^{n} \binom{k}{i} = \sum_{i=0}^{d} \binom{n+1}{i+1}$. Easy computations show that

$\sum_{i=0}^{d} \binom{n+1}{i+1} \leq 2(n+1)^{d+1}$ which can be much smaller than $2^{n+1} - 1$ (for example when d is much smaller than n). Experimentally, we observed that models with small values d were easily solved.

Finally, the number of calls to EnhanceBasis is bounded by the number of nonzeros of the initial basis B, which is bounded by nd (since the number of nonzeros decreases at least by 1 at each call of EnhanceBasis, except for the last call).

4.2   Implementation

We chose to implement algorithms given in Section 3 using the Computer Algebra software Maple, which natively handles long integers and contains many linear algebra routines with exact coefficients. With no surprise, those algorithms can be improved because many useless computations are performed. For example, many useless rank computations are done in Algorithm IsTask. The next section describes the improvements of the algorithms given in Section 3.

4.3 Improvements

Computation and Choice of the Solved Task. Algorithm BasisToSolvedTask stops when it first encounters a solved task. This solved task may change if one pushes t_2 before t_1 in the Algorithm BasisToSolvedTask. This change has no real impact since it speeds up some examples, and slows down others.

We have experimented another strategy consisting in computing the set of all the solved tasks stemming from B instead of stopping at the first solved task. Once this set is computed, one can choose the solved task that leads to the biggest decrease of the number of nonzeros, and only keep solved tasks with $\text{Rank}(A) = m - 1$ if one encounters a solved task with $\text{Rank}(A) = m - 1$ at some point. Indeed, solved tasks with $\text{Rank}(A) = m - 1$ lead to easy computations in Algorithm EnhanceBasisUsingSolvedTask since $p = 1$.

It is not clear whether the strategy of computing all the solved tasks is better or worst than stopping on the first solved task. Indeed, searching all solved tasks is obviously more costly, but choosing a suitable solved task might decrease the number of subsequent calls to Algorithm EnhanceBasis.

Using Reduced Row Echelon Forms. It is possible to request more properties for a task TASK$[A, \Lambda, c, k]$. For example, one can request A to be in reduced row echelon form with no zero rows. Moreover, one can request the rows of Λ to be reduced w.r.t. the matrix A in the following sense: a row b is reduced w.r.t. A if the row contains a zero at the location of the pivots of A (reducing a row Λ_i by A can be done by subtracting multiple of rows of A to Λ_i to get zeros at the positions of the pivots of A).

Those requirements have several advantages, especially in Algorithms 4 and 5. In Algorithm 4, the computation of $\text{Rank}(A)$ is immediate since it equals the number of rows of A. Moreover the condition $\text{Rank}\begin{pmatrix} A \\ \Lambda_i \end{pmatrix} = \text{Rank}(A) + 1$ can be checked immediately: indeed, since Λ_i is reduced w.r.t. A, the condition is true if and only if the row Λ_i is not the zero row. In Algorithm 5, the computation of the matrix K is immediate and can be done by simply rearranging the entries of A in a new matrix K.

Finally, since the rows Λ_i are reduced w.r.t. A, it is easier to detect that two rows Λ_i and Λ_j are equal modulo a linear combination of lines of A. Indeed, if this is the case, both lines are necessarily proportional (and then one can discard one of them).

5 Benchmarks

We have tested our methods on some models taken from the BioModels database [1] (accessed on June 16th, 2014). Among the curated models, we have selected all the models involving only one compartment, with rational and integer stoichiometric coefficients. After rejecting models without conservation laws, we ended up with 214 models.

After computing a basis of linear conservation laws for each of the 214 models, our method detected that 141 bases were already sparsest ones. In the rest of the section, one only considers the remaining 73 models.

In this section, the *CSB* version denotes the non-improved version of the algorithm ComputeSparsestBasis, as described in Section 3, and the *CSB'* version denotes the version based on RREF computations as described in Section 4.3. Timings were measured on a Pentium Xeon 3.40GHz with 32Gb of memory.

Model	Size of B	d	Time (in s) CSB	CSB'	Model	Size of B	d	Time (in s) CSB	CSB'
068	3×8	4	0.15	0.06	475	7×23	14	309.8	10.59
064	4×21	18	28.4	1.99	014	8×86	45	> 3000	235.9
183	4×67	61	505.1	36.50	478	11×33	11	419.4	1.85
086	5×17	12	15.99	3.12	153	11×75	38	> 3000	964.6
336	5×18	7	3.68	0.30	152	11×64	32	> 3000	97.46
237	6×26	17	57.93	0.91	334	13×73	50	> 3000	132.6
431	6×27	15	70.68	10.93	019	15×61	13	> 3000	24.36

Fig. 1. Timing comparison between the basic version and the improved version (with B basis matrices of the models and $d = \max\{\mathcal{N}(B_i), i \in [\![1, m]\!]\}$ as described Section 4.1)

Checking that the 141 bases are indeed sparsest ones takes at most 1s for each model. The improved *CSB'* version takes less than 3000s for 69 models out of the 73 models. The remaining 4 models involve heavier computations: model 332 takes around 4000s, model 175 takes around 1 day, computation of models 205 and 457 were stopped after 2 days. These 4 models involve between 15 and 40 conservation laws (*i.e.* B has between 15 and 40 rows), and between 50 and 200 species (*i.e.* B has between 50 and 200 columns). As shown in [3], finding a sparsest basis for the null space of a matrix is NP-hard, so it is not surprising that some models are challenging.

The *CSB'* version is faster than the basic version, usually by a factor of at least 10. Figure 1 shows some timings for a sample of the 69 models. Note that the timings can be different between models with bases of similar sizes (like models 153 and 152).

For each basis B, one can define the ratio $x = \frac{\mathcal{N}(B')}{\mathcal{N}(B)}$ where B' is a sparsest basis equivalent to B. For a non sparsest basis B, this ratio satisfies $0 < x < 1$. Figure 2 represents the frequency of the 69 models, plus the models 332 and 175 involving heavier computations, as a function of the ratio x.

Appendix B shows a comparison between the number of nonzeros obtained after some usual linear algebra methods and our algorithm.

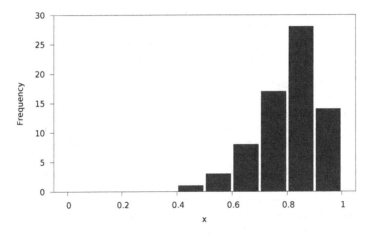

Fig. 2. Number of bases given some proportion $x = \frac{\mathcal{N}(B')}{\mathcal{N}(B)}$

6 Conclusion

We plan to implement our method in Python, for different reasons: it is a free software, it handles large integers natively, and it is easy to interface with other software. One could also implements our method in C (by using the multiprecision arithmetic library GMP [6]) for performance reasons.

Looking back to the problem of getting "good" sets of conservation laws we discussed in Section 1, it is not always possible to have a basis with the less possible negative values and a sparsest basis at the same time. There probably exists a compromise between these two properties, which is left for further work.

Acknowledgements. We would like to thank the reviewers for their helpful comments and Sylvain Soliman for his help on the Nicotine software (http://contraintes.inria.fr/~soliman/nicotine.html).

References

1. Biomodels database, http://www.ebi.ac.uk/biomodels-main/publmodels (accessed June 16, 2014)
2. Berry, M.W., Heath, M.T., Kaneko, I., Lawo, M., Plemmons, R.J., Ward, R.C.: An algorithm to compute a sparse basis of the null space. Numerische Mathematik 47(4), 483–504 (1985)
3. Coleman, T.F., Pothen, A.: The null space problem i. complexity. SIAM J. Algebraic Discrete Methods 7(4), 527–537 (1986)
4. Coleman, T.F., Pothen, A.: The null space problem ii. algorithms. SIAM Journal on Algebraic Discrete Methods 8(4), 544–563 (1987)
5. Fieker, C., Stehlé, D.: Short bases of lattices over number fields. In: Hanrot, G., Morain, F., Thomé, E. (eds.) ANTS-IX. LNCS, vol. 6197, pp. 157–173. Springer, Heidelberg (2010)

6. Granlund, T., the GMP development team: GNU MP: The GNU Multiple Precision Arithmetic Library, http://gmplib.org/
7. Lenstra, A.K., Lenstra, H.W., Lovász, L.: Factoring Polynomials with Rational Coefficients. Afdeling Informatica: IW. Mathematisch Centrum, Afdeling Informatica (1982)
8. Panteleev, M.A., Ovanesov, M.V., Kireev, D.A., Shibeko, A.M., Sinauridze, E.I., Ananyeva, N.M., Butylin, A.A., Saenko, E.L., Ataullakhanov, F.I.: Spatial propagation and localization of blood coagulation are regulated by intrinsic and protein c pathways, respectively. Biophysical Journal 90(5), 1489–1500 (2006)
9. Roman, S.: Advanced Linear Algebra. Graduate Texts in Mathematics. Springer (2007)
10. Sauro, H.M., Ingalls, B.: Conservation analysis in biochemical networks: computational issues for software writers. Biophysical Chemistry 109(1), 1–15 (2004)
11. Schilling, C.H., Letscher, D., Palsson, B.Ø.: Theory for the Systemic Definition of Metabolic Pathways and their use in Interpreting Metabolic Function from a Pathway-Oriented Perspective. Journal of Theoretical Biology 203(3), 229–248 (2000)
12. Schuster, S., Hilgetag, C.: On elementary flux modes in biochemical reaction systems at steady state. Journal of Biological Systems 2(2), 165–182 (1994)
13. Soliman, S.: Invariants and Other Structural Properties of Biochemical Models as a Constraint Satisfaction Problem. Algorithms for Molecular Biology 7(1), 15 (2012)
14. Vallabhajosyula, R.R., Chickarmane, V., Sauro, H.M.: Conservation analysis of large biochemical networks. Bioinformatics 22(3), 346–353 (2006)

A   Proof of Algorithms

Each subsection of this section proves an algorithm, by proving it stops and returns the correct result. When needed, some mathematical properties are proved at the beginning of the subsections.

A.1   ComputeSparsestBasis(B)

Halting. The algorithm halts because EnhanceBasis returns either **false**, B or **true**, B' and this last case cannot happen indefinitely as the number of nonzeros in B' is strictly decreasing and bounded by 0.

Correction. The algorithm halts when $a = $ **false**, *i.e.* when EnhanceBasis detects that B' is a sparsest basis.

A.2   EnhanceBasis(B)

Halting. Trivial.

Correction. BasisToSolvedTask returns either a solved task or $\emptyset$ if no such solved task exists. If $t \neq \emptyset$, EnhanceBasisUsingSolvedTask computes a new basis matrix B' with $\mathcal{N}(B') < \mathcal{N}(B)$, otherwise no solved task exists, which proves that B is a sparsest basis.

A.3   EnhanceBasisUsingSolvedTask(t, B)

Lemma 1 (Finite union of vector subspaces). *Let $\mathbb{K}$ be an infinite field. If E is a $\mathbb{K}$-vector subspace, then every finite union of proper subspaces of E is strictly included in E.*

Proof. See [9].   □

Theorem 2 (Characteristic theorem of solutions of $(\mathcal{S})$). *Consider $A \in \mathbb{Q}^{r \times m}, \Lambda \in \mathbb{Q}^{c \times m}, K \in \mathbb{Q}^{m \times q}$ a matrix representing a basis of $\mathrm{Ker}(A)$ stored columnwise, and the system $(\mathcal{S}) : \begin{cases} Ax = 0 \\ \Lambda x \not\equiv 0 \end{cases}$.*

The following assertions are equivalent:

1. *$(\mathcal{S})$ has nonzero solutions in $\mathbb{Q}^m$,*
2. *$\left(\forall i \in [\![1, c]\!], \mathrm{Rank} \begin{pmatrix} A \\ \Lambda_i \end{pmatrix} = \mathrm{Rank}(A) + 1 \right)$ and $(\mathrm{Rank}(A) \leq m - 1)$,*
3. *$\exists u \in \mathbb{Z}^q \setminus \{0\}, \Lambda K u \not\equiv 0$.*

Proof. $(1) \Rightarrow (3)$: Take a nonzero solution v in $\mathbb{Q}^m$ of $(\mathcal{S})$. Thus, there exists u' in $\mathbb{Q}^q \setminus \{0\}$ such that $v = Ku'$ (since the columns of K are a basis of $\mathrm{Ker}(A)$). Taking $u = \lambda u'$ (with a suitable integer λ such that u belongs to $\mathbb{Z}^q \setminus \{0\}$), one has $Ku = \lambda v$. Since λv is also a nonzero solution of $(\mathcal{S})$, one has $\Lambda K u \not\equiv 0$.

$(3) \Rightarrow (1)$: Take $v = Ku$. Then $Av = 0$ and $\Lambda v \not\equiv 0$, so v is a nonzero solution of $(\mathcal{S})$.

$(1) \Rightarrow (2)$: Consider a nonzero solution v of $(\mathcal{S})$. Since v is nonzero and satisfies $Av = 0$, one has $\mathrm{Ker}(A) \neq \{0\}$ thus $\mathrm{Rank}(A) \leq m - 1$ (according to the rank-nullity theorem). Let us suppose that $\mathrm{Rank} \begin{pmatrix} A \\ \Lambda_i \end{pmatrix} = \mathrm{Rank}(A)$ for some i. Then, Λ_i is a linear combination of rows of A so $\Lambda_i v = 0$, hence v is not a solution of $(\mathcal{S})$, contradiction. We conclude that $\forall i \in [\![1, c]\!], \mathrm{Rank} \begin{pmatrix} A \\ \Lambda_i \end{pmatrix} = \mathrm{Rank}(A) + 1$ and $\mathrm{Rank}(A) \leq m - 1$.

$(2) \Rightarrow (1)$: The set of solutions of $(\mathcal{S})$ is $V = \mathrm{Ker}(A) \setminus \bigcup_{i=1}^{c} \mathrm{Ker}(\Lambda_i) = \mathrm{Ker}(A) \setminus \bigcup_{i=1}^{c} \mathrm{Ker} \begin{pmatrix} A \\ \Lambda_i \end{pmatrix}$. For any $i \in [\![1, c]\!]$, $\mathrm{Ker} \begin{pmatrix} A \\ \Lambda_i \end{pmatrix}$ is a proper vector subspace of $\mathrm{Ker}(A)$ as $\mathrm{Rank} \begin{pmatrix} A \\ \Lambda_i \end{pmatrix} = \mathrm{Rank}(A) + 1$. According to Lemma 1, $\bigcup_{i=1}^{c} \mathrm{Ker} \begin{pmatrix} A \\ \Lambda_i \end{pmatrix}$ is strictly included in $\mathrm{Ker}(A)$. Moreover, since $\mathrm{Rank}(A) \leq m-1$, one has $\mathrm{Ker}(A) \neq \{0\}$ which implies $V \setminus \{0\} \neq \emptyset$.   □

Halting. If $p = 1$, it is trivial that the algorithm halts. If $p \neq 1$, one needs to check that the while loop halts:

- As t is a solved task, there exists a nonzero solution v in $\mathbb{Q}^m$ of $(\mathcal{S})$ and $j \in \mathcal{U}$ such that $v_j \neq 0$ and $c < \mathcal{N}(B_j)$ according to Proposition 1.

- Using the third point of Theorem 2, there exists a vector u in $\mathbb{Z}^q$ such that the vector $v = Ku$ is a nonzero solution of $(\mathcal{S})$. Hence, the while loop will eventually reach such a u since NextVector enumerates all elements of $\mathbb{Z}^q$.

Correction. At line 2, $B' \leftarrow B$. At the end of the algorithm, B'_i is modified and contains more zeros than B_i. Indeed, $\mathcal{N}(B'_i) = c$ and since $i \in \mathcal{U}$, $c < \mathcal{N}(B_i)$. Finally, B' is also a basis matrix since $B'_i = {}^t vB$ with $v_i \neq 0$.

A.4 NextVector

There is some freedom in coding this algorithm (which is not given in this paper), in particular the ordering in which the p-tuples are output. However, to ensure the termination of Algorithm EnhanceBasisUsingSolvedTask, one requires the Algorithm NextVector(u) to iterate all the p-tuples of $\mathbb{Z}^p$. This can be achieved for example by starting from the zero tuple, and enumerating the tuples of $\mathbb{Z}^p$ by increasing 1-norm (where the 1-norm of a tuple is the sum of the absolute values), and by lexicographic order for tuples of the same 1-norm. For example, when $p = 2$, the p-tuples can be enumerated in the following way: $(0,0)$, $(-1,0)$, $(0,-1)$, $(0,1)$, $(1,0)$, $(-2,0)$, $(-1,-1)$, $(-1,1)$, $(0,-2)$, $(0,2)$, $\ldots$

One could also rely on a random number generator, provided it has the property to eventually generate any p-tuple with a nonzero probability (in order to ensure the halting of Algorithm EnhanceBasisUsingSolvedTask).

A.5 BasisToSolvedTask(B)

Halting. Consider a current task t. The algorithm stops if t is solved (*i.e.* $k = n$). Otherwise, t is not a solved task and generates new tasks, obtained from t, with $k+1$ columns processed (instead of k). This last case cannot occur indefinitely.

Correction. Consider a non solved task t inside the while loop. One creates the object t_1 (resp. t_2) corresponding to the cancellation (resp. the non cancellation) of the coefficient number $k+1$ of the linear combinations of the lines of B. The objects t_1 and t_2 are possibly discarded thanks to the function IsTask if they are not tasks (*i.e.* if they cannot be used to increase the number of zeros). Consequently, all cases are considered and the function will return $\emptyset$ if and only if there does not exist any solved task stemming from B.

A.6 IsTask(t, B)

Proposition 2. *Let M be a matrix. Then M has a zero column if and only if $\mathrm{Ker}(M)$ contains at least one vector with exactly one nonzero coefficient.*

Proof. Trivial   □

Corollary 1. *Let M be a matrix. Then M does not have any zero column if and only if $\mathrm{Ker}(M) \setminus \{0\}$ only contains vectors with at least two nonzero coefficients.*

Proposition 3. *Take $A \in \mathbb{Q}^{r \times m}, \Lambda \in \mathbb{Q}^{c \times m}$ and the system $(\mathcal{S}) : \begin{cases} Ax = 0 \\ \Lambda x \neq 0 \end{cases}$. If $(\mathcal{S})$ has nonzero solutions and $\mathrm{Rank}(A) \leq m - 2$, then $(\mathcal{S})$ has nonzero solutions with at least two nonzero coefficients.*

Proof. $\mathrm{Rank}(A) \leq m - 2 \Leftrightarrow \mathrm{Dim}(\mathrm{Ker}(A)) \geq 2$ with the rank-nullity theorem. Consequently, there exist at least two independent nonzero vectors v_1 and v_2 solutions of $Ax = 0$. Consider a nonzero solution v of $(\mathcal{S})$. By a topology argument the vector $v + \varepsilon_1 v_1 + \varepsilon_2 v_2$ is also solution of $(\mathcal{S})$ for any ε_1 and ε_2 satisfying $|\varepsilon_1| < \varepsilon$ and $|\varepsilon_2| < \varepsilon$ (for a suitable small fixed $\varepsilon > 0$). Suppose that $v + \varepsilon_1 v_1 + \varepsilon_2 v_2$ has exactly one nonzero coefficient for any $|\varepsilon_1| < \varepsilon$ and $|\varepsilon_2| < \varepsilon$. That would imply that both v_1 and v_2 have exactly one nonzero coefficient at the same position, which is impossible since v_1 and v_2 are linearly independent. Consequently, V contains at least one nonzero vector with two nonzero coefficients.

Theorem 3. *Consider $A \in \mathbb{Q}^{r \times m}, \Lambda \in \mathbb{Q}^{c \times m}$ and the system $(\mathcal{S}) : \begin{cases} Ax = 0 \\ \Lambda x \neq 0 \end{cases}$.*

The following assertions are equivalent :

1. *$(\mathcal{S})$ has nonzero solutions with at least two nonzero coefficients,*
2. *$(\mathcal{S})$ has nonzero solutions and at least one of the two following conditions is true:*
   - *$\mathrm{Rank}(A) \leq m - 2$,*
   - *A does not have any zero column.*

Proof. $(1) \Rightarrow (2)$: Since $(\mathcal{S})$ has nonzero solutions, then $\mathrm{Rank}(A) \leq m - 1$. If the condition $\mathrm{Rank}(A) \leq m - 2$ is not true, then one has $\mathrm{Rank}(A) = m - 1$. Consequently, $\mathrm{Ker}(A)$ is generated by one vector containing at least two nonzero coefficients. According to Corollary 1, A does not have any zero column.

$(2) \Rightarrow (1)$: Direct consequence of Corollary 1 if $\mathrm{Rank}(A) = m - 1$, or Proposition 3 if $\mathrm{Rank}(A) \leq m - 2$. $\qquad \square$

Definition 2. *A matrix A is **row-unit of index** j if there exists a row of A with only one nonzero coefficient, which is at position j.*

Theorem 4. *Consider a matrix A' in $\mathbb{Q}^{r \times m}$ under reduced row echelon form. Let K be a matrix containing a basis of $\mathrm{Ker}(A')$ stored columnwise. Suppose that $\mathrm{Ker}(A') \neq \{0\}$. For any index j, the following assertions are equivalent:*

1. *A' is row-unit of index j,*
2. *$K_j = 0$.*

Proof. (1) ⇒ (2): There exists a row $A'_i = \begin{pmatrix} 0 \cdots 0\ 1\ 0 \cdots 0 \end{pmatrix}$ where the 1 is at position j, for some i in $[\![1, r]\!]$. Thus, any solution v (in particular all elements of the basis K) of $A'v = 0$ must satisfy $v_j = 0$, hence $K_j = 0$.

(2) ⇒ (1): From $A'K = 0$, one has ${}^tK{}^tA' = 0$. Since $K_j = 0$, the column j of ${}^tK$ is zero, which implies that the canonical vector e_j belongs to $\mathrm{Ker}({}^tK)$ using Proposition 2. Hence, the row $l = {}^te_j = \begin{pmatrix} 0 \cdots 0\ 1\ 0 \cdots 0 \end{pmatrix}$, where the 1 is at position j, is a linear combination of the rows of A'. Since A' is in reduced row echelon form, if the combination l of rows of A' involved strictly more than one row of A', l would at least involve two nonzero coefficients (corresponding to the pivots). Thus, l is a row of A' and A' is row-unit of index j.   □

Halting. Trivial.

Correction

1. If the first condition (line 3) is true, then $(\mathcal{S})$ has no nonzero solutions in $\mathbb{Q}^m$, so t does not satisfy condition **LCP**, and one returns **false**. Otherwise, one has $(\forall i \in [\![1, c]\!], \mathrm{Rank}\begin{pmatrix} A \\ A_i \end{pmatrix} = \mathrm{Rank}(A) + 1)$ and $(\mathrm{Rank}(A) \le m - 1)$ so $(\mathcal{S})$ has nonzero solutions according to Theorem 2.

2. If the second condition (line 6) is true, since $(\mathcal{S})$ admits nonzero solutions and thanks to Theorem 3, $(\mathcal{S})$ does not have solutions with at least two nonzero coefficients, so t does not satisfy **LCP**, and one returns false. Otherwise, t satisfies **LCP**.

3. If the third condition (line 11) is true, then for any $j \in \mathcal{U}$, one has $K_j = 0$ thanks to Theorem 4. Consequently, any solution v of $(\mathcal{S})$ satisfies $v_j = 0$ for all $j \in \mathcal{U}$. Therefore, t does not satisfy **IZP** and one returns false. Otherwise, there exists a $j \in \mathcal{U}$ such that $K_j \neq 0$ thanks to Theorem 4, so there exists a column w of K such that $w_j \neq 0$. It may happen that w_j is not solution of $(\mathcal{S})$. In that case, consider a nonzero solution u of $(\mathcal{S})$ with at least two nonzero coefficients. By a topology argument, the vector $\bar{u} = u + \varepsilon_1 w$ is also solution of $(\mathcal{S})$ for any rational ε_1 satisfying $|\varepsilon_1| < \varepsilon$ (where ε is a small rational). Finally, one can choose a suitable ε_1 to obtain the condition $\bar{u}_j \neq 0$, and $(\mathcal{S})$ satisfies **IZP**. One then returns **true**.

B   Comparison with some Matrix Algorithms

We consider a sample of 10 models taken from the 61 models where computations end in less than 3000s. We present here the number of nonzeros of the initial bases and of the bases obtained after using our algorithm (CSB), the Reduced Row Echelon Form (RREF), the LLL algorithm [7] and the Hermite Normal Form (HNF). When comparing our algorithm with the usual linear algebra algorithms we used, one sees that the sparsest bases are not always reached by these linear algebra algorithms and that they sometimes make it worst.

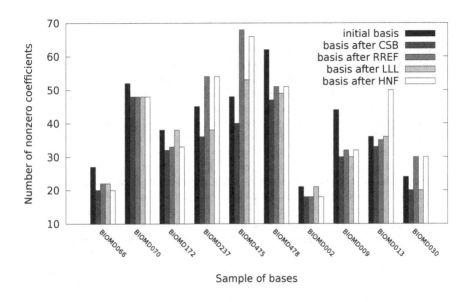

Fig. 3. Number of nonzeros after some matrix algorithms

SAT-Based Metabolics Pathways Analysis
without Compilation

Sabine Peres[1,3], Martin Morterol[1], and Laurent Simon[2]

[1] Univ. Paris-Sud, LRI / CNRS UMR8623
91405 Orsay, France
{speres,morterol}@lri.fr
[2] Univ. Bordeaux, Labri / CNRS UMR5800
33405 Talence, France
lsimon@labri.fr
[3] INRIA Saclay, Orsay

Abstract. Elementary flux modes (*EFMs*) are commonly accepted tools for metabolic network analysis under steady state conditions. They can be defined as the smallest sub-networks enabling the metabolic system to operate in steady state with all irreversible reactions proceeding in the appropriate direction. However, when networks are complex, the number of *EFMs* quickly leads to a combinatorial explosion, preventing from drawing even simple conclusions from their analysis. Since the concept of *EFMs* analysis was introduced in 1994, there has been an important and ongoing effort to develop more efficient algorithms. However, these methods share a common bottleneck: they enumerate all the *EFMs* which make the computation impossible when the metabolic network is large and only few works try to search only *EFMs* with specific properties. As we will show in this paper, enumerating all the *EFMs* is not necessary in many cases and it is possible to directly query the network instead with an appropriate tool. For ensuring a good query time, we will rely on a state of the art SAT solver, working on a propositional encoding of *EFMs*, and enriched with a simple SMT-like solver ensuring *EFMs* consistency with stoichiometric constraints. We illustrate our new framework by providing experimental evidences of almost immediate answer times on a non trivial metabolic network.

1 Introduction

Constraint-based modeling methods allow to predict phenotypes by calculating steady state flux distributions in a metabolic network. The notion of elementary flux mode (*EFM*) is a key concept derived from the analysis of metabolic networks from a pathway-oriented perspective [27]. An *EFM* is defined as the smallest sub-network that enables the metabolic system to operate in steady state with all irreversible reactions proceeding in the appropriate direction [29,28]. Every steady-state flux distribution can be represented as a non negative combination of *EFMs*. Applications of network-based pathway analyses have been presented for predicting functional properties of metabolic networks, measuring different aspects of robustness and flexibility, and even assessing gene regulatory features [35,14]. Actually, *EFMs* can only be enumerated in small to medium-scale metabolic networks because the number of *EFMs* increase exponentially

P. Mendes et al. (Eds.): CMSB 2014, LNBI 8859, pp. 20–31, 2014.

with the network size [21]. The huge number of *EFMs* associated with large biochemical networks prevents from drawing simple conclusions from their analysis. Studies have been carried out on the analysis of sub-networks. Kaleta *et al.* [18] showed that the analysis of small sub-networks can be misleading. To overcome this problem, they introduced the concept of elementary flux patterns that takes into account the steady-state fluxes through a metabolic network at the genome-scale when analyzing pathways in a sub-network. Several approaches have been developed to deal with the combinatorial explosion. ACoM [24,25] which was a first attempt to classify the *EFMs* helped to give biological meaning to the different *EFMs* and to the relatedness between reactions.

Being a computationally demanding task, several approaches to parallel or distributed computation of *EFMs* have been proposed through parallelization techniques [15] or algorithmic reformulations [39,20,36]. To reduce their number, Jol [16] charaterized the flux solution space by determining *EFMs* that are subsequently classified as thermodynamically feasible or infeasible on the basis of experimental metabolome data. To speed up the computation of *EFMs*, gene regulatory information has been taken into account to eliminate mathematically possible *EFMs* [17]. Some analyses on minimal cut set allowed network analyses without enumerating all the *EFMs* [3,38]. Nevertheless, although several improvements have been introduced for computing *EFMs* in large networks, tools are still needed to allow their large-scale analysis and interpretation. Indeed the existing approaches for analyzing the network, despite the insights they have provided, have their capabilities limited because the calculation requires the complete set (or a major part) of *EFMs* in prior given a metabolic network, whose computation is notoriously hard due to the combinatorial explosion when the network size grows.

We were inspired by the seminal paper on *Knowledge Base Compilation Techniques* [10]. Instead of explicitly compute all the *EFMs*, we only answer a set of queries on the network. We propose to develop a method similar to Just in Time Knowledge Compilation [1], based on *SAT solving* to determine network properties. Consequently, using our method of selection *EFMs* under constraints will now allow studying the *EFMs* of large metabolic networks almost instantly.

2 SAT/SMT Encoding of Metabolic Pathways Queries

Since the introduction of so-called "modern" SAT solvers [31,23,13], the practical solving of NP Complete problems (and above) has known a series of performances scale up. Inspired by the central position of the SAT problem in the polynomial hierarchy, the first approaches were relying on a direct SAT translation of the considered problem. However, most of these reductions were addressing academic problems (graphs, puzzles, [32]) until the Blackbox planner [19] was proposed. The idea was to solve planning problems by unrolling time steps and using propositional logic. Thanks to the implementation of zChaff [23], and its lazy data structures, this approach was proved very efficient, even on planning problems encoding huge SAT formulas. A few year after, Bounded Model Checking [7,6] was also successfully proposed. Once again, time steps were unrolled and SAT solvers were called on huge formulas, with success. Most of the progresses observed in practice were due to algorithmic and data structures improvements [13]. Indeed, "Modern" SAT solvers rely on Conflict Driven Clause Learning scheme (CDCL) that deeply changed the backtrack search algorithm proposed before [11]. More recently, SAT solvers were used as SAT oracles for Model Checking [9] or hardest problems [5,40]. Thanks to this technique, the field knew a new scaling up in its practical applications. Problems are nowadays solved with thousands of SAT calls on a set of almost identical successive formulas (this technique is called *incremental SAT solving*).

However, despite this success, some problems are still unreachable for SAT solvers. For instance, enumerating all the models of a formula is not attacked with the same techniques [26]. Typically, each SAT call can only answer Yes or No (in the latter case, some additional information can be gathered like, for instance, the set of initial clauses used to derive the contradiction), and harder questions (minimization and/or enumeration of solutions) are still very hard tasks that even SAT solvers cannot solve efficiently. Another limitation of plain SAT approach is the lack of expressiveness of the propositional logic. One solution for this is to allow more powerful reasoning techniques by using, intuitively, two-levels solvers. One solver is working on an abstraction of the initial problem (typically the SAT solver) and sends abstract (candidates) solutions to a more powerful reasoner (that can, for instance, count or reason with Gaussian elimination) that checks if the abstract solution is indeed a solution of the initial problem. This framework, called DPLL(T) in the literature allows SAT solvers to be used as the corner stone of most of the efficient SMT solvers (Satisfiability Modulo Theory) [4]. The name DPLL(T) comes from the first SAT solvers using backtrack search techniques [12,11]. Often, both solvers are more tightly connected (the Theory solver can also force some propositional variables to take some values during the propagation process), but the general idea remains. The T is the theory used by the more powerful reasoner. In this paper, we will use the CDCL(T) notation, given the fact that SAT solvers used cannot be identified anymore with the DPLL algorithm.

In this paper, we propose to address Metabolic Pathways Analysis based on a SAT solver for quickly finding pathway candidates and a theory solver for ensuring the consistency of stoichiometric constraints on the reactions. We thus have to deal with three levels of problems. The first one is to deal with a SAT encoding of the original metabolic network, and to be able to answer queries efficiently. The second level of problems we will face is to be able to produce minimal pathways only (at the SAT level) before sending them to the theory solver. The third one is to able to produce pathways consistent with stoichiometric constraints. For this, we will use a theory solver with Gaussian elimination, which is more powerful than plain resolution used in SAT solvers.

We will first use a modified version of the SAT solver Glucose in order to enumerate solutions. To minimize them, we will use a solution similar to [34], e.g. a minimization strategy inspired by [22]. Then, we propose to use a very simple theory solver based on matrix kernel computation that will check that candidates at the abstract level are indeed consistent with stoichiometric constraints.

2.1 Previous Work on SAT and Metabolic Pathways Analysis

In a couple of seminal papers [33,34], it was proposed to encode metabolic pathways into SAT problem. Their technique relies on time unrolling, each time step being the execution of one reaction in the network. They propose a computation method to predict gene knockout effects by identifying minimal active pathways using the SAT solver Minisat2. Tiwari *et al.* [37] propose a method using a weighted Max-SAT solver to analyze pathways. They translate reaction laws into soft constraint represented in weighted clauses to compute ordered solutions. However, its ordering is sometimes not acceptable from a biological viewpoint since reaction laws must be held are sometimes violated.

The missing property of these methods is the stoichiometry of the metabolites in reactions which can lead to find non viable pathways and can miss important pathways.

2.2 A Simple Abstraction of the Metabolic Network

We will use in this paper the classical propositional logic notation, with the usual boolean connectors ($\vee, \wedge, \neg$). A propositional variable v can be either true or false (resp. $\top$ or $\bot$). When set, the value of a variable is called its assignment. A literal is a variable or its negation (v or $\neg v$). A clause is a disjunction of literals (for instance $a \vee \neg b \vee c$). A formula in Conjunctive Normal Form (CNF) is a conjunction of clauses. The SAT problem is, given a formula f in CNF, to check if f has a model, *i.e.* there exists an assignment of its variables (the assignment can be partial) that makes f true given the usual semantic of boolean connectors.

The idea in this section is to propose a very simple SAT encoding of our problem, without stoichiometric constraints at this level. We note by $\mathcal{M}$ the set of metabolites M_i, and $\mathcal{R}$ the set of reactions r_i. We will use the set $P(M_i)$ as the set of reactions that produces the metabolite M_i and the set $C(M_i)$ as the set of reactions that consumes M_i. We note each reaction r_i by $r_i : \sum_{j=1}^{|\mathcal{M}|} S(I_{i,j})I_{i,j} \rightarrow \sum_{j=1}^{|\mathcal{M}|} S(O_{i,j})O_{i,j}$. I stands for input metabolites, O for output metabolites. $S(M)$ is the stoichiometric coefficient associated to the metabolite M.

In order to handle reversible reactions, we first have to duplicate each reversible reaction r_i into r_i (left to right reaction) and a new $rrev_i$ (right to left reaction). The SAT encoding is the conjunction of the following set of formulas:

- Producing an output implies at least one of its rules was activated (one clause per produced metabolite):

$$O_{i,j} \rightarrow \bigvee r \in P(O_{i,j}) \tag{1}$$

- The rule r_i was activated implies all its output metabolites were produced:

$$r_i \rightarrow \bigwedge O_{i,j} \tag{2}$$

- The rule r_i was activated implies all its input metabolites were available:

$$r_i \rightarrow \bigwedge I_{i,j} \tag{3}$$

- An internal metabolite I must be consumed in at least one reaction (number of internal metabolites clauses)

$$I \rightarrow \bigvee C(I) \tag{4}$$

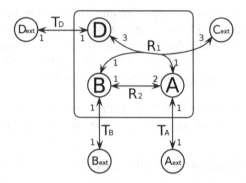

Fig. 1. Small example of a trivial metabolic pathway. The metabolites A, B and D are internal metabolites. $Aext$, $Bext$, $Cext$ and $Dext$ are external metabolites. It describes the following set of reactions : R1 : A + 3 C = B + 3 D, R2 : B = 2 A, TA : Aext = A, TB : Bext = B, TD : Dext = D. Metatool[39] returns 4 reversible *EFMs*: -TB (2 TA) -R2; (-3 TD) -TA R1 R2; (-6 TD) -TB (2 R1) R2; (-3 TD) -TB TA R1.

- In case of reversible reactions, we must ensure that reactions are only activated one-way (number of reversible reactions binary clauses):

$$\neg r_i \lor \neg rrev_i \tag{5}$$

- And, to prevent any empty trivial answer, at least one reaction must be activated

$$\bigvee r_i \bigvee rrev_j \tag{6}$$

It is clear that this encoding does not perfectly match the requirements for enumerating *EFMs*. For instance, there is no reason, with this encoding, that all internal metabolites are consumed by some rule. However, once again, in our architecture, the SAT solver will send candidates solution only to the theory solver.

Example 1. Let us consider here, for clarity, the small example figure 1.

If we apply the SAT encoding, we have:

- A metabolite can be produced only if it is produced by at least one reaction (rule 1) : $\neg A \lor R2 \lor TA \lor R1\_rev$, $\neg B \lor R1 \lor TB \lor R2\_rev$, ...
- An activated reaction produces all its metabolites (rule 2) : $\neg R1 \lor B$, $\neg R1 \lor D$, ...
- A reaction can be activated only if all its reactants are presents (rule 3) : $\neg R1 \lor A$, $\neg R1 \lor Cext$, ...
- A produced internal metabolite must be consumed by at least one reaction (rule 4) : $\neg A \lor R1 \lor R2\_rev \lor TA\_rev$, $\neg B \lor R2 \lor R1\_rev \lor TB\_rev$, ...
- A reversible reaction shouldn't be used at the same time in both ways (rule 5) : $\neg R1 \lor \neg R1\_rev$, $\neg R2 \lor \neg R2\_rev$, ...
- A valid pathway must have at least one reaction (rule 6) : $R1 \lor R1\_rev \lor R2 \lor R2\_rev \lor TA \lor TA\_rev \lor TB \lor TB\_rev \lor TD \lor TD\_rev$

2.3 Ensuring Minimality of Candidates

Before checking the consistency of stoichiometric constraints, we need to send minimal *EFM* candidates only to the theory solver. For this, we will specialize the SAT solver to be able to produce a minimal solution w.r.t. the set of target reactions. In our encoding the set of propositional variables $Target(\mathcal{R})$ encodes the set of target reactions. Thus, finding a minimal set of targeted reactions is exactly the problem of finding an assignment S such that the number of true literal (literals assigned to true in the final solution) $TT = \{x \in var(S) \; s.t. x = \top\}/Target(\mathcal{R})\}$ is minimal. This problem is called "Minimal Model Generation" [22]. $var(S)$ is the set of variables occurring in the assignment (S can be partial over a formula f) and TT stands for "True Target literals". We must ensure its minimality in terms of subset of true literals on TT. More formally, given TT, we have: $\forall TT' \subset TT$ and $\forall \; S' \; s.t. \; x = \top \in S' \; iff \; x \; \in TT'$, at least one clause in f is not satisfied by S'.

For this, we used the method described in [22] (which is also used in [34]). Intuitively, each time a solution is found, multiple SAT calls ensure its minimality by successively trying to remove at least one true literal from the initial solution.

2.4 Validating the Pathway with the Stoichiometry Using SMT

To check if the pathway obtained with SAT is consistent with the stoichiometry of the network, we examine if it is contained in the kernel of the stoichiometric matrix. At this step, the pathway is represented by a set of reactions. If this pathway is consistent with the stoichiometry of the network, there exists a vector with non-null coefficients belonging to the kernel of the sub-matrix of the set of reactions of the pathway. If the solution vector (which represents the pathway) has only strictly positive values or only strictly negative value, it is selected to be a potential solution of the network. If it contains opposite sign values, it is rejected because all the reversible reactions are splitted into forward and backward reactions. If it contains null values (but not all), it is rejected because this pathway is not minimal, taking into account the stoichiometry. If all the values are null, there is no solution, the pathway is not at steady state with the stoichiometry and so, it is not complete. This solution returns to the SAT solver to be completed, if possible.

After minimization in the small example of the figure 1, there are still 14 minimal possible solutions. Each solution must be examined by the SMT solver. It creates a sub-matrix of the stoichiometric matrix which only contains the reactions of the found solution. Then, three cases have to be considered:

- *Incomplete :* there is no vector with non-null coefficients in the kernel. This solution is not at steady state for some metabolites and needs one or more activated reactions to be valid with the stoichiometry of the network. For example, $R1 \land R2 \land TD\_rev$ has been found as a potential solution; the only vector belonging to the kernel which contains only these reactions is the null vector. This solution returns to the SAT solver to be completed, if possible.
- *Impossible :* if the solution is false because the vector of the kernel contains null values (not all) or opposite sign values, it cannot be transformed into a valid one.

For example, the SAT solver found $R1\_rev \wedge R2\_rev \wedge TA\_rev \wedge TD$. The coefficients of this solution in the kernel are $(1\ 1\ -1\ 3)^t$ the stoichiometry imposes to product more metabolite A to be at steady state. This solution is rejected.

- *Valid* : if the solution is a correct pathway (*i.e.* an *EFM*). The coefficients found in the kernel have all their values non null and with the same sign. For example, the kernel vector of $R2 \wedge TA\_rev \wedge TB$ is $(1\ 2\ 1)^t$.

2.5   Optimizing EFM with Our CDCL(T) Architecture

The SAT solver works on an abstraction of the original network. Thus, even a minimized solution will have no guarantee to comply with the stoichiometric constraints. Each time Glucose is producing a solution, it first minimizes it on the target set of reactions. Then, the candidate solution is sent to the Theory solver. Three cases can occur, as described section 2.4: (1) The solution is compliant with stoichiometric constraints and the solution can be considered. Then a clause is added to the SAT solver to block further solutions subsumed by it. (2) The solution is not possible w.r.t. stoichiometric constraints. The solution is discarded and exactly the same clause is added to the SAT solver to prevent further solutions like it. At last (3), the solution can be incomplete with stoichiometric constraints, i.e. it may be possible to find a Kernel to the matrix but at least one reaction must be added to it. Once again, a special clause is added to the SAT solver exactly meaning that.

For each of these special cases, the SAT solver must add a special clause to express the result.

1. The candidate solution $\bigwedge r_i$ is compliant with the stoichiometric constraints. Then a clause $\bigvee \neg r_i$ is added to the clause database.
2. The candidate solution $\bigwedge r_i$ is not compliant. Again, a clause $\bigvee \neg r_i$ is added to the clause database.
3. Now, if the candidate solution $\bigwedge r_i$ is incomplete from a stoichiometry point of view, we cannot add the same clause. Otherwise the matching *EFM*, if it exists, will not be found by the SAT solver (it is a superset of all r_i). To handle this, we add the following clause $\bigvee \neg r_i \bigvee r_j$ where r_j are all the reactions that does not occur in the initial candidate.

In the small example of figure 1, our solver returns the same *EFMs* as Metatool.

3   Experiments on the Yeast Mitochondrial Energetic Metabolism

We implemented our tool on top of Glucose, an award winning SAT solver [2]. We added minimization features, as described in the previous section, and added it the simple Theory module to reason on stoichiometric constraints. Tests were conducted on 64 bits 4 cores workstation Intel I7 at 3GHz, with 16Gb, running under Linux. Note that our tool only uses one core.

We applied this method to the mitochondrial bioenergy metabolism of the yeast *S. cerevisiae* which includes the TCA cycle, transaminases, oxidative phosphorylation, ethanol metabolism and carriers. The model contains 36 reactions, 10 irreversible reactions, 32 internal metabolites and 25 external metabolites. We identified $11,121$ *EFMs* with metatool [39].

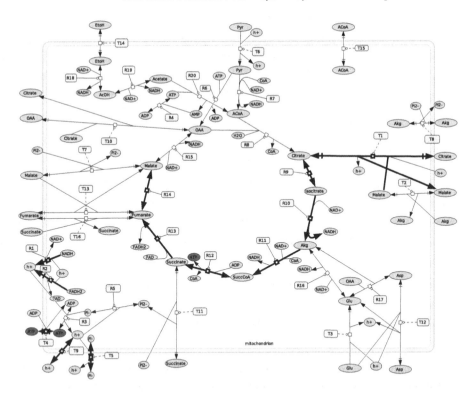

Fig. 2. ATP synthesis of yeast mitochondrial metabolism without *ATP synthase*. The rectangular nodes represent the reactions and the oval nodes represent the metabolites. The bold edges describe the pathway.

3.1 Enumerating the First 100 *EFMs*

As we pointed out, the goal of our method is to be able to directly query any metabolic network without needing to enumerate all *EFMs*. We however want to report here the behavior of our tool in the enumeration of the first 100 *EFMs* on the above example, when no queries are asked. This is not our typical application but reporting this should allow the reader to compare the performances and the behavior of our method with more traditional, compilation-based, methods (7 seconds is needed by `efmtool` to find the 11, 121 *EFMs*).

A few conclusions can be drawn from our analysis reported figure 3. First, 50 *EFMs* are almost instantly found (less than 0.1s), which clearly demonstrates that interactive queries over this kind of network is possible without any compilation. The short computation effort will be transparent for the user. We also noticed that, surprisingly enough, the number of SAT calls is larger than the number of conflicts. Conflicts are essentials in modern SAT solvers. They occur when a clause is falsified by the current partial assignment, when searching for a model, before backtracking. At each conflict, a clause is learnt by the solver, and added to its clauses database. Observing that the number of conflicts is smaller than the number of SAT calls clearly states that each solution is

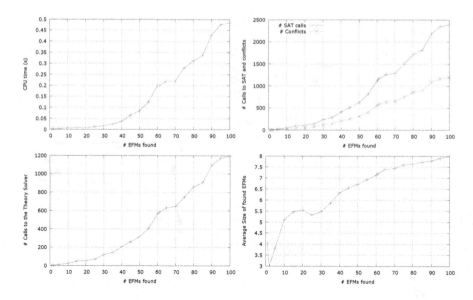

Fig. 3. Behavior of our tool on the first 100 *EFMs* for the yeast mitochondrial metabolism

almost trivially found by the SAT solver at each call. Recall that SAT calls are used either to find a first solution or to minimize it recursively (and thus this observation shows how simple is this example from a SAT point of view). The third sub-figure also shows that the minimization process at the SAT level is relatively efficient, despite its relatively high level of abstraction: each call to the theory solver is done on a minimized formula on the abstract network, and we observed only a few additional SAT calls for minimizing purpose. On average, around 12 calls to the theory solver are needed to find a solution with possible stoichiometric constraints. The last observation is about the size of *EFMs* the solver is able to produce. We clearly observe, on the last figure, that the solver has a natural tendency to produce short *EFMs* first. Adding features that will allow the solver to find most interesting *EFMs* (defined by the user) first is however an ongoing work.

3.2   Querying the Mitochondrial Network

The above example demonstrated that our tool may be used to enumerate all *EFMs*. However, we observed the same limitations of SAT solvers for enumerating solutions (or even minimal solutions). Clearly enough, an important number of solutions is a practical limitation for our tool. Thus, we would like to demonstrate that it is indeed possible to query the above network with no compilations and be able to test/check some hypothesis on it.

The first query is about finding a classical *EFM* in this network. We retrieve classical pathways such that the production of ATP from pyruvate uptake through TCA cycle and respiratory chain in 0.044 CPU time (s) with the following formulae:

$$T6 \wedge R7 \wedge R8 \wedge R9 \wedge R10 \wedge R11 \wedge R12 \wedge R13 \wedge R14 \wedge R15 \wedge R1 \wedge R2 \wedge R3 \wedge T4$$

We also retrieve not well-known pathways. Schwimmer *et al.* [30] showed that ATP production in the TCA cycle (substrate-level phosphorylation) can be important particularly in special conditions. Citrate produced in the glyoxylate cycle and transported into the mitochondria by *Odc1p* (gene encoding an oxodicarboxylate carrier) can enter the TCA cycle and lead to the production of succinate coupled to substrate-level phosphorylation. Either malate or oxaloacetate can then be transported to the cytosol via *Odc1p* and enter the glyoxylate cycle. This gives an increase of flux through TCA cycle giving advantage to the ATP synthesis. We looked for *EFMs* which can represent the experimental data. We searched the *EFMs* which contain respiratory chain (without ATP synthase), *Odc1p* and TCA cycle (between citrate and malate) with the following formulae:

$$\neg R3 \wedge T4 \wedge R1 \wedge R2 \wedge R9 \wedge R10 \wedge R11 \wedge R12 \wedge R13 \wedge R14 \wedge T1\_rev$$

We found 20 *EFMs* in 3.068 CPU time (s) which satisfy this constraint. We selected the *EFMs* which does not contain other supplementary reactions but only transports. So we added the literals $\neg R_i$ with $i \neq 1, 2, 9, 10, 11, 12, 13, 14$. We found one *EFM* (in less than a second) where two more transports were needed to have the steady state: the proton leak and phosphate transport. Figure 2 shows the calculated *EFM*. We do the same with $T10\_rev$ instead of $T1\_rev$. We found 12 *EFMs* in 2.036 CPU time (s) which satisify the constraint. As before, we selected the *EFMs* which does not contain other supplementary reactions but only transports. The same *EFM* has been found with $T10\_rev$ instead of $T1\_rev$. We see here a typical run case of our algorithm. Each query is built and answered without the need of computing all the *EFMs*.

4 Conclusion

If this work is still preliminary in some aspects, it demonstrates that new approaches for manipulating *EFMs* are possible. We showed that a SMT-like architecture (CDCL(T)) can be efficiently used to enumerate *EFMs*, thanks to an efficient SAT solver. This implies a new interaction between the SAT solver and the theory solver, in order to ensure the minimality of each *EFM*. As a main result for our work (this was our initial motivation), it may be not necessary to compute *a priori* all the *EFMs* before reasoning on them.

Our tool was able to recover interesting *EFM* from non trivial metabolic networks. We think it can answer complicated queries on very large metabolic networks, or be used when other methods fail. It also allows for a high degree of flexibility in possible constraints to add to the networks.

References

1. Audemard, G., Lagniez, J., Simon, L.: Just-in-time compilation of knowledge bases. In: 23rd International Joint Conference on Artificial Intelligence(IJCAI 2013), pp. 447–453 (August 2013)
2. Audemard, G., Simon, L.: Predicting learnt clauses quality in modern sat solvers. In: IJCAI (2009)
3. Ballerstein, K., von Kamp, A., Klamt, S., Haus, U.: Minimal cut sets in a metabolic network are elementary modes in a dual network. Bioinformatics 28(3), 381–387 (2012)
4. Barrett, C., Sebastiani, R., Seshia, S.A., Tinelli, C.: Satisfiability Modulo Theories. In: Biere et al (eds.) [8], ch. 26, vol. 185, pp. 825–885 (February 2009)
5. Belov, A., Lynce, I., Marques-Silva, J.: Towards efficient mus extraction. AI Communications 25(2), 97–116 (2012)
6. Biere, A.: Bounded Model Checking. In: Biere, et al. (eds.) [8] ch. 14, vol. 185, pp. 455–481 (2009)
7. Biere, A., Cimatti, A., Clarke, E., Zhu, Y.: Symbolic model checking without bdds (1999)
8. Biere, A., Heule, M., van Maaren, H., Walsh, T. (eds.): Handbook of Satisfiability. Frontiers in Artificial Intelligence and Applications, vol. 185. IOS Press (February 2009)
9. Bradley, A.R.: IC3 and beyond: Incremental, inductive verification. In: Madhusudan, P., Seshia, S.A. (eds.) CAV 2012. LNCS, vol. 7358, pp. 4–4. Springer, Heidelberg (2012)
10. Darwiche, A., Marquis, P.: A knowledge compilation map. J. of AI Research, 229–264 (2002)
11. Davis, M., Logemann, G., Loveland, D.: A machine program for theorem proving. JACM 5, 394–397 (1962)
12. Davis, M., Putnam, H.: A computing procedure for quantification theory. JACM 7, 201–215 (1960)
13. Eén, N., Sörensson, N.: An extensible SAT-solver. In: Giunchiglia, E., Tacchella, A. (eds.) SAT 2003. LNCS, vol. SAT, pp. 502–518. Springer, Heidelberg (2004)
14. Gagneur, J., Klamt, S.: Computation of elementary modes : a unifying framework and the new binary approach. BMC Bioinformatics 5(175) (2004)
15. Jevremovic, D., Trinh, C., Srienc, F., Sosa, C.P., Boley, D.: Parallelization of nullspace algorithm for the computation of metabolic pathways. Parallel Computing 37(6-7), 261–278 (2011)
16. Jol, S.J., Kümmel, A., Terzer, M., Stelling, J., Heinemann, M.: System-level insights into yeast metabolism by thermodynamic analysis of elementary flux modes. PLoS Computational Biology 8(3) (2012)
17. Jungreuthmayer, C., Ruckerbauer, D.E., Zanghellini, J.: Regefmtool: Speeding up elementary flux mode calculation using transcriptional regulatory rules in the form of three-state logic. Biosystems 113(1), 37–39 (2013)
18. Kaleta, C., De Figueiredo, L.F., Schuster, S.: Can the whole be less than the sum of its parts? pathway analysis in genome-scale metabolic networks using elementary flux patterns. Genome Research 19, 1872–1883 (2009)
19. Kautz, H., Selman, B.: Blackbox: A new approach to the application of theorem proving to problem solving. In: Working notes of the Workshop on Planning as Combinatorial Search, Held in Conjunction with AIPS 1998, pp. 58–60 (1998)
20. Klamt, S., Saez-Rodriguez, J., Gilles, E.: Structural and functional analysis of cellular networks with cellnetanalyzer. BMC Systems Biology 1(1), 2 (2007)
21. Klamt, S., Stelling, J.: Combinatorial complexity of pathway anaysis in metabolic networks. Mol. Bio. Rep. 29, 233–236 (2002)

22. Koshimura, M., Nabeshima, H., Fujita, H., Hasegawa, R.: Minimal model generation with respect to an atom set. In: International Workshop on First-Order Theorem Proving (2009)
23. Moskewicz, M., Madigan, C., Zhao, Y., Zhang, L., Malik, S.: Chaff: Engineering an efficient SAT solver. In: Proceedings of DAC, pp. 530–535 (2001)
24. Peres, S., Beurton-Aimar, M., Mazat, J.P.: Pathway classification of tca cycle. IEE Journal for Systems Biology 153(5), 369–371 (2006)
25. Peres, S., Vallée, F., Beurton-Aimar, M., Mazat, J.P.: Acom: a classification method for elementary flux modes based on motif finding. Biosystems 103(3), 410–419 (2011)
26. Sang, T., Bacchus, F., Beame, P., Kautz, H.A., Pitassi, T.: Combining component caching and clause learning for effective model counting. In: Proc. SAT (2004)
27. Schuster, S., Dandekar, T., Fell, D.A.: Detection of elementary modes in biochemical networks : A promising tool for pathway analysis and metabolic engineering. Trends Biotechnol. 17, 53–60 (1999)
28. Schuster, S., Fell, D.A., Dandekar, T.: A general definition of metabolic pathways useful for systematic organization and analysis of complex metabolic networks. Nat. Biotechnol. 18, 326–332 (2000)
29. Schuster, S., Hilgetag, C.: On elementary flux modes in biochemical reaction systems at steady state. Journal of Biological Systems 2(2), 165–182 (1994)
30. Schwimmer, C., Lefebvre-Legendre, L., Rak, M., Devin, A., Slonimski, P., di Rago, J.P., Rigoulet, M.: Increasing mitochondrial substrate-level phosphorylation can rescue respiratory growth of an atp synthase-deficient yeast. J. Biol. Chem. 280(35), 30751–30759 (2005)
31. Silva, J.P.M., Sakallah, K.A.: GRASP - a new search algorithm for satisfiability. In: Proceedings of ICCAD, pp. 220–227 (1996)
32. American Mathematical Society (ed.) Second DIMACS implementation challenge: cliques, coloring and satisfiability, vol. 26 (1996)
33. Soh, T., Inoue, K.: Identifying necessary reactions in metabolic pathways by minimal model generation. In: ECAI, pp. 277–282 (2010)
34. Soh, T., Inoue, K., Baba, T., Takada, T., Shiroishi, T.: Predicting gene knockout effects by minimal pathway enumeration. In: BIOTECHNO 2012 : The Fourth International Conference on Bioinformatics, Biocomputational Systems and Biotechnologies, pp. 11–19 (2012)
35. Stelling, J., Klamt, S., Bettenbrock, K., Schuster, S., Gilles, E.D.: Metabolic network structure determines key aspect of functionnality and regulation. Nature 420, 190–193 (2002)
36. Terzer, M., Stelling, J.: Large-scale computation of elementary flux modes with bit pattern trees. Bioinformatics 24(19), 2229–2235 (2008)
37. Tiwari, A., Talcott, C., Knapp, M., Lincoln, P., Laderoute, K.: Analyzing pathways using SAT-based approaches. In: Anai, H., Horimoto, K., Kutsia, T. (eds.) AB 2007. LNCS, vol. 4545, pp. 155–169. Springer, Heidelberg (2007)
38. von Kamp, A., Klamt, S.: Enumeration of smallest intervention strategies in genome-scale metabolic networks. PLoS Comput Biol 10(1), e1003378 (2014)
39. von Kamp, A., Schuster, S.: Metatool 5.0: Fast and flexible elementary modes analysis. Bioinformatics 22, 1930–1931 (2006)
40. Wieringa, S.: Incremental Satisfiability Solving and its Applications. PhD thesis, Aalto University (2014)

Model Integration and Crosstalk Analysis
of Logical Regulatory Networks

Kirsten Thobe[1,2], Adam Streck[1], Hannes Klarner[1], and Heike Siebert[1,2]

[1] Freie Universität Berlin, Berlin, Germany
[2] International Max Planck Research School for Computational Biology and Scientific
Computing, Max Planck Institute for Molecular Genetics, Berlin, Germany
`Kirsten.Thobe@fu-berlin.de`

Abstract. Methods for model integration have become increasingly
popular for understanding of the interplay between biological processes.
In this work, we introduce an approach for coupling models taking
uncertainties concerning the crosstalk into account. Using constraint-
based modeling and formal verification techniques, a pool of possible
integrated models is generated in agreement with previously validated
behavior of the isolated models as well as additional experimental observations. Correlation- and causality-based analysis allows us to uncover
the importance of particular crosstalk connections for specific functionalities leading to new biological insights and starting points for experimental design. We illustrate our approach studying crosstalk between
the MAPK and mTor signaling pathways.

Keywords: Systems Biology, Logical Modeling, Crosstalk Analysis.

1 Introduction

With rapidly growing technical progress and more accessibility of experimental data, more and more models are built to decipher the mechanisms controlling cellular processes. Often, these models are focused on capturing a specific
local observation. To obtain a more global understanding, techniques for integrating validated models to more comprehensive systems are of interest [14].

In this article, we address this issue in the context of Boolean network modeling, which has been shown to yield meaningful results for systems biology
applications (see e.g. [17] and references therein). Here, finite parameter and
state spaces allow for a top-down modeling approach considering sets of models in agreement with the available data rather than having to exclude viable
models on the basis of unsupported assumptions. This is a key feature in
our model integration procedure allowing us to evaluate different possible
crosstalk connections between models and thus distinguishing our approach
from coupling methods via predefined crosstalk such as in [10].

Starting with two or more validated models describing systems known to
be connected within the organism, we construct the basis for an integrated

P. Mendes et al. (Eds.): CMSB 2014, LNBI 8859, pp. 32–44, 2014.

model by introducing crosstalk interactions between components belonging to different isolated models. These are labeled according to the available data as, e.g., activating or inhibiting, but also possibly as not necessarily having an observable effect. A pool of models in agreement with these constraints is then generated and further reduced to obtain models satisfying a list of desired properties. First, we filter for the models that preserve the validated behavior of the isolated models. Second, we incorporate new experimental data pertaining to the integrated models. This can be implemented using formal verification techniques such as presented in, e.g., [15] and [6]. Analysis of the model pool should then focus on elucidating commonalities and differences between the remaining models in the pool. Here, we propose a statistical analysis focusing on topological characteristics of the networks.

Our approach is illustrated on two signaling pathways, MAPK and mTor, which are known to be connected, but the details are still unclear. Of particular interest is to uncover the crosstalk in cancer cell lines. To tackle this problem we explicitly address how to translate the specificities of such cell lines into constraints to adapt the underlying generic models. Our results, although obtained from strongly simplified models, are in agreement with experimental findings, provide biological insights and may give rise for experimental design.

We start by providing the terminology used within the Boolean formalism and then present our general approach for model integration and crosstalk analysis. The following two sections introduce our statistical pool analysis method and the approach for incorporating genotype information as, e.g., specifics of particular cancer cell lines. The concepts are then illustrated on the MAPK-mTor pathway.

2   Background

Due to limitation of space, a formal description of the formalism is given here and the application is illustrated in Section 6 (for detailed description see [9]). Throughout the paper, we consider Boolean network models where components can only adopt the values 0 or 1. We capture a regulatory network as a directed graph $\mathcal{R} = (V, E, l)$, where $V = \{1, \ldots, n\}$ is a set of components, $E \subseteq V \times V$ is a set of interactions and $l : E \to L$ is an edge labeling where L is a set of well-formed formulas with variables $+$ and $-$.

The dynamical behavior of such a network is described via so-called *logical parameters* that must be explicitly specified. Denote $v^- = \{u \in V \mid (u, v) \in E\}$ the set of *regulators* of $v \in V$. For each $v \in V$ we then need to specify the *parametrization function* $K_v : 2^{v^-} \to \mathbb{B} = \{0, 1\}$. This function defines the value the component v evolves to based on what regulators are currently active.

The parametrization functions need to be consistent with the edge labels given by l [9]. For each $v \in V$, we say that K_v is a solution of the edge labeling *iff* the $l(u, v)$ evaluates to *true* for each $u \in v^-$. Here, we say that the variable $+$ adopts the value *true* for an edge (u, v) *iff* there exists $\omega \subseteq v^-$ such that $K_v(\omega) = 1$ and $K_v(\omega - \{u\}) = 0$. Likewise, $-$ is interpreted as *true* for an edge (u, v) iff there exists $\omega \subseteq v^-$ such that $K_v(\omega) = 0$ and $K_v(\omega - \{u\}) = 1$. The parametrization $K = (K_v)_{v \in V}$ is a solution to l if K_v is a solution to l for each $v \in V$. The set of all K that are solution to l is called the *model pool*, denoted $\mathcal{K}(V, E, l)$.

Having a parametrization K we can describe the complete dynamical behavior of the network $\mathcal{R}$ as a state transition graph (STG) $\mathcal{R}(K) = (S, \rightarrow)$. This is again a directed graph where $S = \prod_{v \in V} \mathbb{B}$ is the set of states and $\rightarrow \subseteq S \times S$ is a transition relation. To define the state transitions, we employ the asynchronous update schedule [16] where only one component can change its value at a time, often yielding more realistic trajectories than synchronous update. For a state $s = (s_1, \ldots, s_v, \ldots, s_n)$ denote $\bar{s}^v = (s_1, \ldots, \neg s_v, \ldots, s_n)$ the state which differs from s in the value of the component v. The transition relation is then obtained as follows: $s \rightarrow \bar{s}^v \iff K_v(\{u \in v^- \mid s_u = 1\}) \neq s_v$. In the STG, a strongly connected set of states that cannot be left by any trajectory is called an *attractor*. In case this set consists of a single state, we call it a *fixpoint* or steady state, otherwise a *cyclic attractor*.

3   Model Integration and Crosstalk Analysis

We developed our approach to model coupling with different requirements in mind. First, the method should allow to decide which characteristics of the original models should be preserved in the integrated model. Second, we wanted to be able to handle uncertainty w.r.t. to the crosstalk connections between the original models. Lastly, the constraints posed by the original characteristics as well as experimental observations pertaining the integrated system should be exploited to obtain a clearer understanding of the crosstalk, possibly linking particular edges to specific functionalities of the integrated system.

We have implemented these ideas in a four step procedure illustrated in Fig. 1. In the following we give a general description of all steps. While here we restrict ourselves to the case of coupling two models, extension to several models is straightforward.

1. Single Model Analysis. We start out with two networks $\mathcal{R}^1 = (V^1, E^1, l^1)$ and $\mathcal{R}^2 = (V^2, E^2, l^2)$ with their parametrizations K^1 and K^2, that we assume to be validated and analyzed w.r.t. some features of interest. For each model, we decide on a set $\mathcal{P}$ of properties to be preserved by an integrated model. Here, both structural properties such involvement of components in feedback circuits and dynamical properties such as attractor characteristics or input-output behavior can be considered. In application, these properties should describe experimentally validated behavior or characteristics of the systems that should not be lost when combining the models.

2. Model Integration. Model integration is accomplished in two steps: first, the regulatory graphs are combined to one network by merging identical components and adding crosstalk edges, and in a second step the model pool comprising all possible parametrizations consistent with this network is generated.

We define the coupled network $\mathcal{R} = (V, E, l)$ in the following way. The component set V is given by $V^1 \cup V^2$, where we assume that vertices that represent the same biological component coincide in both original models, i.e., they are merged within the integrated model. Dependencies and regulations within the single networks are kept and additionally new regulations between components of the uncoupled networks are introduced, so $E = E^1 \cup E^2 \cup E^{new}$ with $E^{new} \subseteq V \times V \setminus ((V^1 \times V^1) \cup (V^2 \times V^2))$ the so-called crosstalk. Lastly, denote l^{new} the labeling of the edges from E^{new}. The labeling $l : E \rightarrow L$ of the integrated network is defined as:

$$
l(u,v) = \begin{cases}
(l^1(u,v)) \vee (l^2(u,v)) & \text{if } (u,v) \in E^1 \cap E^2 \\
l^1(u,v) & \text{if } (u,v) \in E^1 \setminus E^2 \\
l^2(u,v) & \text{if } (u,v) \in E^2 \setminus E^1 \\
l^{new}(u,v) & \text{otherwise}
\end{cases}.
$$

To generate the model pool, we keep the parametrizations K_v of the single models for those components v that are not influenced by new edges resulting from the model integration or crosstalk inclusion. For all other component we consider all parametrizations in agreement with the edge constraints.

In general, the integrated model will have a higher dimensional state space than the single models. In order to interpret the properties $\mathcal{P}$ within this new context, we may have to translate them into this new setting. This might not necessarily be straightforward. For example, consider that an attractor A^1 of the model $\mathcal{R}^1$ should be preserved. One possibility would be to demand that A^1 is the projection of some attractor of the integrated model. A weaker condition would be that A^1 is an attractor of the state transition graph derived from projection from the state transition graph of $\mathcal{R}$. Also it is important to consider whether a property is an observed behavior for the elements of the single model in context of the joint system or might be an artifact of the isolated model. In application, the decision on how to translate the properties might be supported by biological knowledge or reasonable assumptions.

3. Model Pool Refinement. The model pool generated in the second step contains all models consistent with the integrated network $\mathcal{R}$. In general, many of these models will not satisfy the properties $\mathcal{P}$. Therefore, the pool is filtered for these properties using a suitable method. Structural properties can be validated using graph algorithms, which in some cases might also be useful to test dynamical properties encoded in the state transition graph. For the latter, formal verification approaches have also proved useful as shown e.g. in [2]. In our work we describe dynamical properties as formulas in Computation Tree Logic (CTL) [1]. A parametrized regulatory network is then consistent with a property if and only if its transition system is a model of the respective formula, which can be checked efficiently with available tools [15]. For $\mathcal{P}$ comprised of such dynamical properties, we formally create the refined model pool $\mathcal{K}(V, E, l, \mathcal{P})$ where $K \in \mathcal{K}(V, E, l, \mathcal{P}) \iff \forall P \in \mathcal{P} : \mathcal{R}(K) \models P$. More detail on this is provided in Sec. 6.2.

In addition, data available for the integrated model can now also be exploited to refine the pool further using the same methods. Here, experimental observations need to be translated into formal properties as can easily be done for, e.g., time series data [9,15].

4. Model Pool Analysis. Our overall aim in this approach is to gain new insight about the underlying biology by analyzing the model pool. Properties shared by all models can be viewed as strongly supported by the available data while distinguishing features highlight different possibilities of implementing biological functionalities and can be exploited for experimental design.

Depending on the analysis focus different structural and dynamical properties can be used for classification of the pool models, e.g., network connectivity or input-output behavior. The resulting classes can be further analyzed and evaluated. For the crosstalk analysis meaningful information can be gained by, e.g., considering the models with the minimal number of functional crosstalk edges and by correlating crosstalk edges with newly emerging behavior. Again, there are different strategies to investigate such questions, one of which we will introduce in more detail in the next section.

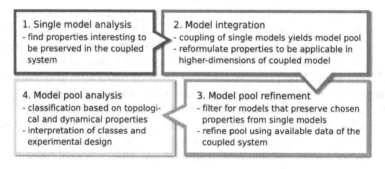

Fig. 1. Overview about the model integration and crosstalk analysis approach

4   Statistical Analysis of Large Model Pools

Even after incorporating numerous constraints, the resulting set $\mathcal{K}$ may be too large for manual analysis. In that case we propose to employ statistical methods that capture the nature of the set. Here, we focus mainly on computing correlations between the state of the system and kinetic parameters of the individual component to evaluate the effect of regulations. Note that the parametrization function describes a causal relationship—the dynamical behavior of a component is implied by the state of the system. For each pair $(u, v) \in E$ and a parametrization K we therefore compute the impact of u on v as the correlation between the current value of u and K_v.

Formally, for each $K \in \mathcal{K}$ we define the impact function $imp_K : E \to [-1, 1] \subset \mathbb{R}$ as $imp_K(u, v) = corr((s_u)_{\{s \in S\}}, (K_v(s))_{\{s \in S\}})$ where $corr$ is the

Pearson product-moment correlation coefficient. Note that $imp_K(u, v) = 0$ is equivalent with $l(u, v) = (\neg +) \wedge (\neg -)$, meaning the edge is non-functional.

This notion can be easily extended to parametrization sets by employing the mean. Formally we create an extended impact function $imp_K : E \rightarrow [-1, 1]$ as $imp_K(u, v) = \frac{\sum_{K \in \mathcal{K}} imp_K(u,v)}{|\mathcal{K}|}$. Lastly we are also interested in how often an edge is functional in the resulting set, which we describe by the *frequency* function $freq_K : E \rightarrow [0, 1]$ defined as $freq_K(u, v) = \frac{|\{K \in \mathcal{K} | imp_K(u,v) \neq 0\}|}{|\mathcal{K}|}$.

5   Incorporating Genotype Information

Often processes in cells with changes in the genotype are of interest, since they show abnormal behavior leading to diseases, such as cancer. Often generic rather than cell line specific models are build. Under the assumption that the more specific are derived from the generic models by adapting node parameter values or edge labels, two scenarios can be directly implemented in our coupling approach.

Component mutations resulting in knock-outs or overexpression, can be modeled by requiring their value to remain constantly at 0 resp. 1 along all considered system trajectories. This can be phrased as additional property for model refinement in the third step in Sec. 3, which is added to each constraint derived from the experimental data used for filtering. This ensures that the observed behavior is tested under the conditions imposed by the mutation.

Mutations can also alter the character of interactions, i.e., affecting the edges in the model. In case a crosstalk edge is targeted by a mutation, we can find this as result of the analysis provided that there is meaningful data. However, if an edge from a single model is lost or a new edge is gained the information must be directly included on the level of the edge constraints in the single models, see Step 1 in Sec. 3.

6   Application on MAPK-mTor Crosstalk

In the following, we illustrate the method on the MAPK and the mTor pathways since they are known to be connected via crosstalk, but the exact information about interactions are sparse and unclear [11]. Mutations in these pathways are very prominent in tumors, motivating research for medical purposes. Several comprehensive logical models are available used for studying input-output behavior [5,12]. Since we aim at a more complex analysis under uncertainty w.r.t. the crosstalk connections, we focus for our illustration on a very much reduced representation of both MAPK and mTor networks which is still able to reproduce the essential pathway behavior.

The single models are built based on literature information. The MAPK model is extracted from Kholodenko et al. [8], the mTor model from Engelman et al. [3]. The resulting interaction graph is shown in Fig. 2 A, along with the logical rules governing the component behavior in B. Here, we only employ the strictest edge labels $+ \wedge \neg-$ for activating or $\neg + \wedge-$ for inhibiting edges, so that the logical rules immediately imply the edge constraints and parametrization of the components resulting in K^{MAPK} and K^{mTor}.

6.1   Single Model Analysis

Both systems exhibit a fixpoint representing a quiescent stable state and a cyclic attractor. Oscillations in agreement with the cyclic attractors were experimentally shown for the components Erk and Akt in [7], the quiescent stable state represents the behavior of the inactive pathway. Both are biologically relevant and the corresponding behavior should be preserved in an integrated model. The following properties thus make up the set $\mathcal{P}$:

MAPK P_1: $\exists$ fixpoint in $\mathcal{S}^{MAPK}$ with (RTK=0, Raf=0, Mek=0, Erk=0)

   P_2: $\exists$ attractor in $\mathcal{S}^{MAPK}$ with RTK=1, s.t. Raf, Mek and Erk oscillate

mTor P_3: $\exists$ fixpoint in $\mathcal{S}^{mTor}$ with (RTK=0, PI3K=0, Akt=0, Tsc=1, mTorC1=0)

   P_4: $\exists$ attractor in $\mathcal{S}^{mTor}$ with RTK=1, s.t. PI3K, Akt, Tsc and mTorC1 oscillate

6.2   Generating and Refining the Model Pool

According to the second step in Sec. 3 we integrate the MAPK and mTor model by combining components and edges of both models. Here, the component RTK featuring in both models is merged to one component, since both pathways are activated by this receptor. Lastly, the crosstalk edges are added. We selected 5 possible crosstalk connections from the literature, which are given in Fig. 2 C. Based on the available biological information, they are labeled with $\neg+$ and $\neg-$, respectively. This translates to edges being either inhibiting resp. activating or not functional.

   The parametrizations of components not targeted by any crosstalk edge are determined by the single models. The other components, namely RTK, Raf, PI3K, Tsc and mTorC1, have new regulatory contexts. For them, we consider all parametrizations in agreement with the edge labels, as defined in Sec. 2, leading to a model pool $\mathcal{K}(V, E, I, \mathcal{P})$ of size 13,266.

   Now, the properties $P_1 - P_4$ observed in the single models need to be transferred to the dimension of the coupled system:

- P_1 and P_3 both characterize the steady state without stimulus and are fused to one property: $FP1$: $\exists$ fixpoint in $\mathcal{S}$ with (RTK=0, Raf=0, Mek=0, Erk=0, PI3K=0, Akt=0, Tsc=1, mTorC1=0).
- P_2 and P_4 both describe cyclic attractors, assumed to be preserved in MAPK and mTor components as 2 distinct attractors in the state space:

   $P_2 \rightarrow Cyc.MAPK$: $\exists$ attractor in $\mathcal{S}$ where Raf, Mek and Erk oscillate,

   $P_4 \rightarrow Cyc.mTor$: $\exists$ attractor in $\mathcal{S}$ where Akt and mTor oscillate.

To obtain the set $\mathcal{K}$ of parametrizations satisfying the properties in $\mathcal{P}$ we use an appropriate model checking tool [9]. The exact specification is given in Tab. 1 B. Here, $FP1$ is chosen to have the strictest verification criterion w.r.t. the choice of initial states (see Tab. 1 D), since the quiescent state should be reached for all states with inactive receptor, while we allow for the cyclic attractors to emerge dependent on the initial value of the internal components. After filtering, only 2263 models remain that all share the validated asymptotic behavior of the single models.

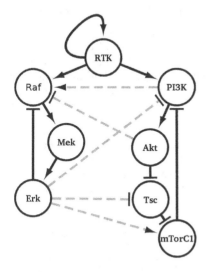

B

MAPK: RTK = RTK,
         Raf = RTK ∧ ¬ Erk,
         Mek = Raf, Erk = Mek
mTor: RTK = RTK, mTorC1 = ¬ Tsc,
         PI3K = RTK ∧ ¬ mTorC1,
         Akt = PI3K, Tsc = ¬ Akt

C

Interaction	Reference
PI3K activates Raf	[18]
Akt inhibits Raf	[11]
Erk inhibits PI3K	[11]
Erk inhibits Tsc	[19]
Erk activates mTorC1	[11]

Fig. 2. Model setup for crosstalk analysis. **A** Network structure of MAPK and mTor in black and crosstalk edges in dashed green lines. **B** Logical rules for regulations of components. **C** List of crosstalk edges added to the single pathways.

To reduce the pool further, experimental data from literature containing information about the integrated system is used. A study from Will et al. investigated the effect of Akt and PI3K inhibitors in connection with MAPK signaling in a breast cancer cell line [18]. The genotype of the cell line needs to be considered when exploiting cancer data. This specific cell line (BT-474) carries an amplification in HER2, which belongs to the RTK family and a mutation in PI3K (PIK3CA) which causes increased levels of activity. As described in Sec. 5, we add the genotype information which amounts to adding the `fixed component` constraint (see Tab. 1 D) to the properties derived from the corresponding experimental observations, fixing RTK and PI3K to value 1 unless explicitly indicated otherwise in the experimental set-up.

In the paper, time series experiments without inhibitor are performed (see Figure S2B in [18]), indicating that the quiescence state shows active Akt (P-Akt) and active Erk (P-Erk). We can translate this observation into CTL formula $FP2$ shown in Tab. 1 C. Moreover, Will et al. performed measurements perturbing with Akt inhibitor MK2206 and PI3K inhibitor BAY 80-694 hypothesizing that PI3K is upstream of MAPK and blocking this kinase should affect Erk activity. We used these western blots (shown in Fig. 2 in [18]) to further refine the model pool. In order to avoid discretization errors, time points with unambiguously active or inactive states are chosen, discretized and collected in Tab. 1 A. The table shows the states of Akt, Erk and P-S6, which is a kinase dependent on mTorC1 and therefore used as its read-out. For the PI3K inhibitor 3 measurements and for the Akt inhibitor 4 measurements are implemented as CTL formulas *BAY* and *MK*, listed in Tab. 1. After filtering, the resulting pool contains 240 models that are in agreement with the properties derived from the single networks and the experimentally observed behavior.

Table 1. Filtering model pool using model checking. **A** Table with discretized western blot data from Will et al. [18] for PI3K inhibitor Bay 80-6946 and Akt inhibitor MK2206. **B** CTL formulas for properties FP1, Cyc.MAPK and Cyc.mTor. **C** CTL formulas derived from western blot data. **D** Description of CTL operators and verification options.

A

	BAY			MK2206			
Time [h]	0	0.5	2	0	0.5	2	8
P-Akt	1	0	0	1	0	0	0
P-S6	1	1	0	1	1	0	1
P-Erk	1	0	1	1	1	1	1

B

```
FP1: CTL: EF(Delta=0&Raf=0&Mek=0&Erk=0&mTorC1=0)
     Initial states: RTK=0; Verification: ForAll
Cyc.MAPK: CTL: EF(AG(EF(deltaErk!=0)))
     Initial states: RTK=1; Verification: ForSome
Cyc.mTor:
     CTL: EF(AG(EF(deltamTorC1!=0)&EF(deltaAkt!=0))))
     Initial states: RTK=1; Verification: ForSome
```

C

```
FP2: CTL: EF(Delta=0&Erk=1&Akt=1); Initial states: RTK=1
     Fixed components: RTK=1,PI3K=1; Verification: ForAll
MK:  CTL: EF(mTorC1=1&Erk=1&EF(mTorC1=0&Erk=1&EF(mTorC1=1&Erk=1)))
     Initial states: mTorC1=1,Erk=1
     Verification: ForAll; Fixed components: Akt=0, RTK=1, PI3K=1
BAY: CTL: EF(Akt=0&mTorC1=1&Erk=0&EF(Akt=0&mTorC1=0&Erk=1))
     Initial states: Akt=1,mTorC1=1,Erk=1
     Verification: ForAll; Fixed components: RTK=1, PI3K=0
```

D

EF(X): is a CTL operator *exists finally*. This states that on some path from an initial state the X holds true at some point.

AG(X): is a CTL operator *all globally*. This states that on every successor of this state, X holds true.

Delta=0: states that no change is possible, i.e. we are in a steady state.

v=b: where $v \in V, b \in \mathbb{B}$ states that value of a component v is set to b.

Initial state: is a list of boolean constraints on the values of the components. A state is considered initial if all the constraints are satisfied.

Verification: specifies the verification strictness. We can either decide that it is sufficient if the property is satisfiable (ForSome), meaning there exists some initial state where it holds. Or we can require that the property is valid (ForAll), meaning it holds in all the initial states.

Fixed component: constrains the listed components to the assigned values for the whole path. This property allows for modeling knock-outs and stimuli.

6.3   Model Pool Analysis

We employ the statistical analysis approach presented in Sec. 4 to investigate the topological characteristics of the integrated models. For identifying important influences and structures in the model pool, the frequency of the edges and correlations between the components is calculated, depicted as edge width and edge color, respectively. In Fig. 3 the statistical analysis of both the refined model pool $\mathcal{K}_{\mathcal{R}}$ resulting after the filtering as well as a reference pool $\mathcal{K}_{ref}$ and the difference between the two is visualized. The reference pool contains all models of the originally generated pool $\mathcal{K}(V, E, l, \mathcal{P})$ that have been discarded in the filtering process. The result for $\mathcal{K}_{\mathcal{R}}$ is shown in graph A, where the crosstalk edge from PI3K to Raf has a frequency of 1. Thus, in every model in the pool this edge is functional. In order to evaluate possible enrichments of other edges, we need information about $\mathcal{K}_{ref}$, shown in C. Here, the frequency and correlation is given by the edge constraints and the arising combinations of parametrizations. Finally, the difference between the filtered and the reference pool is depicted in B. Again, the connection PI3K and Raf is shown to be more prominent and highly correlated comparing the filtered models to all possible models. This result is in line with findings of Will et al., where it was concluded that PI3K is upstream of the MAPK cascade [18]. The crosstalk from Erk to PI3K is less frequent in $\mathcal{K}_{\mathcal{R}}$, which is reasonable since in the data PI3K is a fixed component and therefore this edge cannot be functional here.

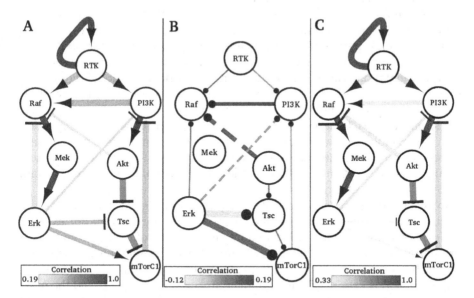

Fig. 3. Visualization of the statistical analysis of the model pool. The edge width represents the frequency of occurrence in the pool and the heads show the sign. **A** illustrates the refined pool $\mathcal{K}_{\mathcal{R}}$, and **C** the reference pool $\mathcal{K}_{ref}$. In **B** the difference graph is shown, where the edge signs has been dropped, solid lines are drawn for connections more frequent on average in $\mathcal{K}_{\mathcal{R}}$ than in $\mathcal{K}_{ref}$ and dashed lines for lower frequency in the refined pool. Figures generated using Cytoscape ($http : //www.cytoscape.org/$).

Moreover, the influence of Erk on Tsc and mTorC1 is strongly enriched in the selected model pool. Further investigation shows that every model in $\mathcal{K}_{\mathcal{R}}$ contains at least one of these edges. They are directly linked to the experimental data showing value changes in mTorC1. Further experiments could clarify which of the two crosstalk edges are actually functional in this setting.

7 Conclusions

Biological processes do not work isolated, but in concert with other cellular mechanisms. For many of these processes exist validated models, but their interactions among each other are often unclear. Here, we present a novel approach for integrating discrete models allowing for uncertainty concerning their connections and subsequent crosstalk analysis. Prior models, candidate crosstalk and constraints derived from validated properties of the original models and new experimental observations give rise to a model pool whose analysis allows to uncover essential features and provides pointers for experimental design, as illustrated with the MAPK-mTor system. Recognizing particular requirements in application, we address, on the modeling side, ways to integrate genotype information and, concerning the analysis, statistical approaches to evaluate the relevance of the crosstalk edges.

Here, we reduce the pool as much as possible and then perform the pool analysis. Another possibility would be to add constraints stepwise and perform analysis after each step. This may allow to link specific network characteristics to the properties and functionalities encoded in each constraint.

One difficulty that needs to be addressed in application is the translation of experimental data into logical constraints for a Boolean model. For checking CTL formulas, for example, the choice of verification type, i.e., whether the property needs to hold for all or only for some initial states, will generally strongly impact the results. Similar problems arise when discretization data. Often, biological knowledge allows for well-supported decisions in these matters. Otherwise, comparative analysis of different interpretations might be useful. In some cases, multi-valued models are more suited to capture certain biological aspects. Our method can easily be extended to this setting.

To make the approach accessible, we plan to develop a tool using, as a starting point, available model-checking software and extending our existing model pool analysis methods. Depending on the nature of the constraints used for filtering, integration of other verification techniques might also prove useful. The biggest challenge to be addressed is certainly scalability of the method, as the problem is exponential in both the number of components and the maximum degree of the network. For smaller models, as shown for the simplified MAPK-mTor system, comprehensive analysis is quite simple. For larger systems, more elaborate methods are needed. We plan to exploit methods yielding significant reduction while preserving meaningful properties, as e.g. in [10], as well as methods identifying core network modules governing the system behavior in subspaces of state space [13]. This will allow for a balance of efficiency and property resolution that can be tailored to the requirements of a given application. Lastly, the size of the problem can be also tackled by incorporating more involved algorithms as recently shown in [4].

References

1. Baier, C., Katoen, J.-P.: Principles of Model Checking. The MIT Press (2008)
2. Bernot, G., Comet, J.-P., Richard, A., Guespin, J.: Application of formal methods to biological regulatory networks: extending Thomas' asynchronous logical approach with temporal logic. Journal of Theoretical Biology 229(3), 339–347 (2004)
3. Courtney, K.D., Corcoran, R.B., Engelman, J.A.: The PI3K pathway as drug target in human cancer. Journal of Clinical Oncology 28(6), 1075–1083 (2010)
4. Gallet, E., Manceny, M., Gall, P.L., Ballarini, P.: Adapting LTL model checking for inferring biological parameters. In: AFADL 2014, p. 46 (2014)
5. Grieco, L., Calzone, L., Bernard-Pierrot, I., Radvanyi, F., Kahn-Perlès, B., Thieffry, D.: Integrative Modelling of the Influence of MAPK Network on Cancer Cell Fate Decision. PLoS Computational Biology 9(10), e1003286 (2013)
6. Guziolowski, C., Videla, S., Eduati, F., Thiele, S., Cokelaer, T., Siegel, A., Saez-Rodriguez, J.: Exhaustively characterizing feasible logic models of a signaling network using Answer Set Programming. Bioinformatics 29(18), 2320–2326 (2013)
7. Hu, H., Goltsov, A., Bown, J.L., Sims, A.H., Langdon, S.P., Harrison, D.J., Faratian, D.: Feedforward and feedback regulation of the MAPK and PI3K oscillatory circuit in breast cancer. Cellular Signalling 25(1), 26–32 (2013)
8. Kholodenko, B.N.: Negative feedback and ultrasensitivity can bring about oscillations in the mitogen-activated protein kinase cascades. European Journal of Biochemistry 267(6), 1583–1588 (2000)
9. Klarner, H., Streck, A., Šafránek, D., Kolčák, J., Siebert, H.: Parameter Identification and Model Ranking of Thomas Networks. In: Gilbert, D., Heiner, M. (eds.) CMSB 2012. LNCS, vol. 7605, pp. 207–226. Springer, Heidelberg (2012)
10. Mendes, N.D., Lang, F., Le Cornec, Y.-S., Mateescu, R., Batt, G., Chaouiya, C.: Composition and abstraction of logical regulatory modules: application to multicellular systems. Bioinformatics 29(6), 749–757 (2013)
11. Mendoza, M.C., Er, E.E., Blenis, J.: The Ras-ERK and PI3K-mTOR pathways: crosstalk and compensation. Trends in Biochemical Sciences 36(6), 320–328 (2011)
12. Samaga, R., Saez-Rodriguez, J., Alexopoulos, L.G., Sorger, P.K., Klamt, S.: The logic of EGFR/ErbB signaling: theoretical properties and analysis of high-throughput data. PLoS Computational Biology 5(8), e1000438 (2009)
13. Siebert, H.: Analysis of discrete bioregulatory networks using symbolic steady states. Bulletin of Mathematical Biology 73(4), 873–898 (2011)
14. Stelling, J., Mendes, P., Tonin, F., Klipp, E., Zecchina, R., Heinemann, M., Przulj, N., Wodke, J., Stoma, S., Kaltenbach, H.: et al. Defining modeling strategies for systems biology. In: Technical report, FutureSysBio Workshop (2011)
15. Streck, A., Kolcák, J., Siebert, H., Šafránek, D.: Esther: Introducing an online platform for parameter identification of boolean networks. In: Gupta, A., Henzinger, T.A. (eds.) CMSB 2013. LNCS, vol. 8130, pp. 257–258. Springer, Heidelberg (2013)
16. Thomas, R.: Regulatory networks seen as asynchronous automata: a logical description. Journal of Theoretical Biology 153(1), 1–23 (1991)
17. Wang, R.-S., Saadatpour, A., Albert, R.: Boolean modeling in systems biology: an overview of methodology and applications. Physical Biology 9(5), 055001 (2012)

18. Will, M., Qin, A.C.R., Toy, W., Yao, Z., Rodrik-Outmezguine, V., Schneider, C., Huang, X., Monian, P., Jiang, X., De Stanchina, E., et al.: Rapid induction of apoptosis by PI3K inhibitors is dependent upon their transient inhibition of RAS-ERK signaling. Cancer discovery CD-13 (2014)
19. Winter, J.N., Jefferson, L.S., Kimball, S.R.: ERK and Akt signaling pathways function through parallel mechanisms to promote mTORC1 signaling. American Journal of Physiology - Cell Physiology 300(5), C1172–C1180 (2011)

A   Author's Contributions

KT designed and implemented the method, constructed and analyzed the biological example, and wrote the manuscript; AS designed, implemented and performed statistical analysis, and participated writing the manuscript; HK participated designing and implementing the method; HS initiated and supervised the development of methods, implementation, analysis, and writing the manuscript.

Improved Parameter Estimation in Kinetic Models: Selection and Tuning of Regularization Methods

Attila Gábor and Julio R. Banga

BioProcess Engineering Group, IIM-CSIC, Vigo, Spain
{attila.gabor,julio}@iim.csic.es

Abstract. Kinetic models are being increasingly used as a systematic framework to understand function in biological systems. Calibration of these nonlinear dynamic models remains challenging due to the nonconvexity and ill-conditioning of the associated inverse problems. Nonconvexity can be dealt with suitable global optimization. Here, we focus on simultaneously dealing with ill-conditioning by making use of proper regularization methods. Regularized calibrations ensure the best trade-offs between bias and variance, thus reducing over-fitting. We present a critical comparison of several methods, and guidelines for properly tuning them. The performance of this procedure and its advantages are illustrated with a well known benchmark problem considering several scenarios of data availability and measurement noise.

Keywords: Dynamic models, parameter estimation, Tikhonov regularization, regularization tuning.

1 Introduction

Dynamic mathematical models (i.e. kinetic models) are central in systems biology as a way to understand the function of biological systems [16], to generate new hypotheses, and to identify possible ways of intervention, especially in metabolic engineering [1]. Recent efforts are focused on the development and exploitation of large-scale kinetic models [28].

Parameter estimation aims to find the unknown parameters of the model which give the best fit to a set of experimental data. Parameter estimation belongs to the class of so called inverse problems, where it is important to include both, a priori (i.e. structural) and a posteriori (i.e. practical) parameter identifiability studies. In this way, parameters which cannot be measured directly will be determined in order to ensure the best fit of the model with the experimental results. This will be done by globally minimizing an objective function which measures the quality of the fit.

Global optimization methods must be used in order to avoid convergence to local solutions [2,3]. However, we also need to deal with the typical ill-conditioning of these problems [14], arising from (i) models with large number of parameters,

P. Mendes et al. (Eds.): CMSB 2014, LNBI 8859, pp. 45–60, 2014.

(ii) experimental data scarcity and (iii) significant measurement errors. As a consequence, we often obtain over-fitting of such kinetic models, i.e. calibrated models with reasonable fits to the available data but poor capability for generalization (low predictive value).

Regularization methods have a rather long history in inverse problems [9] as a way to surmount ill-posedness and ill-conditioning. The regularization process introduces additional information, usually by penalizing model complexity and/or wild behaviour. It also has links with Bayesian estimation in the sense that it can be regarded as a way of introducing prior knowledge about the parameters. It has been mainly used in fields dealing with estimation in distributed parameter systems, such as tomography (with applications in geophysics and medicine) and other image reconstruction techniques. Recently, it has enjoyed success in machine learning, gaining attention from the systems identification area [17].

However, the use of regularization in systems biology has been marginal [8], especially regarding kinetic models. Bansal et. al [4] compared Tikhonov and truncated singular value decomposition regularization for the linear regression model of green fluorescent protein reporter systems to recover transcription signals from noisy intensity measurements. Wang and Wang [31] presented a two stage Bregman regularization method for parameter estimations in metabolic networks. A clear conclusion from these studies is that for nonlinear inverse problems, there is no general recipe for the selection of regularization method and its tuning. Further, it is known that even for linear systems, choosing a method from the plethora of existing techniques is nontrivial [6].

Here we present a critical comparison of a wide range of regularization methods applicable to nonlinear kinetic models. Further, we detail a procedure with guidelines for regularization method selection and tuning. Finally, we use numerical experiments with a challenging benchmark problem to illustrate the usage and benefits of regularization.

2   Parameter Estimation in Dynamic Models

We consider kinetic models given by arbitrary nonlinear ordinary differential equations (ODEs) formulated as

$$\frac{dx(t,\theta)}{dt} = f(u(t), x(t,\theta), \theta), \quad y(x,\theta) = g(x(t,\theta),\theta),$$
$$x(t_0) = x_0(\theta), \quad t \in [t_0, t_f] \ ,$$

(1)

where the dynamics of the states $x \in \mathcal{R}_+^{n_x}$ are determined by the vectorfield $f(\cdot)$, $\theta \in \mathcal{P} \subset \mathcal{R}^{n_\theta}$ is the vector of model parameters (e.g. Hill-coefficients, reaction rate coefficients, Michaelis-Menten parameters, etc.), $u(t) \in \mathcal{R}^{n_u}$ denotes the time dependent stimuli, and $\tilde{y}(t)$ are measured values of the observed variables $y(x(t),\theta) \in \mathcal{R}^{n_y}$. The latter are related with the dynamic states via the observation function $g(x,\theta)$.

The parameter estimation problem is usually formulated as the maximization of the likelihood function. The measurement of the j-th observed quantity, taken at time t_i in the k-th experiment is assumed to be contaminated by random measurement error distributed according to the normal distribution, i.e. $\tilde{y}_{ijk} = y_{ijk}(x(t_i), \theta) + \epsilon_{ijk}$ and the error term $\epsilon_{ijk} \sim \mathcal{N}(0, \sigma_{ijk}^2)$, where σ_{ijk}^2 is the error variance. Then, the maximization of the likelihood function leads to the minimization of the weighted least squares cost function [30].

$$Q_{\mathrm{LS}}(\theta) = \frac{1}{2} \sum_{k=1}^{N_e} \sum_{j=1}^{N_{y,k}} \sum_{i=1}^{N_{t,k,j}} \left(\frac{y_{ijk}(x(t_i,\theta),\theta) - \tilde{y}_{ijk}}{\sigma_{ijk}} \right)^2 = \frac{1}{2} R(\theta)^T R(\theta) , \qquad (2)$$

where N_e is the number of experiments, $N_{y,k}$ is the number of observed compounds in the k-th experiment, $N_{t,k,j}$ is the number of measurement time points of the j-th observed quantity in the k-th experiment, and $R(\theta)$ is the normalized residual vector.

2.1 Optimization Method

It is well known that the cost function (2) can be highly nonlinear and nonconvex in the model parameters, so one should use global optimization in order to avoid local optima. However, the current state of the art in global optimization for this class of problems is still somewhat unsatisfactory. Deterministic global optimization methods [22,18] can guarantee global optimality but their computationally cost increases exponentially with the number of estimated parameters. Thus, stochastic methods [19], or meta-heuristic approaches [26] are better alternatives, given adequate solutions in reasonable time, although at the price of no guarantees.

Here, we have used a global-local hybrid metaheuristic which combines scatter search [25] with the very efficient adaptive nonlinear least squares algorithm NL2SOL [7]. In order to further increase the convergence rate of NL2SOL, the Jacobian of the normalised residual vector is computed based on the solution of the forward sensitivity equations corresponding to (1) via the SUNDIALS CVODES [11] software package.

3 Regularization Methods

Here we consider general family of penalty type regularization methods, which incorporate a term $\Gamma(\theta)$ in the optimization cost function

$$Q_{\mathrm{Reg}}(\theta) = Q_{\mathrm{LS}}(\theta) + \alpha \Gamma(\theta) . \qquad (3)$$

Specific methods differ in the form of the penalty; e.g. for Tikhonov regularization $\Gamma_{\mathrm{T}}(\theta) = ||W\theta||^2$, where $W \in \mathcal{R}^{n_\theta \times n_\theta}$ is a weighting matrix; for Bregman regularization [31] $\Gamma_{\mathrm{B}}(\theta) = \sum_{i=1}^{n_\theta} \theta_i \log(\theta_i)$; for LASSO regularization [29] $\Gamma_{\mathrm{L}}(\theta) = \sum_{i=1}^{n_\theta} |\theta_i|$; and the so-called elastic net [32] combines the Tikhonov and the LASSO regularization.

The Tikhonov regularized optimization problem can be formulated as

$$\underset{\theta}{\text{minimize}} \quad \frac{1}{2}R(\theta)^T R(\theta) + \alpha(W\theta)^T(W\theta) \tag{4}$$

$$\text{subject to } \theta \in \mathcal{P}, \text{ Eqs. (1)}.$$

Since (4) is still a *nonlinear least squares* problem, the above mentioned optimization procedure, with NL2SOL as local method, is still fully applicable. Optimization methods for the LASSO regularization has been reviewed in [27].

3.1   Regularization Parameter (Tuning Methods)

One of the crucial step in the regularization of ill-posed problems is the choice of the regularization parameter α, which balances the model fit and the regularization penalty. Recent studies [6,24] have compared more than twenty parameter choice methods for *linear* inverse problems. In our study, we consider the problem of regularization parameter selection for the nonlinear dynamic problem (4) with the Tikhonov scheme above. However, it should be noted that the methods below are general and applicable for other penalty types, and can also be used in iterative regularization procedures [12]. Note that α is a continuous variable, but below we consider the selection among the set of discrete regularization parameters $\alpha_i = \alpha_{\max} \cdot q^i$, for $0 < q < 1$, and $i = 0, 1, 2 \ldots I$.

Optimal regularization (OR). The optimal regularization minimizes the distance between the estimated parameters and the unknown model parameters, i.e. the estimation error. The expected error in the estimated parameters can be decomposed [6] as

$$\mathbb{E}||\hat{\theta}_\alpha^\epsilon - \theta||^2 = ||\hat{\theta}_\alpha^0 - \theta||^2 + \mathbb{E}||\hat{\theta}_\alpha^0 - \hat{\theta}_\alpha^\epsilon||^2, \tag{5}$$

where $\hat{\theta}_\alpha^\epsilon$ is the estimated parameter vector using noisy measurement data, α is the regularization parameter, θ is the (in general unknown) nominal parameter vector and $\hat{\theta}_\alpha^0$ is the estimated parameters from noise-free data. The first term in the right hand side is the *regularization error*, which accounts for the regularization bias and is a monotonically increasing function of α. The second, variance term is the data noise *propagated error*, which monotonically decreases with increasing α. Therefore, a minimum of the estimation error is expected for a certain α, denoted by α_{opt}. In the discretized framework, if the resolution is fine enough, the problem of finding the optimal regularization parameter is reduced to the selection of the best candidate in the set $\{\alpha_i\}$. It should be noted that OR can only be computed for synthetic problems where the true parameters are known. In other words, the direct computation of (5) is impossible in real problems, since θ and the noise-free data are unknown in practice. The OR results presented below, for the sake of comparison, could be computed because the problems considered are synthetic.

Parameter choice methods. Since in general we do not know the true parameters (as this is obviously the objective of the estimation problem), several parameter choice methods have been developed to find the optimal regularization parameter in an indirect way. Most of the methods have been developed for linear inverse problems (see [6,9,21] and the references therein) or for nonlinear problems in combination with a local (Newton-type) optimization method [12], which cannot handle the nonconvexity of the objective function. In the following part of this section we shortly summarize a selection of existing regularization parameter choice methods that can be used for nonlinear problems in combination with our global metaheuristic optimization approach. In our implementation, the regularized estimates are first obtained on the whole set of α_i for illustrative purposes. However note that most methods can be used in an iterative way, thus reducing the number of regularized solutions required. Several methods require a maximum index I_m of the regularization parameter, such that the optimal index $i_{\text{opt}} \leq I_m$. Details on how the maximum index is computed for those methods are given in Appendix B.

Discrepancy principle (DP)[20]. The discrepancy principle chooses the regularization parameter such that the observed discrepancy between the data and the model prediction is explained by the measurement error, i.e. $||y(\hat{\theta}) - \tilde{y}|| \approx ||\epsilon||$. Since the residuals are normalised (2) and thus each element of the residuals contributes equally to the cost function, the principle chooses the index $n_{\text{DP}} = i$ for which

$$Q_{LS}(\hat{\theta}_{\alpha_i}) \leq \tau N_{\text{data}} \leq Q_{LS}(\hat{\theta}_{\alpha_{i-1}}),$$

where N_{data} is the total number of data and τ is a small tuning parameter of this method. We used $\tau = 1.5$ according to [6] and also 2.0 [9], but did not find significant differences. The results below correspond to $\tau = 1.5$.

Balancing principle (BP1, BP2) [15,6]. The balancing principle chooses the regularization parameter that balances the propagated error bound $||\hat{\sigma}_R(\hat{\theta}_{\alpha_k})||$ and the regularization error. Following [6], the balancing functional is defined as $b(i) = \max\limits_{i < k \leq I_m} \frac{||\hat{\theta}_{\alpha_i} - \hat{\theta}_{\alpha_k}||}{4||\hat{\sigma}_R(\hat{\theta}_{\alpha_k})||}$, where I_m is the maximum regularization index. We considered two submethods: in BP1 the term $||\hat{\sigma}_R(\hat{\theta}_{\alpha_k})||$ was approximated by a local, sensitivity based analysis (A.1) as shown in Appendix B; in case BP2 we used parameter estimates from 4 independent datasets to approximate the standard deviation of the parameters. Then, the smooth balancing functional was computed as $B(n) = \max\limits_{n \leq k \leq I_m} b(k)$. The optimal index ($n_{B1}$ and n_{B2} for the two cases respectively) according to the balancing principle is the first index i such that $B(i) \leq \kappa$, where κ is a tuning parameter. For our test problems $\kappa = 1$.

Hardened-balancing (HB) [5]. This method is a tuning parameter free version of the balancing principle. The smooth balancing functional $B(i)$ is defined as above, but the optimal index is chosen based on the minimisation as $n_{\text{HB}} = \arg\min\limits_{0 \leq k \leq I_m} B(k)\sqrt{||\hat{\sigma}_R(\hat{\theta}_{\alpha_k})||}$, where $||\hat{\sigma}_R(\hat{\theta}_{\alpha_k})||$ is computed as (A.1).

Quasi-optimality criterion (QO) [9]. As the regularization parameter decreases, the corresponding estimated model parameters change. When the regularization parameter is large, the estimated parameters are heavily influenced by the regularization term, while for mildly regularized cases, the fit measure prevails. Quasi-optimality is achieved, when the variability of the estimated parameters is minimized, i.e. the optimal regularization index is defined as $n_{QO} = \arg\min_{0 \leq k \leq I_m} \|\hat{\theta}_{\alpha_k} - \hat{\theta}_{\alpha_{k+1}}\|$. This method showed high sensitivity to the maximum regularization parameter index I_m.

L-curve method (LC1,LC2)[10]. When $\|\hat{\theta}_{\alpha_i}\|$ is plotted against $Q_{LS}(\hat{\theta}_{\alpha_i})$ for $i = 1, 2 \ldots I$, an L-shaped curve is obtained (see Figure 1). The L-curve method chooses the corner point of the curve balancing the propagated error and the regularization error. We considered two variants: method LC1, which identifies the corner by finding the point of the L-curve that has the highest curvature. The corresponding regularization index is n_{LC1}. Method LC2 [23] finds the corner where the tangent of curve is -1, equivalently $n_{LC2} = \arg\min_{0 \leq k \leq I_m} Q_{LS}(\hat{\theta}_{\alpha_k}) \|\hat{\theta}_{\alpha_k}\|$.

Cross validation (CV_{χ^2}, CV_{RSS})[17,24]. When further data is at hand, one can evaluate the performance of the calibrated models with parameters $\hat{\theta}_{\alpha_i}$, $i = 1, 2, \ldots I$ on a second dataset that was not used for the calibration. The performance of the models is measured either by residual sum of squares RSS_{CV} or by the χ^2_{CV} defined in Appendix C. The optimal regularization parameter index is chosen as the index of the estimated parameter vector that performed the best in cross validation, i.e. the index selected by the method CV_{RSS} is $n^{CV}_{RSS} = \arg\min_{0 \leq k \leq I_m} RSS_{CV}(\hat{\theta}_{\alpha_k})$ and by the method CV_{χ^2} is $n^{CV}_{\chi^2} = \arg\min_{0 \leq k \leq I} \chi^2_{CV}(\hat{\theta}_{\alpha_k})$.

4   Numerical Experiments

4.1   Test Problems

We have constructed 45 parameter estimation problems as test cases using the three-step metabolic pathway model [13], described in Appendix A. For a given stimuli, the model was simulated using the parameters in Table A.1 and the computed trajectories were sampled. These parameters and sampled trajectories are called the *nominal parameters* and the *nominal model predictions*, respectively. Then, random noise was added to the samples that generated an experimental dataset. We considered parameter estimation problems with:

  - 3 levels of experimental data (8, 12 and 16 experiments),
  - 3 noise levels per experiment (1, 5 and 10% additive Gaussian noise),
  - 5 realizations of each scenario.

Therefore, the total number of scenarios is 45. For the cross validation based method 8 further sets of data were generated by the same procedure. The corresponding stimuli is indicated by "CV" in Table A.1. This data contains 5% error.

Each test problem is solved for a set of regularization parameters: $I = 24$ regularization parameters were chosen a-priory ranging from 10^3 to 10^{-8} equidistantly on logarithmic scale. Equivalently, $\alpha_i = 10^3 \cdot q^i$, for $i = 1, \ldots I$ and $q = 0.3325$. Altogether, this results $24 \cdot 45 = 1080$ nonlinear, nonconvex estimation problems to be solved. We have not applied any scaling in the regularization, i.e. W is the unity matrix. The set of regularization parameters was chosen this way to give a uniform base for each regularization tuning method and for illustrative purposes. As mentioned above, a careful implementation of each method could reduce the required points.

4.2   Comparison Criteria

Each tuning method selected a regularized parameter estimate $\hat{\theta}_{\alpha_m}$ by solving (4) for the whole set of α_l, $l = 1, \ldots I$ and applying the above tuning procedures. Then, the methods are compared based on well known metrics, such as the residual sum of squares $\mathrm{RSS}(\hat{\theta}_{\alpha_m})$, $\chi^2(\hat{\theta}_{\alpha_m})$ and model prediction error $\mathrm{PE}(\hat{\theta}_{\alpha_m})$ (for details see Appendix C). The inefficiency IE of a tuning method measures the estimation error EE in the chosen regularized estimate compared to the optimal regularized estimate

$$\mathrm{IE}(\hat{\theta}_{\alpha_m}) = \frac{\mathrm{EE}(\hat{\theta}_{\alpha_m})}{\mathrm{EE}(\hat{\theta}_{\alpha_{\mathrm{opt}}})} = \frac{||\theta - \hat{\theta}_{\alpha_m}||}{||\theta - \hat{\theta}_{\alpha_{\mathrm{opt}}}||} \ , \tag{6}$$

where $\hat{\theta}_{\alpha_{\mathrm{opt}}}$ is the parameter estimate based on the optimal regularization parameter α_{opt}, for which the index is $n_{\mathrm{opt}} = \arg\min_{0 \leq l \leq I} ||\theta - \hat{\theta}_{\alpha_l}||$.

5   Results

Figure 1A) shows the trade-off (3) between model fit $Q_{\mathrm{LS}}(\hat{\theta}_\alpha)$ and regularization penalty $\Gamma_{\mathrm{T}}(\hat{\theta}_\alpha)$ for a typical estimation scenario. Large regularization biases the estimation and cause large discrepancy between the model and the measured data (large Q_{LS}). As the regularization parameter decreases, the discrepancy decreases towards a lower limit, but the variability of the estimated parameters (given by the parameter norm) increases drastically. Some methods, such as the L-curve method, try to come up with an optimal trade-off between the two effects by finding the so-called knee-point of the curve (in other words, they treat the problem as a bi-criteria optimization where the L-curve is a Pareto-optimal set).

5.1   Estimation Error and Optimal Regularization

The parameter estimation error (A.6) was calculated for each of the 1080 estimation problem. Figure 2 shows this magnitude for the 5 replicates of a selected scenario (estimation using 8 datasets containing 10% noise). The error curves can be divided into three regions, as the regularization index increases (i.e., the

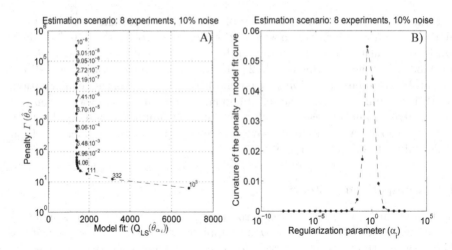

Fig. 1. Results of a regularized estimations. A) Trade-off between model fit and the penalty (3): for a given estimation scenario the regularization penalty $\Gamma_{\mathrm{T}}(\hat{\theta}_{\alpha_i})$ is plotted against the fit of the model $Q_{\mathrm{LS}}(\hat{\theta}_{\alpha_i})$ for each regularization parameter (denoted by the text next to each point), that results in an L-shape curve. Figure B) shows the curvature of the L-shape curve.

regularization parameter decreases). For large regularization parameters, the error in the estimated parameters is dominated by the regularization term (rE domain), while for small regularization parameter the noise propagated error is the main contributor (pE domain). In most cases we found that the propagated error levels off at some value $\mathrm{EE}_{\mathrm{lim}}$ as the regularization parameter reaches a certain limit α_{lim}, which varies with replicates. Below this limit, not only the error in the estimated parameters, but generally the estimated parameters themselves did not change, i.e. $\hat{\theta}_{\alpha_i} \approx \hat{\theta}_{\alpha_{i+1}}$ for all $\alpha_i < \alpha_{lim}$. The theory of inverse problems [9] shows that the regularization parameter must be larger than the smallest eigenvalue of the Hessian of the objective function (2), which can justify our results. Between the rE and pE regions one can find a domain qO, in which the estimation error is smaller than $\mathrm{EE}_{\mathrm{lim}}$. The minimum of the curve ($\mathrm{EE}_{\mathrm{min}}$) is taken at the optimal regularization parameter α_{opt}.

Similar trends and domains can be identified for all the estimation problems, in which the noise level is medium or high. Further, in these cases, the optimal regularization parameter index only slightly varies between 6 and 8. However, in the cases of small measurement noise (1%), the error in the estimated parameters due to the noise propagation is negligible, the pE region is flat and there is not a unique, optimal regularization parameter.

5.2 Performance of the Methods

The statistics described in Section 4.2 were calculated for each scenario and each value of the regularization parameter. The different regularization tuning

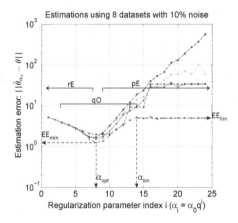

Fig. 2. Parameter estimation error. The estimation error is plotted against the regularization parameters for the 5 replicates of a selected scenario. The notations corresponding to the lowest, green curve: the regularization bias is dominating in domain rE; the noise propagated error is the main contributor to the estimation error in pE. Typically, the propagated error levels off at EE_{lim} as the regularization parameter reaches a certain limit α_{lim}. Any regularization parameter in the qO region gives lower estimation error, than without the regularization ($\alpha = 0$). The optimal regularization parameter α_{opt} corresponds to the minimum of the estimation error curve.

methods (Section 3.1) were used to find the regularization parameter for each scenario. To serve as a reference, we also computed the scores corresponding to the estimations *without* regularization (NR), i.e. $\alpha = 0$. Figure 3 shows the distribution of the inefficiencies given by (6), i.e. the relative parameter estimation error, computed for each regularization method. For the sake of clarity, only the estimation scenarios corresponding to the 10% noise are depicted in the figure. More detailed numerical results for all scenarios can be found in Table A.2 in Appendix D.

From Figure 3 we see that the estimation error grows rapidly as the number of experimental datasets decreases in the non-regularized estimations (NR), i.e. these estimations are greatly affected by ill-conditioning. The same trend can be observed based on the numerical results when the noise level of the data increases, leading to larger inefficiencies of the non-regularized estimations. These results also indicate that, using regularization, such estimation error is reduced up to 2 orders of magnitude. For the more ill-conditioned scenarios (more noise and less data), almost all regularization methods perform better than the non-regularized estimation. For the mildly ill-conditioned cases (more data with less noise), the discrepancy principle (DP) and the L-curve method based on the tangent condition (LC2) perform rather poorly due to over-regularization.

The cross validation based methods (CV_{χ^2}, CV_{RSS}) result in a generally low estimation error and perform the best for the cases when the calibration data is highly contaminated by noise. For the situations where there is no additional data set, the L-curve method based on the maximal curvature detection (LC1), the quasi-optimality (QO) criteria and the balancing principle (BP1) are the best alternatives from estimation error point of view. All these methods performed similarly well for almost all cases. Among them, the LC1 performed also very well in the mildly ill-posed cases. Furthermore, the LC1 method outperformed the NR case in almost all scenarios from the prediction error (PE) point of view, too (see Table A.2).

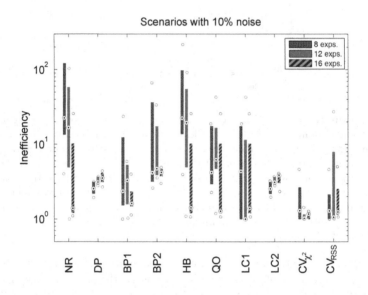

Fig. 3. Inefficiencies of tuning method for three selected scenarios. The inefficiency shows the parameter estimation error normalized by the estimation error with optimal regularization. The color indicates the number of datasets used in the estimations according to the legend. Each column represents the distribution of the 5 measurement error realizations: the circle with the black dot shows the median, the filled area spreads between the 25th and 75th percentiles of the points, the rest of the points are shown individually. NR: non-regularized solution, DP: discrepancy principle, BP: balancing principle, HB: hardened balancing, QO: quasi optimality criteria, LC: L-curve method, CV: cross validation based tuning methods.

6   Conclusions

In this study we considered regularization as a way to improve the calibration of (nonlinear) kinetic models in systems biology, reducing the typical ill-conditioning of these problems. We considered the Tikhonov regularization framework coupled with a global optimization solver. We focused on the specific question of regularization method selection and tuning. We compared several regularization parameter tuning methods, including the discrepancy principle, balancing principle, hardened balancing, quasi optimality criteria, L-curve method and cross validation based methods. The different methods were tuned and tested considering several scenarios of a challenging kinetic model.

Overall, the results obtained indicate that regularization can reduce the parameter estimation error very significantly (up to 2 orders of magnitude for the example considered). The results also indicate that, for the situations where a second data set is available, the cross validation (CV) χ^2 score based method gives the best tuning results. When no further data is available for cross-validation, the L-curve method based on the maximum curvature detection (LC2) is the most robust tuning algorithm.

Acknowledgement. This research was supported by the funding from EU FP7 ITN "NICHE", project no. 289384.

References

1. Almquist, J., Cvijovic, M., Hatzimanikatis, V., Nielsen, J., Jirstrand, M.: Kinetic models in industrial biotechnology - Improving cell factory performance. Metabolic Engineering 1–22 (April 2014)
2. Ashyraliyev, M., Fomekong-Nanfack, Y., Kaandorp, J.A., Blom, J.G.: Systems biology: parameter estimation for biochemical models. FEBS Journal 276(4), 886–902 (2009)
3. Banga, J.R., Balsa-Canto, E.: Parameter estimation and optimal experimental design. Essays in Biochemistry 45, 195–210 (2008)
4. Bansal, L., Chu, Y., Laird, C., Hahn, J.: Regularization of Inverse Problems to Determine Transcription Factor Profiles from Fluorescent Reporter Systems. AIChE Journal 58(12), 3751–3762 (2012)
5. Bauer, F.: Some considerations concerning regularization and parameter choice algorithms. Inverse Problems 23(2), 837–858 (2007)
6. Bauer, F., Lukas, M.A.: Comparingparameter choice methods for regularization of ill-posed problems. Mathematics and Computers in Simulation 81(9), 1795–1841 (2011)
7. Dennis, J.E., Gay, D.M., Welsch, R.E.: An Adaptive Nonlinear Least-Squares Algorithm. ACM Transaction on Mathematical Software 7(3), 348–368 (1981)
8. Engl, H.W., Flamm, C., Kügler, P., Lu, J., Müller, S., Schuster, P., Philipp, K.: Inverse problems in systems biology. Inverse Problems 25(12), 123014 (2009)
9. Engl, H.W., Hanke, M., Neubauer, A.: Regularization of Inverse Problems. Kluwer Academic Publishers (1996)
10. Hansen, P.C., O'Leary, D.P.: The use of the L-Curve in the regularization of discrete ill-posed problems. SIAM Journal on Scientific Computing 14(6), 1487–1503 (1993)
11. Hindmarsh, A.C., Brown, P.N., Grant, K.E., Lee, S.L., Serban, R., Shumaker, D.E., Woodward, C.S.: {SUNDIALS: Suite of Nonlinear and Differential / Algebraic Equation Solvers}. ACM Transaction on Mathematical Software 31(3), 363–396 (2005)
12. Kaltenbacher, B., Neubauer, A., Scherzer, O.: Iterative Regularization Methods for Nonlinear Ill-Posed Problems. Radon Series on Computational and Applied Mathematics. Walter de Gruyter, Berlin, New York (2008)
13. Kitano, H. (ed.): Foundations of Systems Biology. The MIT Press (2001)
14. Kravaris, C., Hahn, J., Chu, Y.: Advances and selected recent developments in state and parameter estimation. Computers & Chemical Engineering 51, 111–123 (2013), http://linkinghub.elsevier.com/retrieve/pii/S0098135412001779

15. Lepskii, O.: On a Problem of Adaptive Estimation in Gaussian White Noise. Theory of Probability & Its Applications 35(3), 454–466 (1991)

16. Link, H., Christodoulou, D., Sauer, U.: Advancing metabolic models with kinetic information. Current Opinion in Biotechnology 29, 8–14 (2014)

17. Ljung, L., Chen, T.: What can regularization offer for estimation of dynamical systems? In: 11th IFAC International Workshop on Adaptation and Learning in Control and Signal Processing. IFAC, vol. 5 (2013)

18. Miró, A., Pozo, C., Guillén-Gosálbez, G., Egea, J.A., Jiménez, L.: Deterministic global optimization algorithm based on outer approximation for the parameter estimation of nonlinear dynamic biological systems. BMC Bioinformatics 13(1), 90 (2012)

19. Moles, C.G., Mendes, P., Banga, J.R.: Parameter Estimation in Biochemical Pathways: A Comparison of Global Optimization Methods. Genome Research 13, 2467–2474 (2003)

20. Morozov, V.A.: Methods for Solving Incorrectly Posed Problems. Springer (1984)

21. Palm, R.: Numerical Comparison of Regularization Algorithms for Solving Ill-Posed Problems. Ph.D. thesis, University of Tartu, Estonia

22. Papamichail, I., Adjiman, C.S.: Global optimization of dynamic systems. Comput. Chem. Eng. 28, 403–415 (2004)

23. Regiska, T.: A Regularization Parameter in Discrete Ill-Posed Problems. SIAM Journal on Scientific Computing 17(3), 740–749 (1996)

24. Reichel, L., Rodriguez, G.: Old and new parameter choice rules for discrete ill-posed problems. Numerical Algorithms 63(1), 65–87 (2012)

25. Rodriguez-Fernandez, M., Egea, J.A., Banga, J.R.: Novel metaheuristic for parameter estimation in nonlinear dynamic biological systems. BMC Bioinformatics 7, 483 (2006)

26. Rodriguez-Fernandez, M., Mendes, P., Banga, J.R.: A hybrid approach for efficient and robust parameter estimation in biochemical pathways. Bio Systems 83(2-3), 248–265 (2006)

27. Schmidt, M., Fung, G., Rosaless, R.: Optimization Methods for l1-Regularization. Tech. rep. (2009), http://www.cs.ubc.ca/cgi-bin/tr/2009/TR-2009-19.pdf

28. Stanford, N.J., Lubitz, T., Smallbone, K., Klipp, E., Mendes, P., Liebermeister, W.: Systematic construction of kinetic models from genome-scale metabolic networks. PloS one 8(11), e79195 (2012)

29. Tibshirani, R.: Regression Shrinkage and Selection via the Lasso. Journal of the Royal Statistical Society: Series B (Statistical Methodology) 58(1), 267–288 (1996)

30. Walter, E., Prorizato, L.: Identification of Parametric Models from experimental data. Springer (1997)

31. Wang, H., Wang, X.C.: Parameter estimation for metabolic networks with two stage Bregman regularization homotopy inversion algorithm. Journal of Theoretical Biology 343, 199–207 (2014)

32. Zou, H., Hastie, T.: Regularization and variable selection via the elastic net. Journal of the Royal Statistical Society: Series B (Statistical Methodology) 67(2), 301–320 (2005), http://doi.wiley.com/10.1111/j.1467-9868.2005.00503.x

Authors Contribution

This paper describes one of the main tasks in the Ph.D. of Attila Gabor. The work presented here describes the methods that he has developed as well as the results that he has obtained. Julio R. Banga is his Ph.D. supervisor.

A Kinetic Model of a Three-Steps Metabolic Pathway – Details

The parameters, initial values and stimuli conditions corresponding to the estimation problems can be found in Table A.1. The ODEs read as:

$$\dot{G}_1 = \frac{V_1}{1 + (\frac{P}{Ki_1})^{ni_1} + (\frac{Ka_1}{S})^{na_1}} - k_1 G_1$$

$$\dot{G}_2 = \frac{V_2}{1 + (\frac{P}{Ki_2})^{ni_2} + (\frac{Ka_2}{M_1})^{na_2}} - k_2 G_2$$

$$\dot{G}_3 = \frac{V_3}{1 + (\frac{P}{Ki_3})^{ni_3} + (\frac{Ka_3}{M_2})^{na_3}} - k_3 G_3$$

$$\dot{E}_1 = \frac{V_4 G_1}{K_4 + G_1} - k_4 E_1$$

$$\dot{E}_2 = \frac{V_5 G_2}{K_5 + G_2} - k_5 E_2$$

$$\dot{E}_3 = \frac{V_6 G_3}{K_6 + G_3} - k_6 E_3$$

$$\dot{M}_1 = \frac{kcat_1 E_1 (\frac{1}{Km_1})(S - M_1)}{1 + \frac{S}{Km_1} + \frac{M1}{Km_2}} - \frac{kcat_2 E_2 \frac{1}{Km_3}(M_1 - M_2)}{1 + \frac{M_1}{Km_3} + \frac{M_2}{Km_4}}$$

$$\dot{M}_2 = \frac{kcat_2 E_2 \frac{1}{Km_3}(M_1 - M_2)}{1 + \frac{M_1}{Km_3} + \frac{M_2}{Km_4}} - \frac{kcat_3 E_3 \frac{1}{Km_5}(M_2 - P)}{1 + \frac{M_2}{Km_5} + \frac{P}{Km_6}}$$

B Finding the Maximal Regularization Index

We see from Figure 2, that the estimation error is levelling off for small regularization parameters, i.e. the regularization parameter does not influence the estimation problem any more. The goal of the maximal index is to find the minimum regularization parameter after which the estimation error levels off. However, the curve is not available in practice, since the nominal parameters is required to compute the estimation error. Alternatively, an estimate of the second term in (5) can be made based on the Hessian of the regularized cost function. Let $\tilde{R}(\theta, \alpha) = [R(\theta)^T \sqrt{\alpha}(W\theta)^T]^T$ be the augmented regularized residual vector (c.f. (4)) and define $F_R : \mathcal{R}^{n_p \times 1} \mapsto \mathcal{R}^{n_p \times n_p}$ as $F_R(\theta, \alpha) \doteq \frac{\partial \tilde{R}(\theta, \alpha)}{\partial \theta}^T \frac{\partial \tilde{R}(\theta, \alpha)}{\partial \theta}$. Note that,

Table A.1. Nominal parameter values, parameter estimation bounds, stimuli (input) values and initial conditions (I.C.) for the dynamic model of the 3-Steps Metabolic Pathway

Param.	Value	$[p_{lb}, p_{ub}]$	Param.	Value	$[p_{lb}, p_{ub}]$	Param.	Value	$[p_{lb}, p_{ub}]$
V1	1	$[10^{-12}, 10^6]$	V3	1	$[10^{-12}, 10^6]$	V6	0.1	$[10^{-12}, 10^6]$
Ki1	1	$[10^{-12}, 10^6]$	Ki3	1	$[10^{-12}, 10^6]$	K6	1	$[10^{-12}, 10^6]$
ni1	2	$[0.1, 1]$	ni3	2	$[0.1, 1]$	k_6	0.1	$[10^{-12}, 10^6]$
Ka1	1	$[10^{-12}, 10^6]$	Ka3	1	$[10^{-12}, 10^6]$	kcat1	1	$[10^{-12}, 10^6]$
na1	2	$[0.1, 1]$	na3	2	$[0.1, 1]$	Km1	1	$[10^{-12}, 10^6]$
k_1	1	$[10^{-12}, 10^6]$	k_3	1	$[10^{-12}, 10^6]$	Km2	1	$[10^{-12}, 10^6]$
V2	1	$[10^{-12}, 10^6]$	V4	0.1	$[10^{-12}, 10^6]$	kcat2	1	$[10^{-12}, 10^6]$
Ki2	1	$[10^{-12}, 10^6]$	K4	1	$[10^{-12}, 10^6]$	Km3	1	$[10^{-12}, 10^6]$
ni2	2	$[0.1, 1]$	k_4	0.1	$[10^{-12}, 10^6]$	Km4	1	$[10^{-12}, 10^6]$
Ka2	1	$[10^{-12}, 10^6]$	V5	0.1	$[10^{-12}, 10^6]$	kcat3	1	$[10^{-12}, 10^6]$
na2	2	$[0.1, 1]$	K5	1	$[10^{-12}, 10^6]$	Km5	1	$[10^{-12}, 10^6]$
k_2	1	$[10^{-12}, 10^6]$	k_5	0.1	$[10^{-12}, 10^6]$	Km6	1	$[10^{-12}, 10^6]$

Inputs:	[S]	[P]		[S]	[P]		[S]	[P]
exp. #1	0.1	0.050	exp. #7	10	0.368	exp. #13	10	0.050
exp. #2	0.1	1.0	exp. #8	10	1.0	exp. #14	0.1	0.368
exp. #3	0.464	0.136	exp. #9	0.1	0.136	exp. #15	0.464	0.050
exp. #4	0.464	1.0	exp. #10	0.464	0.368	exp. #16	2.15	0.136
exp. #5	2.15	0.05	exp. #11	2.15	1.000			
exp. #6	2.15	0.368	exp. #12	10	0.136			
CV. #1	1.0	0.02	CV. #4	4.0	0.02	CV. #7	8.0	0.02
CV. #2	1.0	0.2	CV. #5	4.0	0.2	CV. #8	8.0	0.2
CV. #3	1.0	0.8	CV. #6	4.0	0.8	CV. #9	8.0	0.8

Sampling time points:	equidistantly 21 points on [0s 120s].

States	I.C.	States	I.C.	States	I.C.
G1	0.6667	E1	0.4	M1	1.419
G2	0.5725	E2	0.3641	M2	0.9346
G3	0.4176	E3	0.2946		

as $\alpha \to 0$, $F_R(\hat{\theta}_\alpha, \alpha)$ becomes the observed Fisher Information matrix (FIM). The inverse of the FIM (if exists) is the Cramer-Rao lower bound (CRLB) of the covariance matrix of the parameters [30]. Although, the FIM is practically non-invertible when the estimation is highly ill-posed, the inverse of F_R always exists for sufficiently large $\alpha > 0$ and invertible weighting matrix W. Thus, the α-dependent regularized CRLB is estimated by $\text{CRLB}_R(\hat{\theta}_{\alpha_l}) = F_R(\hat{\theta}_{\alpha_l}, \alpha_l)^{-1}$. The regularized variance is therefore bounded by

$$\sigma_R^2(\hat{\theta}_{\alpha_l}) \geq \text{diag}(\text{CRLB}_R(\hat{\theta}_{\alpha_l})) = \hat{\sigma}_R^2(\hat{\theta}_{\alpha_l}). \tag{A.1}$$

The maximum regularization parameter corresponds to the index

$$I_m = \max_{1 \leq i \leq I}(i \mid \|\hat{\sigma}_R(\hat{\theta}_{\alpha_i})\| < 0.9 \max_{1 \leq k \leq I}(\|\hat{\sigma}_R(\hat{\theta}_{\alpha_k})\|)), \tag{A.2}$$

where 0.9 is a tuning parameter that tries to avoid the small numerical disturbances.

C   Computational Details of Comparison Criteria

Some quantities, such as the residual sum of squares

$$\text{RSS}(\hat{\theta}_{\alpha_l}) = \sum_{k=1}^{N_e} \sum_{j=1}^{N_{y,k}} \sum_{i=1}^{N_{t,k,j}} \left(y_{ijk}\left(x(t_i, \hat{\theta}_{\alpha_l}), \hat{\theta}_{\alpha_l} \right) - \hat{y}_{ijk} \right)^2 \tag{A.3}$$

and

$$\chi^2(\hat{\theta}_{\alpha_l}) = \sum_{k=1}^{N_e} \sum_{j=1}^{N_{y,k}} \sum_{i=1}^{N_{t,k,j}} \frac{\left(y_{ijk}\left(x(t_i, \hat{\theta}_{\alpha_l}), \hat{\theta}_{\alpha_l} \right) - \hat{y}_{ijk} \right)^2}{\sigma_{ijk}^2} \tag{A.4}$$

can be easily computed from the model prediction and the data. They measure the explanatory potential of the model with estimated parameter $\hat{\theta}_{\alpha_l}$.

Some quantities, as the *nominal parameters* θ and nominal model prediction (i.e. the measurement error free concentrations), are not known in practice. However, a synthetic framework let us compute these values and we can compare the regularization methods based on these quantities. The prediction error defined as

$$\text{PE}(\hat{\theta}_{\alpha_l}) = \sum_{k=1}^{N_e} \sum_{j=1}^{N_{y,k}} \sum_{i=1}^{N_{t,k,j}} (y_{ijk}(x(t_i, \hat{\theta}_{\alpha_l}), \hat{\theta}_{\alpha_l}) - y_{ijk}(x(t_i, \theta), \theta))^2 \tag{A.5}$$

measures the distance of the model prediction $y(x, \hat{\theta})$ and the noise-free underlying data y, that is unknown in practical applications. A model that tends to over-fit the data, i.e. fits also the noise in the data, likely to generate a good fit to the estimation data (small RSS value), but performs worst according to the PE.

The accuracy of the estimated parameters is measured by the estimation error:

$$\text{EE}(\hat{\theta}_{\alpha_l}) = ||\hat{\theta}_{\alpha_l} - \theta|| , \tag{A.6}$$

which is the 2-norm measure of the deviation of the estimated parameters from the nominal parameters.

D   Detailed Numerical Results

Table A.2 contains the averaged statistics corresponding to each tuning method.

Table A.2. Performance of the Parameter Choice Methods. Each statistics is obtained by taking the average of the five replicates. N_{exps} number of experiments used for the estimations, **N.**: amplitude of the noise in %. **NR**: non-regularized solution, **DP**: discrepancy principle, **B**: balancing principle, **HB**: hardened balancing, **QO**: quasi optimality criteria, **LC**: L-curve method, **CV**: cross validation based

					Inefficiency averages: $\langle \mathrm{IE}(\hat{\theta}_{\alpha_m}) \rangle$					
N_e	N.	NR	DP	B1	B2	HB	QO	LC1	LC2	CV_{χ^2}	CV_{RSS}
8	1	1.34	6.09	3.50	12.41	1.06	1.06	1.33	3.50	1.65	1.08
	5	1.34	2.98	1.46	4.12	1.25	1.23	1.25	2.43	1.06	1.12
	10	85.31	2.61	7.46	20.60	63.52	9.10	8.44	2.61	1.99	1.85
12	1	1.07	6.44	3.60	14.48	1.12	1.12	1.07	3.60	2.59	1.23
	5	1.37	3.55	1.66	4.92	1.29	1.31	1.10	3.55	1.23	1.23
	10	34.05	3.24	3.42	11.57	32.25	12.81	9.39	3.24	1.11	6.39
16	1	1.24	8.75	2.84	12.73	1.25	1.24	1.24	2.84	2.50	1.24
	5	1.59	4.17	1.90	6.20	1.56	1.60	1.60	4.17	1.16	1.33
	10	6.91	3.61	2.02	4.14	6.91	6.88	6.91	3.53	1.12	2.05

					χ^2 averages: $\langle \chi^2(\hat{\theta}_{\alpha_m}) \rangle$					
N_e	N.	NR	DP1	B1	B2	HB	QO	LC1	LC2	CV_{χ^2}	CV_{RSS}
8	1	1254.3	1635.4	1342.8	5597.5	1256.7	1256.7	1254.3	1342.8	1267.5	1256.2
	5	1305.5	1740.1	1324.2	6444.3	1305.5	1305.6	1305.6	1452.8	1307.1	1308.4
	10	1297.0	1767.5	1297.6	4543.6	1297.0	1486.9	1298.7	1767.5	1300.3	1301.5
12	1	1909.3	2215.6	1971.4	6474.4	1910.2	1910.2	1909.3	1971.4	1940.6	1911.8
	5	2041.2	2565.4	2064.4	6095.7	2041.4	2041.8	2041.6	2565.4	2042.5	2041.8
	10	1988.2	2477.8	1990.6	5137.9	1988.2	1988.3	1989.7	2477.8	1996.2	1994.7
16	1	2676.7	3845.5	2730.6	6997.3	2677.3	2677.3	2676.7	2730.6	2722.1	2676.6
	5	2593.2	3120.8	2622.4	6857.5	2593.4	2593.7	2593.2	3120.8	2599.1	2605.2
	10	2636.3	3514.7	2645.3	6374.5	2636.2	2636.2	2636.2	3344.3	2639.3	2637.4

					RSS averages: $\langle \mathrm{RSS}(\hat{\theta}_{\alpha_m}) \rangle$					
N_e	N.	NR	DP1	B1	B2	HB	QO	LC1	LC2	CV_{χ^2}	CV_{RSS}
8	1	0.15	0.21	0.17	0.78	0.15	0.15	0.15	0.17	0.16	0.15
	5	4.45	6.11	4.53	15.49	4.45	4.45	4.45	4.99	4.45	4.47
	10	15.11	20.45	15.14	33.08	15.11	17.66	15.16	20.45	15.17	15.20
12	1	0.24	0.28	0.25	0.80	0.24	0.24	0.24	0.25	0.24	0.24
	5	6.25	7.74	6.32	16.33	6.25	6.26	6.26	7.74	6.25	6.26
	10	22.39	28.15	22.47	41.76	22.39	22.39	22.41	28.15	22.47	22.46
16	1	0.31	0.47	0.32	0.86	0.31	0.31	0.31	0.32	0.32	0.31
	5	7.24	8.93	7.35	18.62	7.24	7.24	7.24	8.93	7.26	7.29
	10	31.32	41.37	31.51	58.06	31.32	31.33	31.32	40.03	31.42	31.37

					PE averages: $\langle \mathrm{PE}(\hat{\theta}_{\alpha_m}) \rangle$					
N_e	N.	NR	DP1	B1	B2	HB	QO	LC1	LC2	CV_{χ^2}	CV_{RSS}
8	1	0.069	0.244	0.126	0.799	0.069	0.069	0.069	0.126	0.080	0.070
	5	0.392	1.361	0.472	3.299	0.390	0.390	0.389	0.852	0.390	0.398
	10	0.729	2.383	0.718	3.520	0.731	1.446	0.703	2.383	0.719	0.716
12	1	0.069	0.215	0.115	0.760	0.071	0.071	0.069	0.115	0.091	0.074
	5	0.437	1.266	0.466	3.158	0.431	0.429	0.429	1.266	0.416	0.423
	10	0.696	2.353	0.723	3.765	0.698	0.698	0.645	2.353	0.624	0.657
16	1	0.080	0.401	0.111	0.742	0.082	0.081	0.080	0.111	0.103	0.080
	5	0.376	1.215	0.382	3.315	0.368	0.366	0.376	1.215	0.321	0.318
	10	0.770	3.041	0.852	4.762	0.770	0.765	0.771	2.809	0.788	0.775

Uncertainty Analysis for Non-identifiable Dynamical Systems: Profile Likelihoods, Bootstrapping and More

Fabian Fröhlich[1,2], Fabian J. Theis[1,2], and Jan Hasenauer[1,2]

[1] Institute of Computational Biology, Helmholtz Zentrum München,
85764 Neuherberg, Germany
{fabian.froehlich,fabian.theis,jan.hasenauer}@helmholtz-muenchen.de
[2] Department of Mathematics, Technische Universität München,
85748 Garching, Germany

Abstract. Dynamical systems are widely used to describe the behaviour of biological systems. When estimating parameters of dynamical systems, noise and limited availability of measurements can lead to uncertainties. These uncertainties have to be studied to understand the limitations and the predictive power of a model. Several methods for uncertainty analysis are available. In this paper we analysed and compared bootstrapping, profile likelihood, Fisher information matrix, and multi-start based approaches for uncertainty analysis. The analysis was carried out on two models which contain structurally non-identifiable parameters. We showed that bootstrapping, multi-start optimisation, and Fisher information matrix based approaches yield misleading results for parameters which are structurally non-identifiable. We provide a simple and intuitive explanation for this, using geometric arguments.

Keywords: parameter estimation, uncertainty analysis, bootstrapping, profile likelihood, identifiability.

1 Introduction

In systems and computational biology, mechanistic models are used to advance our understanding of a process of interest. In order to test whether the model can adequately reproduce measured behaviour, it is necessary to fit the model parameters to the measurement data. This process of inferring model parameters is usually termed parameter estimation.

In general, it is not possible to measure every biochemical component. Furthermore, measurements are noise corrupted. These two factors can result in a non-negligible uncertainty of the estimated parameters [1]. These parameter uncertainties have to be studied, to determine limitations of the model and its predictive power. Moreover, the resulting uncertainties in parameter estimates are building blocks for subsequent investigation, such as model predictions or experimental design [2,3].

P. Mendes et al. (Eds.): CMSB 2014, LNBI 8859, pp. 61–72, 2014.
© Springer International Publishing Switzerland 2014

The model parameters are commonly estimated using maximum likelihood and maximum a posteriori estimators. For the analysis of the uncertainty of such estimators, several methods have been established: asymptotic analysis, bootstrapping, profile likelihoods and Bayesian statistics.

Asymptotic analysis based on the Fisher information matrix (FIM) is related to a local approximation of the objective function at the current optimum [4]. Bootstrapping exploits data resampling and estimation using the resampled data to construct confidence intervals. [5]. The profile likelihood approach approximates the extent of super-level-sets of the likelihood function by constrained optimisation [4]. Some researchers also exploit uncertainty estimates derived from multiple optimiser starts [6]. However, these uncertainty estimates have no statistical foundation. There also exist Bayesian methods for uncertainty analysis, such as marginal densities [7] or profile posteriors [8]. Nonetheless, in the scope of this paper, we will only consider the frequentist perspective of uncertainty analysis, as the comparison between the frequentist perspective and the Bayesian perspective has already been covered in a recent study [8].

Several papers are available which show that bootstrapping, profile likelihood, and FIM derived confidence intervals are rather similar, given that parameters are identifiable [5,9,10]. Similar studies, in the presence of structural non-identifiability (cf. Definition 1), are however missing.

In this paper, we will compare the performance of several methods for uncertainty analysis based on their prediction of confidence intervals. The comparison will be carried out on two examples which both contain structurally non-identifiable parameters. Based on these two models, we will conclude that most methods for uncertainty analysis yield misleading results for parameters that are structurally non-identifiable.

This paper is organised as follows: In Section 2, we will formulate the general problem considered in parameter estimation and introduce methods for parameter estimation, as well as uncertainty analysis. In Section 3, we will compare the previously introduced methods for uncertainty analysis based on two models where structural non-identifiability of parameters is present. In Section 4, we will summarise the results and respective conclusions and give some advice on how to cope with the presented shortcomings of methods in practice.

2   Methods

In the following sections, we will describe the general problem of parameter estimation and methods for parameter estimation. Subsequently, we will introduce several methods for uncertainty analysis, namely the **Fisher information matrix approximation (FIM)**, the **Bootstrapping approach (BS)**, the **Profile Likelihood approach (PL)**, and the **Multi-start approach (MS)**. Eventually, we will present the concept of structural non-identifiability.

2.1   Problem Formulation

We consider ordinary differential equation models, in which the time-dependence of states x of the biological system are captured by a set of differential equations and the observable components y are described by a mapping of these states:

$$y(t; \theta) = h(x(t; \theta), \theta) \text{ with } x \text{ solution to } \dot{x} = f(x, \theta), \quad (1)$$

where f is the right hand side of the differential equation of the dynamical system, x is the state vector of the dynamical system and h is the function that maps states of the system to the observable components. Under the assumption of additive, independent noise, the measurement of the i-th observable component $\bar{y}_i$ at time point t_k is

$$\bar{y}_i(t_k) = y_i(t_k; \theta) + \varepsilon_{ik} \quad i = 1, \ldots, n_y \ k = 1, \ldots, n_t, \quad (2)$$

where n_y is the number of observable components, n_t is the number of observed time points and $\varepsilon_{ik} \sim \mathcal{N}(0, \sigma_{ik}^2)$ is the measurement noise.

The maximum likelihood estimate of the parameters is obtained from measurement data $\mathcal{D} = \{(t_k, \bar{\mathbf{y}}(t_k))\}_{k=1}^{n_t}$, by minimising the negative log-likelihood

$$J_{\mathcal{D}}(\theta) = -\log P(\mathcal{D}|\theta) = \frac{1}{2} \sum_{i=1}^{n_y} \sum_{k=1}^{n_t} \left(\log\left(2\pi\sigma_{ik}^2\right) + \left(\frac{\bar{y}_i(t_k) - y_i(t_k; \theta)}{\sigma_{ik}} \right)^2 \right). \quad (3)$$

The likelihood of θ is equal to the conditional probability $P(\mathcal{D}|\theta)$ to observe the data $\mathcal{D}$, given the parameter vector θ. In the following, the optimiser of the objective function for the respective measurement data will be denoted as

$$\hat{\theta}^{\mathcal{D}} = \arg\min_{\theta} J_{\mathcal{D}}(\theta). \quad (4)$$

In practice, the optimisation problem is often reformulated by transforming the parameters [11]. This can improve the numerical properties of optimisation algorithms, as well as its convergence.

2.2   Parameter Estimation

A multitude of different optimisation algorithms are available for the minimisation of the function $J_{\mathcal{D}}(\theta)$. Most commonly, global optimisation schemes like simulated annealing or particle swarm are used [12,13,14,15]. A recent study also suggests a good performance of multi-start local optimisation, when provided with high-quality gradients [11]. In this paper we employ the particle swarm method [15], as well as multi-start local optimisation. For the local optimisation scheme, the gradient $\nabla J_{\mathcal{D}}(\theta)$ is computed using sensitivity equations.

2.3   Uncertainty Analysis

In the frequentist perspective, the uncertainty of a parameter estimate is usually described in terms of confidence intervals. Confidence intervals to the level δ, will contain the true parameter θ^* in $(\delta * 100)\%$ of the times, given that they are computed from realisations of the true parameter θ^*.

To facilitate the direct analysis of the extent of confidence intervals across a multitude of values for δ we will study the function

$$R_{\theta_i}(c) = \min_{\delta}\{\delta : c \in \mathrm{CI}_{i,\delta}\}, \tag{5}$$

which describes the minimal confidence level δ such that the parameter $\theta_i = c$ is contained in the respective confidence interval $\mathrm{CI}_{i,\delta}$.

Profile Likelihood Approach (PL). The confidence interval to the confidence level δ for a parameter θ_i can be interpreted as the sub-level-set to the level $\log(\delta)$ of the objective function $J_{\mathcal{D}}(\theta)$. The extent of these sub-level-sets can be determined using the profile likelihood ratio

$$R_{\theta_i}^{\mathrm{PL}}(c) = \exp\left(\min_{\theta_{j\neq i}} J_{\mathcal{D}}(\theta) - J_{\mathcal{D}}(\hat{\theta}^{\mathcal{D}})\right) \; s.t. \; \theta_i = c, \tag{6}$$

which internally uses the profile likelihood [1]

$$PL_{\theta_i}(c) = \min_{\theta_{j\neq i}} J_{\mathcal{D}}(\theta). \tag{7}$$

Fisher Information Matrix Approximation (FIM). The FIM approximation relies on a local quadratic approximation of $J_{\mathcal{D}}(\theta)$, which yields the approximation

$$R_{\theta_i}^{\mathrm{FIM}}(c) = \exp\left(-\frac{(\hat{\theta}_i^{\mathcal{D}} - c)^2}{2(F^{-1})_{ii}}\right) \tag{8}$$

where F is the Fisher information matrix [16].

Bootstrapping Approach (BS). For the bootstrapping approach, the model is fitted to the true experimental data, yielding a parameter estimate $\hat{\theta}^{\mathcal{D}}$. Subsequently $n_{\mathcal{D}}$ datasets $\mathcal{D}_k$ are generated by simulating the system $y(\cdot, \hat{\theta}^{\mathcal{D}})$ and corrupting the simulation results with measurement noise. The parameter estimation is repeated for these artificial datasets by minimising $J_{\mathcal{D}_k}(\theta)$, yielding $\hat{\theta}^{\mathcal{D}_k}$. The uncertainty of parameter estimates is then derived from the spread of the estimates $\hat{\theta}^{\mathcal{D}_k}$.

Apparently, different parameter estimation methods can be used to determine the bootstraps. In this study we consider:

- a single **local** optimisation scheme initialised at $\hat{\theta}_{\mathrm{MS}}^{\mathcal{D}}$ to obtain $\{\hat{\theta}_{\mathrm{LG}}^{\mathcal{D}_k}\}_{k=1}^{n_{\mathcal{D}}}$ (BS-LG)
- a **multi-start local** optimisation scheme to obtain $\{\hat{\theta}_{\mathrm{MS}}^{\mathcal{D}_k}\}_{k=1}^{n_{\mathcal{D}}}$ (BS-MS)
- a **particle swarm** based optimisation scheme to obtain $\{\hat{\theta}_{\mathrm{PSO}}^{\mathcal{D}_k}\}_{k=1}^{n_{\mathcal{D}}}$. (BS-PSO)

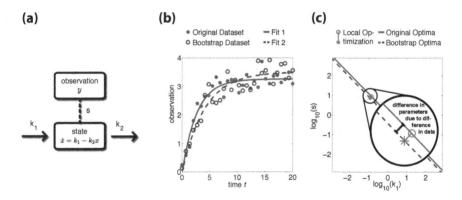

Fig. 1. Illustration of the bootstrapping process. (a) Schematic illustration of the model for the turnover reaction introduced in Section **3.1 (b)** Depiction of original dataset $\mathcal{D}_1$ and bootstrapped dataset $\mathcal{D}_2$. Data-points $y_k^{\mathcal{D}_i}$ are shown as red ($i = 1$) and blue ($i = 2$) circles. Simulations $y(t; \hat{\theta}_{\mathrm{MS}}^{\mathcal{D}_i})$ of the system for corresponding optimal parameters are shown as solid lines of respective colour. **(c)** The continua of optimal parameters corresponding to $\mathcal{D}_i$ are plotted as red ($i = 1$) and blue ($i = 2$) solid lines. The optimal parameter $\hat{\theta}_{\mathrm{MS}}^{\mathcal{D}_1}$ is shown as teal circle. The respective optimal parameter $\hat{\theta}_{\mathrm{LG}}^{\mathcal{D}_2}$ that was obtained by optimisation initiated at $\hat{\theta}_{\mathrm{MS}}^{\mathcal{D}_1}$ is shown as teal star. The difference in the location of the continuum of optimal points induced by bootstrapping is illustrated in black.

The normalised histograms over these three sets yield three approximations $R_{\theta_i}^{\mathrm{BS\text{-}LG}}$, $R_{\theta_i}^{\mathrm{BS\text{-}MS}}$, and $R_{\theta_i}^{\mathrm{BS\text{-}PSO}}$ to $R_{\theta_i}(c)$. The histograms were normalised, such that the height of the bin with highest frequency is equal to 1.

Multi-start Approach (MS). Another approach, which is often used by practitioners, is to use the normalised histogram $R_{\theta_i}^{\mathrm{MS}}(c)$ over the best optima found by multi-start optimisation. At a first glance, MS approaches for uncertainty analysis may seem similar to bootstrapping approaches. However, for a problem where all parameters are identifiable and the optimiser converges always to the global optimum, the MS approach would suggest no parameter uncertainty.

2.4 Structural Identifiability

The aforementioned methods are used to assess the parameter uncertainties, given a certain set of experimental data $\mathcal{D}$. In addition, also the structural identifiability or non-identifiability might be assessed:

Definition 1 (Structural Identifiability [17]). *A parameter θ_i, $i = 1, \ldots, n_\theta$ is structurally identifiable, if for almost any θ',*

$$y(\cdot, \theta) = y(\cdot, \theta') \Rightarrow \theta_i = \theta_i'. \tag{9}$$

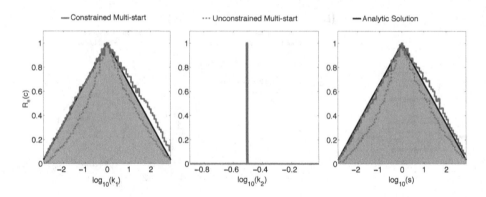

Fig. 2. Comparison of constrained and unconstrained multi-start local optimisation. Histograms for results of multi-start local optimisation (10^5 starting points) in logarithmic parameters using the MATLAB functions `fminunc` and `fmincon` (constrained to $[-3, 3]^3$). Initial points are uniformly drawn from $[-3, 3]^3$. The analytical solution for gradient descent method is shown in black.

Except for pathological cases, structural non-identifiability of a parameter will lead to infinitely extended sets of parameters on which y is invariant. In fact, in most cases the invariant set will also be infinitely extended. It is evident that $R_{\theta_i}(c)$ must be constant on the invariant set.

The assessment of structural identifiability and non-identifiability for large scale system is challenging and can yield inconclusive results [17]. In the following, we will study the practical identifiability of system. In particular, we will study the uncertainty intervals predicted by different methods, investigate whether the methods properly reflect the structural non-identifiabilities and investigate how the results depend on parameter transformations.

3   Results

In this section, we will compare the previously introduced methods for uncertainty analysis based on two example models. The first model is relatively simple and allows for an in-depth analysis and discussion of the results. The second model is more complex and describes a signal transduction pathway. For the second model, experimental data is available.

3.1   Example 1: Turnover Reaction

The first example we consider is a turnover reaction with synthesis rate k_1 and degradation rate k_2. The one observable of the system is $y = sx$. The initial value for the state x is 0. This yields the following system of equations:

$$\dot{x} = k_1 - k_2 x \quad x(0) = 0$$
$$y = sx \, . \tag{10}$$

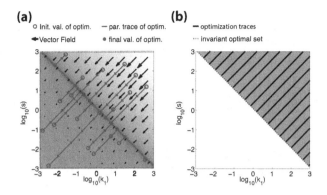

Fig. 3. Illustration of traces of local optimisers and arising histogram of multi-start optimisation results (a) Optimiser traces (○—●) from initial point (○) to final point (●) along the vector field ∇J (→). **(b)** The one half of the lines γ_k are shown in black. The triangular shape shown in Fig. 2 can already be anticipated in this figure.

The analytical solution to this system is $y = \frac{s}{k_1}k_2(1 - \exp(-k_2 t))$. Thus, the parameter k_1 and s are structurally non-identifiable.

For this model, we generated the data $\mathcal{D}$ by simulating the system with parameters $k_1 = 0.75$, $k_2 = 0.25$, $s = 1$ for 30 equi-spaced time points t_k in the interval $[0, 30]$ and adding i.i.d. noise $\epsilon_k \sim \mathcal{N}(0, 1)$.

We used multi-start local optimisation with $n_M = 10^5$ multi-starts in logarithmic coordinates in the hypercube $\Omega = [-3, 3]^3$. The multi-start optimisation was carried out both as constrained optimisation problem (MATLAB function fmincon with default parameters), with parameters restricted to the hypercube Ω, as well as an unconstrained optimisation problem (MATLAB function fminunc with default parameters). For both algorithms the convergence of the algorithm to a local minimum was verified.

A multi-start local optimisation yields an ensemble $\{\hat{\theta}^{\mathcal{D}}_{(k)}\}_{k=1}^{n_M}$ of parameter estimates. The distribution of parameter estimates is depicted in Fig. 2, for the case where constrained and unconstrained optimisation is used. We find that for parameter k_2 all estimates are identical, which can be expected as this parameter is structurally identifiable. For k_1 and s a triangular histogram shape is observed. This is surprising as the parameters are non-identifiable and a flat distribution should be expected.

The triangular shape of the histogram can be explained by studying the emergence of the distribution in more detail. Initial guesses for the parameters are drawn uniformly from the domain Ω and a local optimisation is performed. It is well known that the optimised parameters evolve roughly along the gradient of the objective function $\dot{\theta} = -\nabla_\theta J_\mathcal{D}(\hat{\theta})$ (gradient descent) or a rescaled version of that $\dot{\theta} = -H^{-1}(J)(\hat{\theta})\nabla_\theta J_\mathcal{D}(\hat{\theta})$ (steepest descent). The optimisers stop as soon as a point with negligible gradient, i.e. the global optimum, a local optimum or

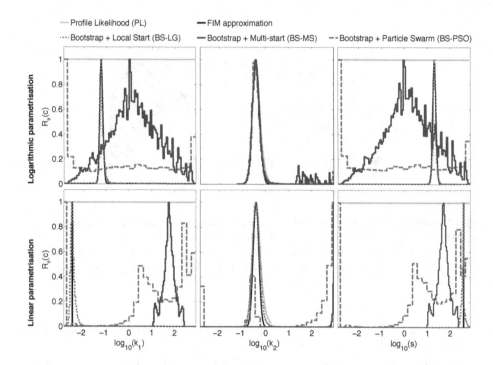

Fig. 4. Comparison of uncertainty analysis using different methods for the turnover model across logarithmic and linear parametrisation. The number of bootstraps was 10^5 for the BS-LG method and 10^4 for the BS-MS and BS-PSO methods. Number of multi-starts was 10 for the BS-MS method. Bin size is chosen according to optimal bin size from Scott's Rule [18].

a point on a non-identifiable manifold, is reached. Accordingly, the optimisation can be interpreted as an optimisation-method-dependent projection of starting points on the subset of points with zero gradient. In addition to the dependence on the optimisation method, the distribution of optimal parameters depends on the choice of the parameter domain Ω. For the considered example, the vector field in the k_1 - s plane (cf. Fig. 3), we find that if we draw random starting points from a square and follow the vector field, the resulting distribution corresponds to the observed triangular. The histograms for constrained and unconstrained optimisation are slightly different, as the projection is altered.

This simple example illustrates that multi-start local optimisation results are insufficient for uncertainty analysis, as they depend on the selected optimisation method and the parameter domain. Furthermore, for identifiable parameters and perfect convergence of the optimiser, a delta-distribution will be observed which also does not reflect the uncertainty correctly.

In a second step, bootstrapping methods using different optimisation schemes have been studied for logarithmic as well as linear parametrisation. For BS-MS and BS-LG we used the MATLAB function `fmincon` with default parameters was used. For BS-PSO we used the MATLAB function `pso`, also with default parameters. The results were compared with profile likelihoods and FIM derived approximations and are shown in Fig. 4.

For the identifiable parameter k_2 all four methods yield similar predictions for the uncertainty. In contrast, for the structurally non-identifiable parameters k_1 and s there are distinct differences. The PLs are flat and suggest structural non-identifiability of k_1 and s. In the case of logarithmic parameterisation, roughly the same is true for BS-PSO, while for linear parameterisation BS-PSO indicates identifiability. This might be due to differences in the distribution of the starting points and indicates the sensitivity of this method to the choice of the parameterisation. BS-MS in logarithmic parameterisation yields a triangular shape, indicating a large uncertainty, but no structural non-identifiability. For the linear parameterisation, the distribution is even tighter.

The most alarming result is revealed by comparing FIM and BS-LG derived uncertainties. In the literature, this comparison seems to be often used to validate the results of the uncertainty analysis [5]. However, we find that although both results agree, they are misleading as they indicate a small uncertainty for structurally non-identifiable problems.

In conclusion, we find already for this simple problem that results of bootstrapping approaches and MS methods depend significantly on the parameterisation, the parameter domain and the optimisation method. FIM-based methods only provide a local approximation. PL is seemingly the only approach to yield reasonable results.

3.2 Example 2: JAK/STAT Pathway

To analyse whether the same problems occur for a more realistic example, we consider the central module of the JAK-STAT signalling pathway [19]. The experimental data presented in [19] has already been studied in great detail and it is well established that the parameters $x_1(0)$, s_1 and s_2 are structurally non-identifiable [1,16,2,3,7,20]. A schematic of the model kinetics, as well as the experimental data is shown in Fig. 5. In this study we employed the `D2D-Toolbox`, which implements state-of-the-art simulation and estimation methods [11].

Figure 5 shows the resulting uncertainty analysis from BS-LG, the FIM approximation, and the PL approach. For the structurally identifiable parameters p_1, p_2, and p_3 all three methods yield similar uncertainty estimates.

For the three structurally non-identifiable parameters $x_1(0)$, s_1, and s_2 the profile likelihood ratio is flat up to some deviations at the border of the shown intervals. The FIM approximation for all three parameter is quite narrow, which indicates identifiability of the respective parameters.

The histograms for $x(0)$, s_1, and s_2 are bimodal. This is surprising and we expect that this is caused by the numerics of the problem. We can conclude that the results of the bootstrapping approach again do not properly reflect the uncertainty of structurally non-identifiable parameters.

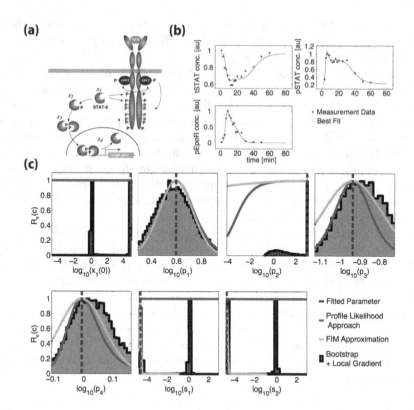

Fig. 5. Model schematic, data and uncertainty analysis for the JAK-STAT model. (a) Schematic illustration of the states and reactions described in the JAK-STAT pathway. Figure taken from [19]. (b) Experimental measurement and respective best fit for the JAK-STAT model. Experimental data is shown as black stars, the optimal fit is shown as solid black line. (c) Uncertainty Analysis for the JAK-STAT model. The number of bootstraps was 10^4. For $x_1(0)$, s_1, p_2, and s_2 there is evident difference in the approximation of $R_{\theta_i}(c)$ between different methods. For p_1, p_2 and p_3 all methods yield comparable results.

4   Discussion

In this paper we investigated the effect of structural non-identifiabilities on the performance of frequentist methods for uncertainty analysis. We reviewed multi-start, bootstrapping, FIM, as well as the profile likelihood based methods for uncertainty analysis.

In Section 3.1, we considered a model, which is simple, but for which two of the three parameters are structurally non-identifiable. Despite the simplicity of the model, none of the methods for uncertainty analysis could indicate the structural non-identifiability of parameters, except for the profile likelihood approach. FIM, and BS-LG approaches even yielded finite confidence intervals and that parameters are identifiable. Similarly, for the model considered in Section

3.2, we observe that the profile likelihood approach also is the only method that properly identifies structural non-identifiability of parameters.

We provided detailed explanations for the emergence of the observed effects which allow for the generalisation of the results to other models and implementations. According to our results, a preceding investigation for structural non-identifiable parameters [17] is advisable in practice, as the emphasised shortcomings of the studied methods could otherwise give rise to misleading results regarding parameter identifiability and uncertainty when using FIM, multi-start and bootstrapping based uncertainty analysis.

Acknowledgements. This work was supported by a German Research Foundation (DFG) Fellowship through the Graduate School of Quantitative Biosciences Munich (QBM; F.F.), the Federal Ministry of Education and Research (BMBF) within the Virtual Liver project (Grant No. 0315766; J.H.), and the European Union within the ERC grant 'LatentCauses' (F.J.T.).

References

1. Raue, A., Kreutz, C., Maiwald, T., Bachmann, J., Schilling, M., Klingmüller, U., Timmer, J.: Structural and Practical Identifiability Analysis of Partially Observed Dynamical Models by Exploiting the Profile Likelihood.. Bioinformatics 25(15), 1923–1929 (2009)
2. Vanlier, J., Tiemann, C.A., Hilbers, P.A.J., van Riel, N.A.W.: A Bayesian Approach to Targeted Experiment Design. Bioinformatics 28(8), 1136–1142 (2012)
3. Vanlier, J., Tiemann, C.A., Hilbers, P.A.J., van Riel, N.A.W.: An Integrated Strategy for Prediction Uncertainty Analysis. Bioinformatics (Oxford, England) 28(8), 1130–1135 (2012)
4. Kreutz, C., Raue, A., Kaschek, D., Timmer, J.: Profile Likelihood in Systems Biology. The FEBS Journal 280(11), 2564–2571 (2013)
5. Joshi, M., Seidel-Morgenstern, A., Kremling, A.: Exploiting the Bootstrap Method for Quantifying Parameter Confidence Intervals in Dynamical Systems. Metabolic Engineering 8(5), 447–455 (2006)
6. Kallenberger, S.M., Beaudouin, J., Claus, J., Fischer, C., Sorger, P.K., Legewie, S., Eils, R.: Intra- and Interdimeric Caspase-8 Self-Cleavage Controls Strength and Timing of CD95-Induced Apoptosis. Science Signaling 7(316) (2014)
7. Schmidl, D., Czado, C., Hug, S., Theis, F.J.: A Vine-copula Based Adaptive MCMC Sampler for Efficient Inference of Dynamical Systems. Bayesian Analysis 8(1), 1–22 (2013)
8. Raue, A., Kreutz, C., Theis, F.J., Timmer, J.: Joining Forces of Bayesian and Frequentist Methodology: A Study for Inference in the Presence of Non-Identifiability. Philosophical Transactions of the Royal Society A: Mathematical, Physical and Engineering Sciences 371(1984) (2013)
9. Hock, S., Hasenauer, J., Theis, F.J.: Modeling of 2D Diffusion Processes based on Microscopy Data: Parameter Estimation and Practical Identifiability Analysis. BMC Bioinformatics 14(suppl.7) (2013)
10. Balsa-Canto, E., Alonso, A.A., Banga, J.R.: An Iterative Identification Procedure for Dynamic Modeling of Biochemical Networks. BMC Systems Biology 4, 11 (2010)

11. Raue, A., Schilling, M., Bachmann, J., Matteson, A., Schelke, M., Kaschek, D., Hug, S., Kreutz, C., Harms, B.D., Theis, F.J., Klingmüller, U., Timmer, J.: Lessons Learned From Quantitative Dynamical Modeling in Systems Biology. PLoS ONE 8(9), e74335 (2013)
12. Gonzalez, O.R., Küper, C., Jung, K., Naval, P.C., Mendoza, E.: Parameter Estimation Using Simulated Annealing for S-System Models of Biochemical Networks. Bioinformatics 23(4), 480–486 (2007)
13. Prata, D.M., Schwaab, M., Lima, E.L., Pinto, J.C.: Nonlinear Dynamic Data Reconciliation and Parameter Estimation through Particle Swarm Optimization: Application for an Industrial Polypropylene Reactor. Chemical Engineering Science 64(18), 3953–3967 (2009)
14. Moles, C.G., Mendes, P., Banga, J.R.: Parameter Estimation in Biochemical Pathways: A Comparison of Global Optimization Methods. Genome Research 13(11), 2467–2474 (2003)
15. Vaz, A., Vicente, L.: A Particle Swarm Pattern Search Method for Bound Constrained Global Optimization. Journal of Global Optimization, 1–6 (2007)
16. Raue, A., Kreutz, C., Maiwald, T., Klingmuller, U., Timmer, J.: Addressing Parameter Identifiability by Model-Based Experimentation. IET Systems Biology 5(2), 120–130 (2011)
17. Chis, O.T., Banga, J.R., Balsa-Canto, E.: Structural Identifiability of Systems Biology Models: A Critical Comparison of Methods. PLoS ONE 6(11), e27755 (2011)
18. Ramsay, P.H., Scott, D.W.: Multivariate Density Estimation, Theory, Practice, and Visualization. Wiley (1993)
19. Swameye, I., Muller, T.G., Timmer, J., Sandra, O., Klingmuller, U.: Identification of Nucleocytoplasmic Cycling as a Remote Sensor in Cellular Signaling by Databased Modeling. PNAS 100(3), 1028–1033 (2003)
20. Balsa-Canto, E., Alvarez-Alonso, A., Banga, J.R.: Computational Procedures for Optimal Experimental Design in Biological Systems. IET Systems Biology 2(4), 163–172 (2008)

Radial Basis Function Approximations of Bayesian Parameter Posterior Densities for Uncertainty Analysis

Fabian Fröhlich[1,2], Sabrina Hross[1,2], Fabian J. Theis[1,2], and Jan Hasenauer[1,2]

[1] Institute of Computational Biology, Helmholtz Zentrum München,
85764 Neuherberg, Germany
{fabian.froehlich,sabrina.hock,fabian.theis,
jan.hasenauer}@helmholtz-muenchen.de
[2] Department of Mathematics, Technische Universität München,
85748 Garching, Germany

Abstract. Dynamical models are widely used in systems biology to describe biological processes ranging from single cell transcription of genes to the tissue scale formation of gradients for cell guidance. One of the key issues for this class of models is the estimation of kinetic parameters from given measurement data, the so called parameter estimation. Measurement noise and the limited amount of data, give rise to uncertainty in estimates which can be captured in a probability density over the parameter space. Unfortunately, studying this probability density, using e.g. Markov chain Monte-Carlo, is often computationally demanding as it requires the repeated simulation of the underlying model. In the case of highly complex models, such as PDE models, this can render the study intractable. In this paper, we will present novel methods for analysis of such probability densities using networks of radial basis functions. We employed lattice generation algorithms, adaptive interacting particle sampling schemes as well as classical sampling schemes for the generation of approximation nodes coupled to the respective weighting scheme and compared their efficiency on different application examples. Our analysis showed that the novel method can yield an expected L_2 approximation error in marginals that is several orders of magnitude lower compared to classical approximations. This allows for a drastic reduction of the number of model evaluations. This facilitates the analysis of uncertainty for problems with high computational complexity. Finally, we successfully applied our method to a complex partial differential equation model for guided cell migration of dendritic cells.

1 Introduction

In computational biology, parameter estimation is of crucial importance to build predictive models. In Bayesian parameter estimation, parameters are considered to be distributed according to a multi-variate probability density. This interpretation emanates from the application of Bayes' Rule:

$$p(\theta|\mathcal{D}) = \frac{p(\mathcal{D}|\theta)p(\theta)}{p(D)} . \tag{1}$$

P. Mendes et al. (Eds.): CMSB 2014, LNBI 8859, pp. 73–85, 2014.

Here, $p(\theta|\mathcal{D})$ is the **posterior** density, a multivariate probability density of the parameter vector θ, given some dataset $\mathcal{D}$. $p(\mathcal{D}|\theta)$ is the **likelihood**, which describes the probability of observing the dataset $\mathcal{D}$ given the parameter vector θ. $p(\theta)$ is the **prior**, which encapsulates information about the parameter which is available before experiments are carried out. $p(\mathcal{D})$ is the **evidence** which normalises the posterior.

To study the uncertainty of individual parameters, the one dimensional marginal densities

$$p(\theta_i|\mathcal{D}) = \int \ldots \int p(\theta|\mathcal{D})d\theta_1 \ldots d\theta_{i-1}d\theta_{i+1} \ldots d\theta_{n_\theta} \tag{2}$$

are analysed. In many applications, however, no closed form of the likelihood or the posterior can be deduced. In those applications, the computation of the evidence $p(\mathcal{D}) = \int p(\theta|\mathcal{D})d\theta$ requires the numerical multivariate integration over all parameters. However, it is usually sufficient to only consider the non-normalised posterior $q(\theta|\mathcal{D}) = p(\mathcal{D}|\theta)p(\theta)$, which does not require computation of integrals.

Still, for the computation of marginal densities, a multivariate integration is necessary. For high-dimensional integration, Markov chain Monte Carlo (MCMC) generally is the method of choice [1]. There have been many developments for efficient computation of marginals using MCMC [2,3]. For example, the MCMC method can be combined with the Variational Bayes approach, which assumes that the posterior factorises over a partition of latent variables [4].

For posterior densities with heavy tails or non-linear correlation structure, MCMC methods can have slow convergence rates [5]. Moreover, the derivation of the factorisation of the posterior can be intractable for many densities. Both of these issues frequently arise when estimating parameters in dynamical systems [3]. The evaluation of the likelihood, and thus also of the posterior, requires the solution of the dynamical system which often is not available in analytical form. Hence, one has to rely on numerical solutions to the dynamical systems which can take several seconds up to minutes when considering PDEs. A good approximation to the marginal density might require millions of samples and thus also millions of simulations of the dynamical system. Therefore, the computation of marginal densities can easily be intractable, even for dynamical systems with a small number of parameters.

In this paper we present a novel method for the approximation of posterior densities using radial basis functions. Radial basis functions are commonly employed in scattered data approximation of multivariate functions, solvers for partial differential equations [6,7] and global optimisation algorithms [8], as they yield good approximations even for small numbers of approximation nodes. In the following, we show how this method can be exploited for uncertainty analysis for parameter estimation. In particular we consider the problem of approximation node generation and compared sampling, lattice and interacting particle based methods. Furthermore, we discuss the advantages and disadvantages of the different methods. Using an example from image-based systems biology, we show that for computationally demanding problems, which for instance require the simulation of PDEs, the proposed scheme is significantly more efficient than classical Monte-Carlo methods.

2   Methods

In the following section, we will first describe the deduction of the multivariate probability densities in Bayesian parameter estimation. Subsequently, we will describe three different methods to approximate the multivariate densities. Eventually we will discuss the approximation of marginal densities, based on marginals of the approximation.

2.1   Bayesian Inference

Recalling Equation (1), the parameter density is the product of likelihood, prior and a normalisation factor. The likelihood $p(\mathcal{D}|\theta)$ describes the probability of observing measurement data $\mathcal{D} = \{(t_k, \bar{y}(t_k))\}_{k=1}^{n_t}$, consisting of n_t time-points t_k and respective n_y-dimensional vector of observations $\bar{y}(t_k)$, given the parameter vector θ:

$$p(\mathcal{D}|\theta) = \prod_{i=1}^{n_y} \prod_{k=1}^{n_t} \sqrt{\frac{1}{2\pi\sigma_{ik}^2}} \exp\left(-\frac{1}{2}\left(\frac{\bar{y}_i(t_k) - y_i(t_k;\theta)}{\sigma_{ik}}\right)^2\right). \tag{3}$$

Here we assumed that $y_i(t_k;\theta)$ is the simulation of the underlying model for species i, which describes the measurement data up to some additive normally distributed measurement noise with variance σ_{ik}^2 at time-point t_k :

$$\bar{y}_i(t_k) \sim \mathcal{N}(y_i(t_k;\theta), \sigma_{ik}^2). \tag{4}$$

Bayesian inference is especially challenging for dynamical systems, as the observables are given by the map of the solution to differential equations

$$y(t;\theta) = h(u(t;\theta), \theta), \tag{5}$$

where u is the solution to some dynamical system with parameters θ, which could be a ordinary or partial differential equation and h is the function that maps states u of the system to the observable components.

2.2   Radial Basis Function Approximation

One of the most common tools for the approximation of probability densities is kernel density estimation. For kernel density estimation, the probability density $p(\theta|\mathcal{D})$ is approximated by a equally weighted convex combination of kernel functions Φ_j. By introducing weighting factors w_j, the approximand φ can be written in a more general form:

$$\varphi^{w,\Phi,N}(\theta) = \frac{1}{\sum_j w_j} \sum_{j=1}^{N} w_j \Phi_j(\theta). \tag{6}$$

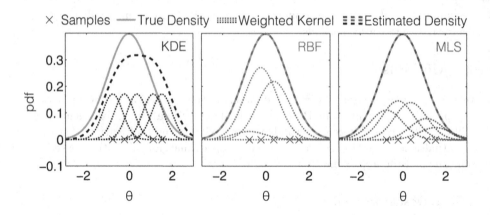

Fig. 1. Illustration of estimated densities from Kernel Density Estimation (KDE), Radial Basis Function interpolation (RBF), and Moving Least Squares Approximation. The approximated density is a univariate normal density with mean 0 and variance 1. The estimation was carried out using 5 random samples from the normal density.

These weighting factors introduce new degrees of freedom which can be used to incorporate additional information on the density such as values of point-wise evaluation.

In the following, we will discuss three different choices for the weighting factors w_j:

- **Kernel density estimation** (KDE): $w_j = 1$
- **Radial Basis Function interpolation** (RBF): $w_j = w_j(\theta^{(j)}, \Phi)$
- **Moving least squares approximation** (MLS): $w_j = \varrho q(\theta^{(j)} | \mathcal{D})$

The approximation of a standard normal density using these kernels is schematically illustrated in Fig. 1.

In this paper, we will only consider Gaussian kernels with mean $\theta^{(j)}$ and covariance matrix $2\varepsilon M$:

$$\Phi_j(\theta) = \sqrt{\frac{1}{\pi^{n_\theta} \det(\varepsilon M)}} \exp\left(-\frac{1}{\varepsilon}(\theta - \theta^{(j)})^T M^{-1}(\theta - \theta^{(j)})\right), \qquad (7)$$

in which $\theta^{(j)}$, $j = 1, \ldots N$, are the approximation nodes, $M \in \mathbb{R}_{sym}^{N \times N} \succeq 0$ describes the correlation structure, $\varepsilon \in \mathbb{R}_+$ is the kernel bandwidth parameter and n_θ is the number of parameters. Although a multitude of alternatives are available, this choice for the kernel allows for analytical formulas for the marginals of φ and efficient evaluations schemes [9]. Moreover, Gaussian kernels are radial basis functions and hence positive definite kernels, which ensures uniqueness of interpolations and allows for efficient numerical interpolation algorithms [6]. However, all methods presented in the following easily translate to other kernel functions, such as the non-smooth Wendland Kernels which should be employed when considering a non-smooth posterior [6].

For Gaussian kernels, the one-dimensional marginals are given by

$$\varphi^{w,\Phi,N}(\theta_i) = \frac{\sum_{j=1}^{N} w_j \sqrt{\frac{1}{\pi^{n_\theta} \det(\varepsilon M)}} \exp\left(-\frac{1}{\varepsilon}(\theta - \theta^{(j)})^T M^{-1}(\theta - \theta^{(j)})\right)}{\sum_j w_j}. \tag{8}$$

The convergence rate for such kernel approximations is usually quantified in terms of the asymptotic mean integrated square error (AMISE)

$$\text{AMISE}_i = \int_{-\infty}^{+\infty} \mathbb{E}(\varphi^{w,\Phi,N}(\theta_i) - p(\theta_i|\mathcal{D}))^2 d\theta. \tag{9}$$

KDE: Kernel density estimation is the common method for estimating probability densities and has been employed for decades [10,11]. As the weight w_j is always 1, no additional computations are necessary. The free parameter ε is usually determined using Scott's rule [11]. For the convergence of the KDE, it is necessary that the approximation nodes $\Theta = \{\theta^{(j)}\}_{j=1}^{N}$ are samples of the density $p(\theta|\mathcal{D})$ which is approximated. The theoretical lower convergence bound for all non-negative KDE kernels is

$$\text{AMISE}^* = \mathcal{O}(N^{-\frac{4}{5}}), \tag{10}$$

which is slower than linear convergence. This lower bound is independent of the dimensionality of θ. However, this is only this bound is only asymptotic and for realistic regimes of N, dimensionality dependent effects and will often dominate.

MLS: In contrast to KDE, MLS exploits the available function values $q(\theta^{(j)}|\mathcal{D})$ at the interpolation nodes. Indeed, the MLS approximation is obtained by minimising a locally weighted distance function [6]. Accordingly, the weights are given by the scaled value of the density at that approximation point, $w_j = \varrho q(\theta^{(j)}|\mathcal{D})$. The scaling parameters ε and ϱ are obtained by minimising $\sum_{j=1}^{N} \left(\varphi^{w,\Phi,N}(\theta^{(j)}) - q(\theta|\mathcal{D})\right)^2$, while M is in general provided. The convergence of the MLS approximation has high regularity requirements on the distribution of points in Θ. In the next section we will discuss options for such regular sets of points.

RBF: Radial basis function interpolation is a common tool employed in scattered data approximation [6,7]. Here the weights are computed based on the interpolation conditions:

$$\underbrace{\begin{bmatrix} \Phi_1(\theta^{(1)}) & \cdots & \Phi_N(\theta^{(1)}) \\ \vdots & \ddots & \vdots \\ \Phi_1(\theta^{(N)}) & \cdots & \Phi_N(\theta^{(N)}) \end{bmatrix}}_{A_{\Theta,\Phi}} \begin{bmatrix} w_1 \\ \vdots \\ w_N \end{bmatrix} = \begin{bmatrix} q(\theta^{(1)}|\mathcal{D}) \\ \vdots \\ q(\theta^{(N)}|\mathcal{D}) \end{bmatrix}. \tag{11}$$

Thus, a linear system of equations must be solved to compute the weights w_j. It can be shown that the condition number of the interpolation matrix $A_{\Theta,\Phi}$

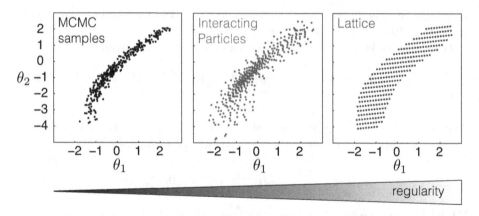

Fig. 2. Illustration of different node construction schemes. The methods are ordered by the regularity in the generated samples. All samples are generated for the density given by (14).

depends on the mesh distance $d_\Theta = \frac{1}{2}\min_{y \neq x} \|x - y\|$ with $x, y \in \Theta$. For the selection of the parameter ε, leave-one-out cross-validation (LOOCV) is the most commonly used approach [6]. For RBF interpolation, the interpolation error depends on the local fill distance $h_\rho(x) := \max_{y \in B(x,\rho)} \min_{\theta \in \Theta} \|y - \theta\| \leq h_0$ of the set Θ [12]. Therefore, the performance of RBF can be tremendously increased by imposing certain regularity based on d_Θ and h_ρ, although RBF interpolation will theoretically work on any set Θ. We will discuss the exact details of these regularity conditions in the next section.

2.3 Construction of Approximation Nodes

As discussed in the previous section, the different approximation schemes for $p(\theta|\mathcal{D})$ have different requirements on the regularity of Θ. In the following we will present methods for the construction of Θ that comply with these requirements. The resulting sets Θ are schematically depicted in Fig. 2.

Sampling: For the convergences of KDEs, the set of approximation nodes Θ has to be a statistically representative sample from the posterior distribution of $p(\theta|\mathcal{D})$. This can be achieved by constructing a Markov chain which has the distribution associated with $p(\theta|\mathcal{D})$ as equilibrium distribution. Markov chain methods enjoy great popularity as method of choice for numerical computation of high-dimensional integrals. Despite numerous improvements [2,3] of the original methods, the number of function evaluations required to obtain a converged Markov chain for multi-modal and heavy-tailed posterior distributions is often large [5].

Lattice: The MLS approach requires regular sets Θ. One possibility to obtain such regular sets are lattices. Lattices are integer linear combinations of basis vectors and in general infinite and thus must be restricted to finite subsets for all practical purposes. It is reasonable to assume that super-level-sets of the density functions are a reasonable choice, as approximation of the density should be carried out in areas of high mass [13]. For the generation of such lattices we used the algorithms presented in [13] and also used the suggestion to motivate the level of super-level-sets using chi-square based confidence levels [13]. In the case of non-identifiable parameters, further restrictions of the domain of interest are necessary for the validity of the method. Theoretically, there exists an infinite number of different lattices as there is no restriction on the choice of basis vectors. However, [13] describes the optimality of A^* root lattices as minimiser of the mesh distance d_Θ and the global maximiser of the local fill distance h_ρ for dimensions up to three. Hence these lattices also constitute a viable choice for RBF interpolation. However, the number of lattice points grows exponentially with the number of dimensions, which limits their applicability. Sparse grids could also be a valid approach for lattice generation [14], however no efficient methods for the restriction to super-level-sets are available.

Interacting Particles: Also for the RBF method, lattices are a viable choice. RBF interpolation, however, allows for more flexibility in the choice of Θ and should also benefit from local refinements in the point density [7]. In general, the generation of locally finer lattices is difficult and a more elegant solution can be obtained by using interacting particle methods [15].

Such an interacting particle system is given by an energy function

$$E(\Theta) = \sum_{p=1}^{N}\sum_{q=1}^{N} V\left(\left\|\theta^{(q)} - \theta^{(p)}\right\|\right), \tag{12}$$

where $V(r)$ is the potential which defines the interacting forces between particles. Here, $\underset{\Theta}{\operatorname{argmin}} E(\Theta)$ is the ground state of the particle system. By locally rescaling the potential of the function by the inverse of the density function

$$\tilde{D}(\theta) = \frac{D_0}{\sqrt{1 + \|\nabla p(\theta|\mathcal{D})\|}}, \tag{13}$$

a higher density in θ can be achieved in areas where the non-normalised density $p(\theta|\mathcal{D})$ exhibits large fluctuations [16].

In summary, there are three commonly used density approximation methods and three methods for the generation of approximation nodes. In the following, we will study the performance of the four most reasonable combinations: (i) KDE with sampling; (ii) MLS with lattices; and (ii) RBF with lattices and interacting particles.

3 Results

The evaluate the performance of the different methods, we consider two complementary examples. Firstly, we study a 2-dimensional numerical example for which analytical solutions are available. Secondly, the parameter inference for the PDE model from imaging data is considered.

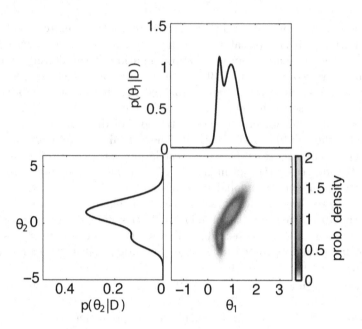

Fig. 3. Posterior density and its marginals for Example 1. The posterior density is plotted in the lower right. The respective marginal densities are plotted above and to the left of the posterior. Both marginal densities and the posterior density are bimodal.

3.1   Example 1: Numerical Example

In the first example, we assume that the posterior distribution $P(\theta|\mathcal{D})$ of the dynamical system is described by the sums of two normal distributions,

$$P(\theta|\mathcal{D}) = (\frac{4}{5}\mathcal{N}(\mu^{(1)}, \Sigma^{(1)}) + \frac{1}{5}\mathcal{N}(\mu^{(2)}, \Sigma^{(2)}) \tag{14}$$

with covariance matrices and means

$$\Sigma^{(1)} = \begin{bmatrix} 0.1 & 0.25 \\ 0.25 & 1 \end{bmatrix}, \; \Sigma^{(2)} = \begin{bmatrix} 0.01 & -0.01 \\ -0.01 & 0.5 \end{bmatrix}, \; \mu^{(1)} = \begin{bmatrix} 1 \\ 1 \end{bmatrix}, \; \mu^{(2)} = \begin{bmatrix} 0.5 \\ -1.5 \end{bmatrix}. \tag{15}$$

The posterior density as well as the respective marginals are plotted in Fig. 3.

For this example the marginal densities can be computed explicitly, which allows for exact error analysis. We carried out two different comparisons. First we compared KDE on samples, MLS on lattice and RBF on lattice (Fig. 4 (left)). Secondly, we also considered RBF with interacting particles with different particle numbers (Fig. 4 (right)). The MCMC samples were generated using the DRAM toolbox [2]. For the generation and restriction of the lattice and the particle method the implementation previously described methods [13] have been used.

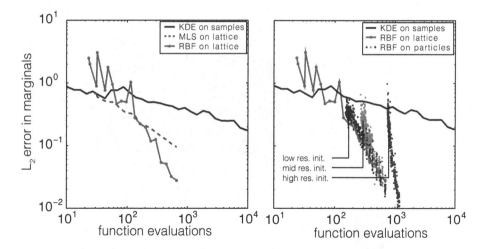

Fig. 4. Comparison of the approximation error in marginals for the KDE, the MLS and the RBF approach. The L_2 error in marginal approximation is plotted for the KDE method using MCMC samples is plotted for several sample numbers of a single Markov chain and compared to the MLS and RBF approach on the lattice. Lattices with different numbers of nodes were obtained by rescaling of the basis vectors. For the RBF method coupled to the particle method, the initialisations were generated using lattices with varying resolution.

The number of function evaluations for the lattice based approaches is determined by the scaling of the basis for the lattice generation. By decreasing the scaling factor of the basis vector, the distance between points in the lattice decreases and the total number of points and thus the total number of function evaluations increases. For low numbers of samples, KDE on MCMC samples and MLS and RBF on lattice yield similar results. However, for higher sample numbers both MLS, as well as RBF yield significantly lower errors. The error which is attained by the MLS approach at about 10^3 samples is obtained by the KDE approach between 10^4 and 10^5 samples. For the same number of samples, the RBF approach already yields an error which is approximately one magnitude lower than both the error from the KDE and the MLS approach.

To avoid additional function evaluations for the interaction particle method, the gradient $\nabla p(\theta|\mathcal{D})$ was approximated using the MLS approach. The number of function evaluation is therefore the number of points in the initial configuration plus the final number of particles. The number of particles is determined by the characteristic length of the inter-particle forces given by D_0 in (13).

Despite the required number of function evaluations for the initialisations, the combination of RBF approximation and interacting particles methods provides slightly better performance, for low and medium resolution, than the lattice-based approach. The fluctuations are a result of stochasticity in the interacting particle scheme [16]. This suggests that, for this example, the resolution of the approximation to $\nabla p(\theta|\mathcal{D})$ has only a negligible influence on the resulting error of the marginal approximation.

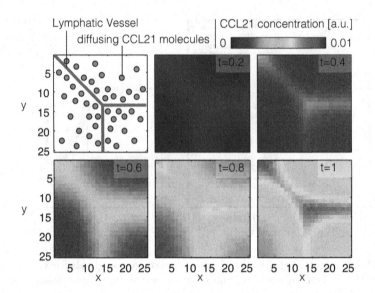

Fig. 5. Schematic of the model and synthetic data for the chemokine gradient model. The upper left plot shows the schematic of the underlying geometry for the generated synthetic data. The synthetic data for the measurement of CCL21 concentration at various time-points is shown in the other subplots.

Summarising, we could show that for this example both the MLS and the RBF approximation schemes coupled to the respective sampling schemes yielded lower errors in marginals compared to the KDE approximation scheme.

3.2   Example 2: PDE Model for Chemokine Gradient Formation

In this section we consider a PDE model describing the formation of gradients of the cytokine CCL21 around lymphatic vessels [17]. Such gradients are, among other processes, responsible for the migration of dendritic cells towards the lymphatic vessels. The subsequent dendritic cell traffic through the lymphoid vessels towards the lymph nodes is a key process in adaptive immune response.

The model considered here utilises a reaction diffusion equation to describe the secretion of CCL21 by the lymphoid vessels and the subsequent diffusion. The secreted CCL21 is then stabilised by complex formation with heparan sulfates in the surrounding tissue, which is modelled by an additional ODE. A schematic of the model as well as the utilised synthetic data is shown in Fig. 5. Of biological interest are the five unknown parameters: diffusion D, secretion α and degradation rate γ of CCL21 as well as the association k_1 and dissociation constant k_{-1} of the complex. For the detailed model, parameter inference and uncertainty analysis for those parameters we refer to the original publication [17]. In the original work the posterior probability was not calculate as it requires the simulation of the discretised PDE with several thousand state variables and is nearly infeasible with traditional methods.

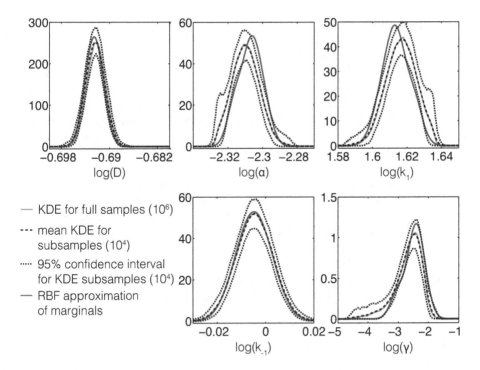

Fig. 6. Comparison of marginal densities computed using KDE and RBF approaches. The RBF approach was employed using a lattice of about 10^4 points. For the KDE approach a total of 10^6 MCMC samples were generated and percentile based confidence intervals calculated using a sliding window bootstrapping scheme with 10^3 bootstraps and a window size of 10^4.

In the following we applied the KDE and RBF approach to this problem to approximate the posterior probability and the resulting estimated marginal densities are shown in Fig 6. First 10^6 MCMC samples were generated using the DRAM toolbox [2], as these samples passed all convergence tests and its approximated marginal was considered as reference solution. Second, we generated approximation nodes for the RBF approach based on 10^4 lattice points. The resulting marginal is a reasonable approximation as it is close to the reference solution (see Fig. 6). Last, we used a sliding window approach to generate subsamples of size 10^4 from the original MCMC. Based on those samples we derived KDE based marginals comparable to the RBF approach.

If we compare both approaches we see that the 95% confidence intervals of the KDE-derived marginal show a rather large uncertainty although a highly efficient adaptive sampling scheme has been employed. In comparison the RBF approximation is often closer to the reference solution. Furthermore, the sample generation took several weeks compared to several days for the RBF lattice points. This suggests that that for this problem, RBF approximation can indeed reduce the number of required function evaluations.

4   Discussion

The problem of posterior densities approximation is omnipresent in Bayesian parameter estimation. One of the common approaches is KDE in combination with MCMC sampling. For some simple examples also analytical approximations are available. In this manuscript we suggest two alternative methods introduced in the context of scattered data approximation, namely MLS and RBF approximation. As these methods require a higher regularity of the approximation nodes, we also introduced lattices and interacting particle methods for node generation.

The resulting methods have been evaluated based on examples. For a 2-dimensional example we showed superior convergence properties for the RBF and MLS approach over the KDE approach. For a computationally demanding PDE model with 5 parameters, for which the MCMC sampling took several weeks, we showed that RBF methods provide a valid alternative which can provide better approximations with smaller samples sizes. This PDE example suggests that the method is suited for problems with a low dimensional parameter space but high computational complexity. In this case, the RBF and MLS approach could significantly reduce the number of required simulations of the systems. As this is one of the first application of MLS and RBF approximations in the context of parameter estimation [18], further studies are necessary to determine the full potential of the methods.

Acknowledgements. We thank Frank Filbir for fruitful discussions on the subject of this paper. This work was supported by a DFG Fellowship through the Graduate School of Quantitative Biosciences Munich (QBM; F.F.), the Federal Ministry of Education and Research (BMBF) within the Virtual Liver project (Grant No. 0315766; J.H.), and the European Union within the ERC grant 'LatentCauses' (F.J.T.).

References

1. Robert, C., Casella, G.: Monte Carlo Statistical Methods. Springer, New York (2004)
2. Haario, H., Laine, M., Mira, A., Saksman, E.: DRAM: Efficient Adaptive MCMC. Statistics and Computing 16(4), 339–354 (2006)
3. Schmidl, D., Czado, C., Hug, S., Theis, F.J.: A Vine-copula Based Adaptive MCMC Sampler for Efficient Inference of Dynamical Systems. Bayesian Analysis 8(1), 1–22 (2013)
4. MacKay, D.: Information Theory, Inference and Learning Algorithms (2003)
5. Jarner, S., Roberts, G.: Convergence of Heavy-tailed Monte Carlo Markov Chain Algorithms. Scandinavian Journal of Statistics 34(1994), 781–815 (2007)
6. Fasshauer, G.: Meshfree Approximation Methods with MATLAB. World Scientific (2007)
7. Sarra, S.A., Kansa, E.J.: Multiquadric Radial Basis Function Approximation Methods for the Numerical Solution of Partial Differential Equations. Advances in Computational Mechanics 2 (2009)

8. Müller, J.: Surrogate Model Algorithms for Computationally Expensive Black-Box Global Optimization Problems. PhD thesis (2012)

9. Potts, D., Steidl, G.: Fast Summation at Nonequispaced Knots by NFFT. SIAM Journal on Scientific Computing 24(6), 2013–2037 (2003)

10. Silverman, B.: Density Estimation for Statistics and Data Analysis. Chapman & Hall/CRC (1986)

11. Ramsay, P.H., Scott, D.W.: Multivariate Density Estimation, Theory, Practice, and Visualization. Wiley (1993)

12. Wu, Z., Schaback, R.: Local Error Estimates for Radial Basis Function Interpolation of Scattered Data. IMA Journal of Numerical Analysis 13(1), 1–15 (1993)

13. Fröhlich, F.: Approximation and Analysis of Probability Densities using Radial Basis Functions. Master's thesis, Technische Universität München, Germany

14. Bungartz, H.J., Griebel, M.: Sparse grids. Acta Numerica 13, 147 (2004)

15. Torquato, S.: Reformulation of the Covering and Quantizer Problems as Ground States of Interacting Particles. Physical Review E 82(5), 1–52 (2010)

16. Reboux, S., Schrader, B., Sbalzarini, I.F.: A Self-Organizing Lagrangian Particle Method for Adaptive Resolution Advection-Diffusion Simulations. Journal of Computational Physics 231(9), 3623–3646 (2012)

17. Hock, S., Hasenauer, J., Theis, F.J.: Modeling of 2D Diffusion Processes based on Microscopy Data: Parameter Estimation and Practical Identifiability Analysis. BMC Bioinformatics 14(suppl.7) (2013)

18. Blizniouk, N., Ruppert, D., Shoemaker, C., Regis, R.: Bayesian Calibration of Computationally Expensive Models using Optimization and Radial Basis Function Approximation. Journal of Computational and Graphical Statistics 17(2) (2008)

Precise Parameter Synthesis
for Stochastic Biochemical Systems

Milan Češka[1,2], Frits Dannenberg[1], Marta Kwiatkowska[1], and Nicola Paoletti[1]

[1] Department of Computer Science, University of Oxford, UK
[2] Faculty of Informatics, Masaryk University, Czech Republic

Abstract. We consider the problem of synthesising rate parameters for stochastic biochemical networks so that a given time-bounded CSL property is guaranteed to hold, or, in the case of quantitative properties, the probability of satisfying the property is maximised/minimised. We develop algorithms based on the computation of lower and upper bounds of the probability, in conjunction with refinement and sampling, which yield answers that are precise to within an arbitrarily small tolerance value. Our methods are efficient and improve on existing approximate techniques that employ discretisation and refinement. We evaluate the usefulness of the methods by synthesising rates for two biologically motivated case studies, including the reliability analysis of a DNA walker.

1 Introduction

Biochemical reaction networks are a convenient formalism for modelling a multitude of biological systems, including molecular signalling pathways, logic gates built from DNA and DNA walker circuits. For low molecule counts, and under the well-mixed and fixed volume assumption, the prevailing approach is to model such networks using continuous-time Markov chains (CTMCs) [11]. Stochastic model checking [17], e.g. using PRISM [18], can then be employed to analyse the behaviour of the models against temporal logic properties expressed in CSL (Continuous Stochastic Logic) [2]. For example, one can establish the reliability and performance of DNA walker circuits by means of properties such as "what is the probability that the walker reaches the correct final anchorage within 10 min?". Since DNA circuits can implement biosensors and diagnostic systems, ensuring appropriate levels of reliability is crucial to guarantee the safety of deploying molecular devices in healthcare applications.

Stochastic model checking, however, assumes that the model is fully specified, including the kinetic rates. In view of experimental measurement error, these are rarely given precisely, but rather as intervals of values. The *parameter synthesis problem*, studied for CTMCs in [13], assumes a formula and a model whose rates are given as functions of model parameters, and aims to compute the parameter valuations that guarantee the satisfaction of the formula. This allows one, for example, to determine the ranges of parameter values for a given level of reliability and performance, which can provide important feedback to the designers of biosensors and similar molecular devices, and thus significantly extends the power of stochastic model checking.

P. Mendes et al. (Eds.): CMSB 2014, LNBI 8859, pp. 86–98, 2014.
© Springer International Publishing Switzerland 2014

In [13], the parameter synthesis problem was solved for CTMCs approximately, and only for probabilistic time-bounded reachability. In this paper, we address the parameter synthesis problem for stochastic biochemical reaction networks for the full time-bounded fragment of the (branching-time) logic CSL [2]. We formulate two variants: *threshold synthesis*, which inputs a CSL formula and a probability threshold and identifies the parameter valuations which meet the threshold, and *max synthesis*, where the maximum probability of satisfying the property and the maximizing set of parameter valuations are returned.

We develop efficient synthesis algorithms that yield answers with arbitrary precision. The algorithms exploit the recently published parameter exploration technique that computes safe approximations to the lower and upper bounds for the probability to satisfy a CSL property over a fixed parameter space [6]. In contrast to the exploration technique, our algorithms automatically derive the satisfying parameter regions through iterative decomposition of the parameter space based on refining the preliminary answer with additional decompositions up to a given problem-specific tolerance value. We also show that significant computational speed-up is achieved by enhancing the max synthesis algorithm by sampling the property at specific points in the parameter space. We demonstrate the usefulness of the method through two case studies: the SIR epidemic model [16], where we synthesize infection and recovery rates that maximize the probability of disease extinction, and the DNA walker circuit [9], where we derive the rates that ensure a predefined level of reliability.

Related Work. Parameter synthesis has been studied for discrete-time Markovian models in [12,7]. The approach applies to unbounded temporal properties and is based on constructing a rational function by performing state elimination [12]. For CTMCs and bounded reachability specifications, the problem can be reduced to the analysis of the polynomial function describing the reachability probability of a given target state [13]. The main limitation here is the high degree of the polynomials, which is determined by the number of uniformization steps. Therefore, in contrast to our method, only an approximate solution is obtained using discretization of parameter ranges. When considering linear-time specifications, specific restrictions can be placed on the rate function to result in a smooth satisfaction function (i.e. having derivatives of all orders). In that case, the function can be approximated using statistical methods which leverage the smoothness [5]. A concept similar to smoothness, uniform continuity, can be used to obtain an unbiased statistical estimator for the satisfaction function [14]. Both methods approximate parameter synthesis using confidence intervals. Inference of parameter values in probabilistic models from time-series measurements is a well studied area of research [1,4], but different from the problem we consider. Interval CTMCs, where transition rates are given as intervals, have been employed to obtain a three-valued abstraction for CTMCs [15]. In contrast to parametric models we work with, the transition rates in interval CTMCs are chosen nondeterministically and schedulers are introduced to compute lower and upper probability bounds.

2 Background

We state preliminary definitions relevant to the study of Parametric Continuous Time Markov Chains [13,6] that permit formal analysis of probabilistic models with uncertain parameters [20].

A *Continuous Time Markov Chain* (CTMC) is a tuple $\mathcal{C} = (S, \pi_0, \mathbf{R})$ where S is a finite set of states, $\pi_0 : S \to \mathbb{R}_{\geq 0}$ is the initial distribution and $\mathbf{R} : S \times S \to \mathbb{R}_{\geq 0}$ is the rate matrix. A transition between states $s, s' \in S$ can occur only if $\mathbf{R}(s, s') > 0$ and in that case the probability of triggering the transition within t time units equals $1 - e^{-t\mathbf{R}(s,s')}$. The time spent in s, before a reaction occurs, is exponentially distributed with rate $E(s) = \sum_{s' \in S} \mathbf{R}(s, s')$, and when the transition occurs the probability of moving to state s' is given by $\frac{\mathbf{R}(s,s')}{E(s)}$. Let $\mathbf{E}$ be a $S \times S$ diagonal matrix such that $\mathbf{E}(s_i, s_i) = E(s_i)$, and define the generating matrix by setting $\mathbf{Q} = \mathbf{R} - \mathbf{E}$. Then a vector $\pi_t : S \to \mathbb{R}_{\geq 0}$ of the transient probabilities at time t is given by $\frac{d\pi_t}{dt} = \pi_t \mathbf{Q}$ such that $\pi_t = \pi_0 e^{\mathbf{Q}t}$. Using standard uniformisation the transient probability at time t is obtained as a sum of state distributions after i discrete-stochastic steps, weighted by the probability of observing i steps in a Poisson process. Let $\mathbf{P} = \mathbf{I} + \frac{1}{q}\mathbf{Q}$ be the uniformised matrix, where $q \geq \max\{E(s) - \mathbf{R}(s,s) \mid s \in S\}$ is called the uniformisation rate. The transient probabilities π_t are computed as $\pi_t = \pi_0 \sum_{i=0}^{k_\epsilon} \gamma_{i,qt} \mathbf{P}^i$ where $\gamma_{i,qt} = e^{-qt}\frac{(qt)^i}{i!}$ denotes the i-th Poisson probability for a process with rate qt, and k_ϵ satisfies the convergence bound $\sum_0^{k_\epsilon} \gamma_{i,qt} \geq 1 - \epsilon$ for some $\epsilon > 0$. The Poisson terms and summation bound can be efficiently computed using an algorithm due to Fox and Glynn [10].

We assume a set K of model parameters. The domain of each parameter $k \in K$ is given by a closed real interval describing the range of possible values, i.e, $[k^\perp, k^\top]$. The *parameter space* $\mathcal{P}$ induced by K is defined as the Cartesian product of the individual intervals: $\mathcal{P} = \times_{k \in K}[k^\perp, k^\top]$. A *parameter point* $p \in \mathcal{P}$ is a valuation of each parameter k. Subsets of the parameter space are also referred to as *parameter regions* or *subspaces*. $\mathbb{R}[K]$ denotes the set of polynomials over the reals $\mathbb{R}$ with variables $k \in K$.

Parametric Continuous Time Markov Chains (pCTMCs) [13] extend the notion of CTMCs by allowing transition rates to depend on model parameters. Formally, a *p*CTMC over a set K of parameters is a triple $\mathcal{C} = (S, \pi_0, \mathbf{R})$ where s and π_0 are as above, and in this case $\mathbf{R} : S \times S \to \mathbb{R}[K]$ is the parametric rate matrix. Given a pCTMC $\mathcal{C}$ and a parameter space $\mathcal{P}$, we denote with $\mathcal{C}_\mathcal{P}$ the (possibly uncountable) set $\{\mathcal{C}_p \mid p \in \mathcal{P}\}$ where $\mathcal{C}_p = (S, \pi, \mathbf{R}_p)$ is the instantiated CTMC obtained by replacing the parameters in $\mathbf{R}$ with their evaluation in p.

We consider the time-bounded fragment of CSL [2] to specify behavioural properties, with the following syntax. A state formula Φ is given as $\Phi ::= \text{true} \mid a \mid \neg\Phi \mid \Phi \wedge \Phi \mid P_{\sim r}[\phi] \mid P_{=?}[\phi]$, where ϕ is a path formula whose syntax is $\phi ::= \mathbf{X}\,\Phi \mid \Phi\,\mathbf{U}^I\,\Phi$, a is an atomic proposition, $\sim \in \{<, \leq, \geq, >\}$, $r \in [0, 1]$ is a probability threshold and I is a bounded interval. Using $P_{=?}[\phi]$ we specify properties which evaluate to the probability that ϕ is satisfied. The synthesis methods presented in this paper can be directly adapted to the time-bounded fragment of CSL with the reward operator [17], but, for the sake of simplicity, here we present our methods only for the probabilistic operator P.

Let ϕ be a CSL path formula and $\mathcal{C}_\mathcal{P}$ be a pCTMC over a space $\mathcal{P}$. We denote with $\Lambda : \mathcal{P} \to [0,1]$ the *satisfaction function* such that $\Lambda(p) = P_{=?}[\phi]$, that is, $\Lambda(p)$ is the probability of ϕ being satisfied over the CTMC $\mathcal{C}_p$. Note that the path formula ϕ may contain nested probabilistic operators, and therefore the satisfaction function is, in general, not continuous.

Biochemical reaction networks provide a convenient formalism for describing various biological processes as a system of well-mixed reactive species in a volume of fixed size. A CTMC semantics can be derived whose states hold the number of molecules for each species, and transitions correspond to reactions that consume and produce molecules. Bounds on species counts can be imposed to obtain a finite-state model. The rate matrix is defined as $\mathbf{R}(s_i, s_j) \stackrel{def}{=} \sum_{r \in \mathsf{reac}(s_i, s_j)} f_r(K, s_i)$ where $\mathsf{reac}(s_i, s_j)$ denotes all the reactions changing state s_i into s_j and f_r is the *stochastic rate function* of reaction r over parameters $k \in K$. In this paper, we assume *multivariate polynomial* rate functions that include, among others, mass-action kinetics where $k \in K$ are kinetic rate parameters.

3   Problem Definition

We consider pCTMC models of biochemical reaction networks that can be parametric in the rate constants and in the initial state. We introduce two parameter synthesis problems for this class of models: the *threshold synthesis* problem that, given a threshold $\sim r$ and a CSL path formula ϕ, asks for the parameter region where the probability of ϕ meets $\sim r$; and the *max synthesis* problem that determines the parameter region where the probability of the input formula attains its maximum, together with an estimation of that maximum. In the remainder of the paper, we omit the min synthesis problem that is defined and solved in a symmetric way to the max case.

In contrast to previous approaches that support only specific kinds of properties (e.g. reachability as in [13]), we consider the full time-bounded fragment of CSL with rewards, thus enabling generic and more expressive synthesis requirements. Moreover, the variants of the synthesis problem that we define correspond to qualitative and quantitative CSL formulas, which are of the form $P_{\geq r}[\phi]$ and $P_{=?}[\phi]$, respectively. Solutions to the threshold problem admit parameter points left undecided, while, in the max synthesis problem, the set of maximizing parameters is contained in the synthesis region. Our approach supports arbitrarily precise solutions through an input tolerance that limits the volume of the undecided region (in the threshold case) and of the synthesis region (in the max case). To the best of our knowledge, no other synthesis methods for CTMCs exist that provide guaranteed error bounds.

Problem 1 (Threshold Synthesis). Let $\mathcal{C}_\mathcal{P}$ be a pCTMC over a parameter space $\mathcal{P}$, $\Phi = P_{\geq r}[\phi]$ with $r \in [0,1]$ be a CSL formula and $\varepsilon > 0$ a volume tolerance. The *threshold synthesis* problem is finding a partition $\{T, U, F\}$ of $\mathcal{P}$, such that:

1. $\forall p \in T.\ \Lambda(p) \geq r$; and
2. $\forall p \in F.\ \Lambda(p) < r$; and
3. $\mathrm{vol}(U)/\mathrm{vol}(\mathcal{P}) \leq \varepsilon$

where Λ is the satisfaction function of ϕ on $\mathcal{C}_\mathcal{P}$; and $\mathrm{vol}(A) = \int_A 1 d\mu$ is the volume of A.

Problem 2 (Max Synthesis). Let $\mathcal{C}_\mathcal{P}$ be a pCTMC over a parameter space $\mathcal{P}$, $\Phi = P_{=?}[\phi]$ be a CSL formula and $\epsilon > 0$ a probability tolerance. The *max synthesis* problem is finding a partition $\{T, F\}$ of $\mathcal{P}$ and probability bounds $\Lambda^\perp$, $\Lambda^\top$ such that:

1. $\Lambda^\perp - \Lambda^\top \leq \epsilon$;
2. $\forall p \in T.\ \Lambda^\perp \leq \Lambda(p) \leq \Lambda^\top$; and
3. $\exists p \in T.\ \forall p' \in F.\ \Lambda(p) > \Lambda(p')$.

where Λ is the satisfaction function of ϕ on $\mathcal{C}_\mathcal{P}$.

Note that we need to consider a probability tolerance to control the inaccuracy of the max probability, and in turn of region T. Indeed, constraining only the volume of T gives no guarantees on the precision of the maximizing region.

4   Computing Lower and Upper Probability Bounds

This section presents a generalization of the parameter exploration procedure originally introduced in [6]. The procedure takes a pCTMC $\mathcal{C}_\mathcal{P}$ and a CSL path formula ϕ, and provides safe under- and over-approximations for the minimal and maximal probability that $\mathcal{C}_\mathcal{P}$ satisfies ϕ, that is, lower and upper bounds $\Lambda_{\min}$ and $\Lambda_{\max}$ satisfying $\Lambda_{\min} \leq \min_{p \in \mathcal{P}} \Lambda(p)$ and $\Lambda_{\max} \geq \max_{p \in \mathcal{P}} \Lambda(p)$. The accuracy of these approximations is improved by partitioning the parameter space $\mathcal{P}$ into subspaces and re-computing the corresponding bounds, which forms the basis of the synthesis algorithms that we discus in the next section. For now we focus on obtaining approximations $\Lambda_{\min}, \Lambda_{\max}$ for a fixed parameter space $\mathcal{P}$. The model-checking problem for any time-bounded CSL formula reduces to the computation of transient probabilities [3], and a similar reduction is applicable to the computation of lower and upper bounds. Following [6], to correctly handle nested probabilistic operators, under- and over-approximations of the satisfying sets of states in the nested formula are computed.

We now re-state the transient probabilities as given by standard uniformisation and include the dependency on the model parameters in our notation, so that $\pi_{t,p} = \pi_0 \sum_{i=0}^{k_\epsilon} \gamma_{i,qt} \mathbf{P}_p^i = \sum_{i=0}^{k_\epsilon} \gamma_{i,qt} \tau_{i,p}$ where $\mathbf{P}_p$ is the uniformised rate matrix obtained from the rate matrix $\mathbf{R}_p$ and $\tau_{k,p} = \pi_0 \mathbf{P}_p^k$ is the probability evolution in the discretized process. Observe that, if some functions $\pi_i^{\min}$ and $\pi_i^{\max}$ can be obtained such that for any step i,

$$\tau_i^{\min} \leq \min_{p \in \mathcal{P}} \tau_{i,p} \text{ and } \tau_i^{\max} \geq \max_{p \in \mathcal{P}} \tau_{i,p} \tag{1}$$

then robust approximations $\pi_t^{min} = \sum_{i=0}^{k_\epsilon} \gamma_{i,qt} \tau_i^{min}$ and $\pi_t^{max} = \sum_{i=0}^{k_\epsilon} \gamma_{i,qt} \tau_i^{max}$ provide the bounds Λ_{min} and Λ_{max} that we seek. As usual, the vector ordering in Equation 1 holds element-wise. If we assume that some functions f_k^{min}, f_k^{max} exist such that $f_k^{min}(\tau_i^{min}) = \tau_{i+k}^{min}$ and $f_k^{max}(\tau_i^{max}) = \tau_{i+k}^{max}$ then recursively the terms in Equation 1 for all i are obtained, given that the first k terms for $\tau_i^{min}, \tau_i^{max}$ are known. We now note that the functions

$$f_k^{min}(\tau_i^{min}) = \min_{p \in \mathcal{P}} \tau_i^{min} \mathbf{P}_p^k \text{ and } f_k^{max}(\tau_i^{max}) = \max_{p \in \mathcal{P}} \tau_i^{max} \mathbf{P}_p^k \qquad (2)$$

can be under- and over-approximated using analytical methods when the parametric rate matrix $\mathbf{R}_p$ employs low-degree multivariate polynomial expressions. Provided that $\mathbf{R}_p(s_i, s_j)$ is a polynomial of at most degree d over the parameter space, the degree of $\tau_{k,p}(s) = \pi_0 \mathbf{P}_p^k(s)$ is at most kd.

An analytical treatment for the case $k = 1$ and $d = 1$ is given in [6]. Here, we derive an effective method to obtain approximations using $k = 1$ for multivariate polynomials where each variable has degree at most 1; full details are included in [21]. More advanced methods can be used, provided that the under- and over-approximations for Equation 2 are sound. Note that the solution $\pi_{t,p}(s)$ itself can be expressed as a polynomial of degree at most $k_\epsilon d$. A direct attempt to bound the polynomial expression of $\pi_{t,p}(s)$ is difficult due to the large number of uniformisation steps, k_ϵ, and previous approaches in parameter synthesis have provided an approximate solution by sampling the value of $\pi_{t,p}$ over a grid in $\mathcal{P}$ [13], rather than bounding the polynomial itself as in our approach. The computational complexity depends on the chosen rate function and the bounding method for the functions in Equation 2, but for our settings it has the same asymptotic complexity as standard uniformisation. Two approximation errors are introduced when we compute π_t^{max} (or π_t^{min}). Firstly, the probabilities $\tau_i^{max}, \tau_{i+k}^{max}, \tau_{i+2k}^{max}, \cdots$ are locally maximized, so that different parameter valuations are allowed at each step and for each state. Secondly, the error of over-approximating $f_k^{max}(\tau_i^{max})$ accumulates in τ_i^{max} at every iteration.

5   Refinement-Based Parameter Synthesis

We present algorithms to solve Problems 1 and 2, based on the computation of probability bounds introduced in Section 4 and iterative parameter space refinement. In the max synthesis case we employ parameter sampling to enhance the synthesis procedure.

Threshold Synthesis. Algorithm 1 describes the method to solve the threshold synthesis problem with input formula $\Phi = P_{\geq r}[\phi]$. The idea, also illustrated in Figure 1, is to iteratively refine the undecided parameter subspace U (line 3) until the termination condition is met (line 14). At each step, we obtain a partition R of U. For each subspace $\mathcal{R} \in R$, the algorithm computes bounds $\Lambda_{min}^{\mathcal{R}}$ and $\Lambda_{max}^{\mathcal{R}}$ on the maximal and minimal probability that $\mathcal{C}_{\mathcal{R}}$ satisfies ϕ (line 5). We then evaluate if $\Lambda_{min}^{\mathcal{R}}$ is above the threshold r, in which case the satisfaction of Φ is guaranteed for the whole region $\mathcal{R}$ and thus it is added to T. Otherwise,

Algorithm 1. Threshold Synthesis

Require: pCTMC $\mathcal{C}_\mathcal{P}$ over parameter space $\mathcal{P}$, CSL formula
$\quad \Phi = P_{\geq r}[\phi]$ and volume tolerance $\varepsilon > 0$
Ensure: T, U and F as in Problem 1
1: $T \leftarrow \emptyset, F \leftarrow \emptyset, U \leftarrow \mathcal{P}$
2: **repeat**
3: $\quad R \leftarrow$ decompose(U), $U \leftarrow \emptyset$
4: $\quad$ **for each** $\mathcal{R} \in R$ **do**
5: $\quad\quad (\Lambda_{min}^{\mathcal{R}}, \Lambda_{max}^{\mathcal{R}}) \leftarrow$ computeBounds($\mathcal{C_R}, \phi$)
6: $\quad\quad$ **if** $\Lambda_{min}^{\mathcal{R}} \geq r$ **then**
7: $\quad\quad\quad T \leftarrow T \cup \mathcal{R}$
8: $\quad\quad$ **else if** $\Lambda_{max}^{\mathcal{R}} < r$ **then**
9: $\quad\quad\quad F \leftarrow F \cup \mathcal{R}$
10: $\quad\quad$ **else**
11: $\quad\quad\quad U \leftarrow U \cup \mathcal{R}$
12: **until** vol(U)/vol($\mathcal{P}$) $\leq \varepsilon$ $\qquad \triangleright$ where vol(A) = $\int_A 1 d\mu$

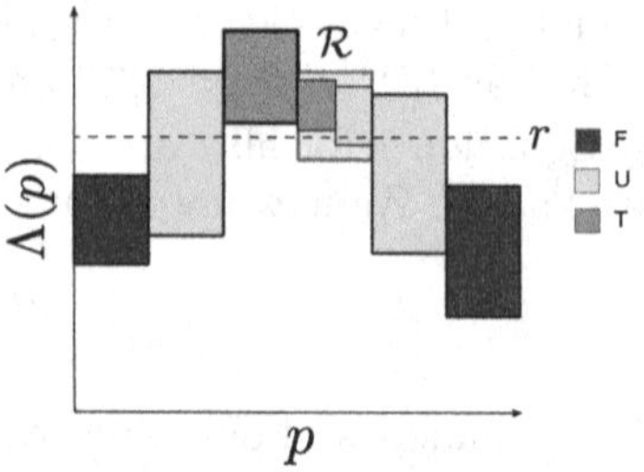

Fig. 1. Refinement in threshold synthesis with $\geq r$. Parameter values are on the x-axis, probabilities on the y-axis. Each box describes a parameter region (width), and its probability bounds (height). The refinement of $\mathcal{R}$ yields regions in T and in U.

the algorithm tests whether $\mathcal{R}$ can be added to the set F by checking if $\Lambda_{max}^{\mathcal{R}}$ is below the threshold r. If $\mathcal{R}$ is neither in T nor in F, it forms an undecided subspace that is added to the set U. The algorithm terminates when the volume of the undecided subspace is negligible with respect to the volume of the entire parameter space, i.e. vol(U)/vol($\mathcal{P}$) $\leq \varepsilon$, where ε is the input tolerance. Otherwise, the algorithm continues to the next iteration, where U is further refined.

Since, for a finite formula ϕ, only a finite number of refinement steps is needed to meet the desired tolerance, the algorithm always terminates. The initial decomposition of the parameter space is guided by a prior sampling of probability values. For more details see [21].

Max Synthesis. Algorithm 2 is used to solve the max synthesis problem, which returns the set T containing the parameter valuations that maximize $\Phi = P_{=?}[\phi]$ and the set F not yielding the maximum value of Φ. Let R be a partition of T. For each subspace $\mathcal{R} \in R$, the algorithm computes bounds $\Lambda_{min}^{\mathcal{R}}$ and $\Lambda_{max}^{\mathcal{R}}$ on the maximal and minimal probability that $\mathcal{C_R}$ satisfies Φ (line 5). The algorithm then rules out subspaces that are guaranteed to be included in F, by deriving an under-approximation (MLB) to the maximum satisfaction probability (line 7). If $\Lambda_{max}^{\mathcal{R}}$ is below the under-approximation, the subspace $\mathcal{R}$ can be safely added to the set F (line 9). Otherwise, it is added to the set T. The bound MLB is derived as follows. In the naive approach, the algorithm uses the maximum over the least bounds in the partition of T, that is, MLB = max$\{\Lambda_{min}^{\mathcal{R}} \mid \mathcal{R} \in R\}$. Let $\overline{\mathcal{R}}$ be the region with highest lower bound. The sampling-based approach improves on this by sampling a set of parameters $\{p_1, p_2, \ldots\} \subseteq \overline{\mathcal{R}}$ and taking the highest value of $\Lambda(p)$, that is, MLB = max$\{\Lambda(p_i) \mid p_i \in \{p_1, p_2, \ldots\}\}$. Although regular CSL model checking is nearly as expensive as the computation of the bounds for a pCTMC, the bound obtained by the sampling method excludes more boxes (see Fig. 2), which in turn leads to fewer refinements in the next iteration. In this

Algorithm 2. Max Synthesis

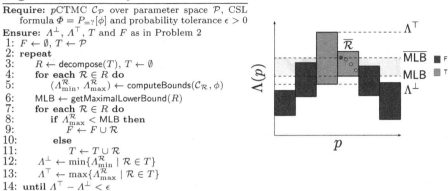

Require: pCTMC $\mathcal{C}_\mathcal{P}$ over parameter space $\mathcal{P}$, CSL
   formula $\Phi = P_{=?}[\phi]$ and probability tolerance $\epsilon > 0$
Ensure: $\Lambda^\perp$, $\Lambda^\top$, T and F as in Problem 2
1: $F \leftarrow \emptyset, T \leftarrow \mathcal{P}$
2: **repeat**
3:     $R \leftarrow$ decompose(T), $T \leftarrow \emptyset$
4:     **for each** $\mathcal{R} \in R$ **do**
5:         $(\Lambda^\mathcal{R}_{\min}, \Lambda^\mathcal{R}_{\max}) \leftarrow$ computeBounds($\mathcal{C}_\mathcal{R}, \phi$)
6:     MLB $\leftarrow$ getMaximalLowerBound(R)
7:     **for each** $\mathcal{R} \in R$ **do**
8:         **if** $\Lambda^\mathcal{R}_{\max} <$ MLB **then**
9:             $F \leftarrow F \cup \mathcal{R}$
10:        **else**
11:            $T \leftarrow T \cup \mathcal{R}$
12:    $\Lambda^\perp \leftarrow \min\{\Lambda^\mathcal{R}_{\min} \mid \mathcal{R} \in T\}$
13:    $\Lambda^\top \leftarrow \max\{\Lambda^\mathcal{R}_{\max} \mid \mathcal{R} \in T\}$
14: **until** $\Lambda^\top - \Lambda^\perp < \epsilon$

Fig. 2. Refinement in max synthesis. The two outermost regions (in red) cannot contain the maximum, as their upper bound is below the maximum lower bound (MLB) found at region $\overline{\mathcal{R}}$. The maximum lower bound is improved by sampling several points $p \in \overline{\mathcal{R}}$ and taking the highest value ($\overline{\text{MLB}}$) of the satisfaction function $\Lambda(p)$. The yellow area highlights the improvement.

case we perform additional checks, discussed in [21], for detecting and discarding regions containing points of jump discontinuity that might prevent the algorithm from reaching the target accuracy and thus from terminating.

The overall time complexity of the synthesis algorithms is directly determined by the number of subspaces that need to be analysed to obtain the desired precision. This number depends on the number of unknown parameters, the shape of the satisfaction function and the type of synthesis. In practice, the algorithms scale exponentially in the number of parameters and linearly in the volume of the parameter space.

6   Results

We demonstrate the applicability and efficiency of the developed algorithms on two case studies.

6.1   Epidemic Model

The SIR model [16] describes the epidemic dynamics in a closed population of susceptible (S), infected (I) and recovered (R) individuals. In the model, a susceptible individual is infected after a contact with an infected individual with rate k_i. Infected individuals recover with rate k_r, after which they are immune to the infection. We can describe this process with the following biochemical reaction model with mass action kinetics: $i : S + I \xrightarrow{k_i} I + I, r : I \xrightarrow{k_r} R$. We

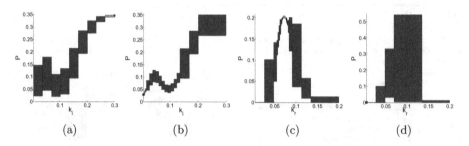

(a)   (b)   (c)   (d)

Fig. 3. Solution to max (a,c) and min (b,d) synthesis using sampling-based refinement for $P_{=?}[(I > 0)U^{[100,120]}(I = 0)]$. Probability tolerance $\epsilon = 1\%$ (a,c) and $\epsilon = 0.1\%$ (b,d).

Table 1. The computation of the synthesis problems for $P_{=?}[(I > 0)U^{[100,120]}(I = 0)]$ using probability tolerance $\epsilon = 1\%$ (problems 1,3,5,6) and $\epsilon = 0.1\%$ (problems 2,4). The sampling-based refinement is used except for problem 5. The minimal bounding box of T is reported in problems 5 and 6. Λ^* denotes $\Lambda^{\perp}$ (problems 1,3,5,6) and $\Lambda^{\top}$ (problems 2,4).

Problem	k_i	k_r	Runtime	Subspaces	$\Lambda^*[\%]$	T
1. Max	$[0.005, 0.3]$	0.05	16.5 s	9	33.94	$[0.267, 0.3]$
2. Min	$[0.005, 0.3]$	0.05	49.5 s	21	2.91	$[0.005, 0.0054]$
3. Max	0.12	$[0.005, 0.2]$	99.7 s	57	19.94	$[0.071, 0.076]$
4. Min	0.12	$[0.005, 0.2]$	10.4 s	5	0.005	$[0.005, 0.026]$
5. Max	$[0.005, 0.3]$	$[0.005, 0.2]$	3.6 h	5817	35.01	$[0.217, 0.272] \times [0.053, 0.059]$
6. Max	$[0.005, 0.3]$	$[0.005, 0.2]$	6.2 h	10249	34.77	$[0.209, 0.29] \times [0.051, 0.061]$

represent the model as a pCTMC with k_i and k_r as parameters, and initial populations $S = 95$, $I = 5$, $R = 0$. We consider the time-bounded CSL formula $\Phi = P_{=?}[(I > 0)U^{[100,120]}(I = 0)]$, which asks for the probability that the infection lasts for at least 100 time units, and dies out before 120 time units. Model parameters and the property are taken from [5], where the authors estimate the satisfaction function for Φ following a Bayesian approach[1].

Figure 3 and Table 1 (problems 1-4) illustrate the solutions using sampling-based refinement for max and min synthesis problems over one-dimensional parameter spaces. We report that, in order to meet the desired probability tolerance, problems 2 (Fig. 3b) and 3 (Fig. 3c) require a high number of refinement steps due to two local extrema close to the minimizing region and due to a bell-shaped Λ with the maximizing region at the top, respectively. Our precise results for problem 1 (Fig. 3a) improve on the estimation in [5], where in the equivalent experiment the max probability is imprecisely registered at smaller k_i values.

We also compare the solutions to the max synthesis problem over the two-dimensional parameter space obtained by applying Alg. 2 with sampling (Fig. 4a, problem 5 in Table 1) and without (Fig. 4b, problem 6 in Table 1). In the

---

[1] In [5], a linear-time specification equivalent to Φ is given.

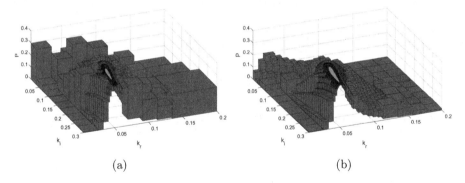

(a)   (b)

Fig. 4. Solutions to max synthesis with sampling-based refinement (a) and without sampling (b) for $P_{=?}[(I > 0)U^{[100,120]}(I = 0)]$ using probability tolerance $\epsilon = 1\%$.

former case, a more precise T region is obtained (with a volume 2.04 times smaller than in the approach without sampling), thus giving a more accurate approximation of the max probability. Sampling also allows us to rule out earlier those parameter regions that are outside the final solution, thus avoiding unnecessary decompositions and improving the runtime (1.72 times faster than in the approach without sampling). This is visible by the coarser approximations of probabilities in the F region.

6.2 DNA Walkers

We revisit models of a DNA walker, a man-made molecular motor that traverses a track of anchorages and can take directions at junctions in the track [22], which can be used to create circuits that evaluate Boolean functions. PRISM models of the walker stepping behaviour were developed previously [9] based on rate estimates in the experimental work. The walker model is modified here to allow uncertainty in the stepping rate, and we consider its behaviour over a single-junction circuit. Given a distance d between the walker-anchorage complex and an uncut anchorage, and d_a being the distance between consecutive anchorages, the stepping rate k is defined as: $k = k_s$ when $d \leq 1.5d_a$, $k = c \cdot k_s/50$ when $1.5d_a < d \leq 2.5d_a$, $k = c \cdot k_s/100$ when $2.5d_a < d \leq 24$nm and $k = 0$, otherwise. The base stepping rate $k_s \in [0.005, 0.020]$ is now defined as an interval, as opposed to the original value of 0.009. We have also added factor c for steps between anchorages that are not directly adjacent, but we will assume $c = 1$ for now. The base stepping rate may depend on buffer conditions and temperature, and we want to verify the robustness of the walker with respect to the uncertainty in the value of k_s.

We compute the minimal probability of the walker making it onto the correct final anchorage (min synthesis for the property $P_{=?}[F^{[T,T]}$ finish-correct$]$) and the maximum probability of the walker making it onto the incorrect anchorage

Table 2. The computation of min-synthesis for $P_{=?}[F^{[T,T]}$ finish-correct] and max-synthesis for $P_{=?}[F^{[T,T]}$ finish-incorrect] using $k_s \in [0.005, 0.020], c = 1$ and probability tolerance $\epsilon = 1\%$. The runtime and subspaces are listed only for the min-synthesis (the results for the max-synthesis are similar).

			Runtime		Subspaces	
Time bound	Min. correct	Max. incorrect	$\emptyset$	Sampling	$\emptyset$	Sampling
$T = 15$	1.68%	5.94%	0.55 s	0.51 s	22	11
$T = 30$	14.86%	10.15%	1.43 s	1.35 s	35	15
$T = 45$	33.10%	12.25%	3.53 s	2.14 s	61	21
$T = 200$	79.21%	16.47%	213.57s	88.97 s	909	329

(max synthesis for the property $P_{=?}[F^{[T,T]}$ finish-incorrect]). We list the probabilities at $T = 15, 30, 45, 200$ minutes in Table 2. For time $T = 30, 45, 200$, we note that the walker is robust, as the minimal guaranteed probability for the correct outcome is greater than the maximum possible probability for the incorrect outcome. For time $T = 15$ this is not the case.

We also consider a property that provides bounds on the *ratio* between the walker finishing on the correct versus the incorrect anchorage. The rates $c \cdot k_s/50$ and $c \cdot k_s/100$ correspond to the walker stepping onto anchorages that are not directly adjacent, which affects the probability for the walker to end up on the unintended final anchorage. For higher values of c, we expect the walker to end up in the unintended final anchorage more often. Now we add uncertainty on the value of c, so that $c \in [0.25, 4]$, and define the performance related property $P_{\geq 0.4}[F^{[30,30]}$ finish-correct] $\wedge P_{\leq 0.08}[F^{[30,30]}$ finish-incorrect], that is, the probability of the walker to make it onto the correct anchorage is at least 40% by time $T = 30$ min, while the probability for it to make it onto the incorrect an-

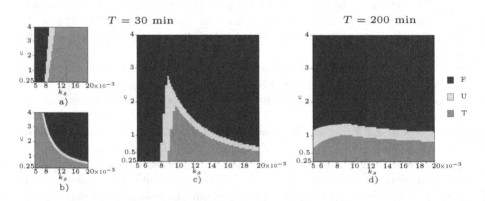

Fig. 5. The computation and results of the threshold synthesis for different formulae, using volume tolerance $\varepsilon = 10\%$. a) $\Phi_1 = P_{\geq 0.4}[F^{[30,30]}$ finish-correct], runtime 443.5 s, 2692 subspaces. b) $\Phi_2 = P_{\leq 0.08}[F^{[30,30]}$ finish-incorrect], runtime 132.3 s, 807 subspaces. c) $\Phi_1 \wedge \Phi_2$. d) $P_{\geq 0.8}[F^{[200,200]}$ finish-correct] $\wedge P_{\leq 0.16}[F^{[200,200]}$ finish-incorrect], runtime 12.3 h, 47229 subspaces.

chorage is no greater than 8%. In other words, we require a correct signal of at least 40% and a correct-to-incorrect ratio of at least 5 by time $T = 30$ min. We define a similar property at time $T = 200$ min, this time requiring a signal of at least 80%: $P_{\geq 0.8}[F^{[200,200]}$ finish-correct$] \wedge P_{\leq 0.16}[F^{[200,200]}$ finish-incorrect$]$. The synthesized ranges of k_s and c where the properties hold are shown in Fig. 5.

7 Conclusion

We have developed efficient algorithms for synthesising rate parameters for biochemical networks so that a given reliability or performance requirement, expressed as a time-bounded CSL formula, is guaranteed to be satisfied. The techniques are based on the computation of lower and upper probability bounds of [6] in conjunction with region refinement and sampling. The high computational costs observed in our case studies can be reduced by parallel processing of individual subspaces, or by utilizing advanced uniformisation techniques [19,8]. We plan to include the synthesis algorithms in the `param` module of the `PRISM` model checker [7,18].

Acknowledgements. We would like to thank Luboš Brim for helpful input and fruitful discussions. This work has been partially supported by the ERC Advanced Grant VERIWARE, Microsoft Research Studentship (Frits Dannenberg) and the Ministry of Education, Youth, and Sport project No. CZ.1.07/2.3.00/30.0009 - Employment of Newly Graduated Doctors of Science for Scientific Excellence (Milan Češka).

References

1. Andreychenko, A., Mikeev, L., Spieler, D., Wolf, V.: Parameter Identification for Markov Models of Biochemical Reactions. In: Gopalakrishnan, G., Qadeer, S. (eds.) CAV 2011. LNCS, vol. 6806, pp. 83–98. Springer, Heidelberg (2011)
2. Aziz, A., Sanwal, K., Singhal, V., Brayton, R.: Verifying Continuous Time Markov Chains. In: Alur, R., Henzinger, T.A. (eds.) CAV 1996. LNCS, vol. 1102, pp. 269–276. Springer, Heidelberg (1996)
3. Baier, C., Haverkort, B., Hermanns, H., Katoen, J.: Model-Checking Algorithms for Continuous-Time Markov Chains. IEEE Trans. on Soft. Eng. 29(6), 524–541 (2003)
4. Bortolussi, L., Sanguinetti, G.: Learning and designing stochastic processes from logical constraints. In: Joshi, K., Siegle, M., Stoelinga, M., D'Argenio, P.R. (eds.) QEST 2013. LNCS, vol. 8054, pp. 89–105. Springer, Heidelberg (2013)
5. Bortolussi, L., Sanguinetti, G.: Smoothed model checking for uncertain continuous time markov chains. CoRR ArXiv, 1402.1450 (2014)
6. Brim, L., Češka, M., Dražan, S., Šafránek, D.: Exploring parameter space of stochastic biochemical systems using quantitative model checking. In: Sharygina, N., Veith, H. (eds.) CAV 2013. LNCS, vol. 8044, pp. 107–123. Springer, Heidelberg (2013)

7. Chen, T., Hahn, E.M., Han, T., Kwiatkowska, M., Qu, H., Zhang, L.: Model repair for Markov decision processes. In: Theoretical Aspects of Software Engineering (TASE), pp. 85–92. IEEE (2013)

8. Dannenberg, F., Hahn, E.M., Kwiatkowska, M.: Computing cumulative rewards using fast adaptive uniformisation. In: Gupta, A., Henzinger, T.A. (eds.) CMSB 2013. LNCS, vol. 8130, pp. 33–49. Springer, Heidelberg (2013)

9. Dannenberg, F., Kwiatkowska, M., Thachuk, C., Turberfield, A.: DNA walker circuits: Computational potential, design, and verification. Natural Computing (to appear, 2014)

10. Fox, B.L., Glynn, P.W.: Computing Poisson Probabilities. CACM 31(4), 440–445 (1988)

11. Gillespie, D.T.: Exact Stochastic Simulation of Coupled Chemical Reactions. Journal of Physical Chemistry 81(25), 2340–2381 (1977)

12. Hahn, E.M., Hermanns, H., Zhang, L.: Probabilistic reachability for parametric Markov models. International Journal on Software Tools for Technology Transfer (STTT) 13(1), 3–19 (2011)

13. Han, T., Katoen, J., Mereacre, A.: Approximate parameter synthesis for probabilistic time-bounded reachability. In: Real-Time Systems Symposium (RTSS), pp. 173–182. IEEE (2008)

14. Jha, S.K., Langmead, C.J.: Synthesis and infeasibility analysis for stochastic models of biochemical systems using statistical model checking and abstraction refinement. Theor. Comput. Sci. 412(21), 2162–2187 (2011)

15. Katoen, J.-P., Klink, D., Leucker, M., Wolf, V.: Three-valued abstraction for continuous-time markov chains. In: Damm, W., Hermanns, H. (eds.) CAV 2007. LNCS, vol. 4590, pp. 311–324. Springer, Heidelberg (2007)

16. Kermack, W.O., McKendrick, A.G.: Contributions to the mathematical theory of epidemics. ii. the problem of endemicity. Proceedings of the Royal Society of London. Series A 138(834), 55–83 (1932)

17. Kwiatkowska, M., Norman, G., Parker, D.: Stochastic Model Checking. In: Bernardo, M., Hillston, J. (eds.) SFM 2007. LNCS, vol. 4486, pp. 220–270. Springer, Heidelberg (2007)

18. Kwiatkowska, M., Norman, G., Parker, D.: PRISM 4.0: Verification of Probabilistic Real-Time Systems. In: Gopalakrishnan, G., Qadeer, S. (eds.) CAV 2011. LNCS, vol. 6806, pp. 585–591. Springer, Heidelberg (2011)

19. Mateescu, M., Wolf, V., Didier, F., Henzinger, T.A.: Fast Adaptive Uniformization of the Chemical Master Equation. IET Systems Biology 4(6), 441–452 (2010)

20. Sen, K., Viswanathan, M., Agha, G.: Model-checking markov chains in the presence of uncertainties. In: Hermanns, H., Palsberg, J. (eds.) TACAS 2006. LNCS, vol. 3920, pp. 394–410. Springer, Heidelberg (2006)

21. Češka, M., Dannenberg, F., Kwiatkowska, M., Paoletti, N.: Precise parameter synthesis for stochastic biochemical systems. Technical Report CS-RR-14-08, Department of Computer Science, University of Oxford (2014)

22. Wickham, S.F.J., Bath, J., Katsuda, Y., Endo, M., Hidaka, K., Sugiyama, H., Turberfield, A.J.: A DNA-based molecular motor that can navigate a network of tracks. Nature Nanotechnology 7, 169–173 (2012)

Parameter Synthesis for Cardiac Cell Hybrid Models Using δ-Decisions

Bing Liu[1], Soonho Kong[1], Sicun Gao[1], Paolo Zuliani[2], and Edmund M. Clarke[1]

[1] Computer Science Department, Carnegie Mellon University, USA
[2] School of Computing Science, Newcastle University, UK

Abstract. A central problem in systems biology is to identify parameter values such that a biological model satisfies some behavioral constraints (*e.g.*, time series). In this paper we focus on parameter synthesis for hybrid (continuous/discrete) models, as many biological systems can possess multiple operational modes with specific continuous dynamics in each mode. These biological systems are naturally modeled as hybrid automata, most often with nonlinear continuous dynamics. However, hybrid automata are notoriously hard to analyze — even simple reachability for hybrid systems with linear differential dynamics is an undecidable problem. In this paper we present a parameter synthesis framework based on δ-complete decision procedures that sidesteps undecidability. We demonstrate our method on two highly nonlinear hybrid models of the cardiac cell action potential. The results show that our parameter synthesis framework is convenient and efficient, and it enabled us to select a suitable model to study and identify crucial parameter ranges related to cardiac disorders.

1 Introduction

Computational modeling and analysis methods are playing a crucial role in understanding the complex dynamics of biological systems [1]. In this paper we address the parameter synthesis problem for hybrid models of biological systems. This problem amounts to finding sets of parameter values for which a model satisfies some precise behavioral constraints, such as time series or reachability properties. We focus on hybrid continuous/discrete models, since one of the key aspects of many biological systems is their differing behavior in various 'discrete' modes. For example, it is well-known that the five stages of the cell cycle are driven by the activation of different signaling pathways. Hence, hybrid system models are often used in systems biology (see, *e.g.*, [2–9]).

Hybrid systems combine discrete control computation with continuous-time evolution. The state space of a hybrid system is defined by a finite set of continuous variables and modes. In each mode, the continuous evolution (*flow*) of the system is usually given by the solution of ordinary differential equations (ODEs). At any given time a hybrid system dwells in one of its modes and each variable evolves accordingly to the flow in the mode. Jump conditions control the switch to another mode, possibly followed by a 'reset' of the continuous variables. Thus, the temporal dynamics of a hybrid system is piecewise continuous.

P. Mendes et al. (Eds.): CMSB 2014, LNBI 8859, pp. 99–113, 2014.

Hybrid models of biological systems often involve many parameters such as rate constants of biochemical reactions, initial conditions, and threshold values in jump conditions. Generally, only a few rate constants will be available or can be measured experimentally — in the latter case the rate constants are obtained by fitting the model to experimental observations. Furthermore, it is also crucial to figure out what initial conditions or jump conditions may lead to an unsafe state of the system, especially when studying hybrid systems used to inform clinical therapy [10]. All such questions fall within the *parameter synthesis* problem, which is extremely difficult for hybrid systems. Even simple reachability questions for hybrid systems with linear (differential) dynamics are undecidable [11]. Therefore, the parameter synthesis problem needs to be relaxed in a sound manner in order to solve it algorithmically — this is the approach we shall follow.

In this paper, we tackle the parameter synthesis problem using δ-complete procedures [12] for deciding first-order formula with arbitrary computable real functions, including solutions of Lipschitz-continuous ODEs [13]. Such procedures may return answers with one-sided δ-bounded errors, thereby overcoming undecidability issues (note that the maximum allowable error δ is an arbitrarily small positive rational). In our approach we describe the set of states of interest as a first-order logic formula and perform bounded model checking [14] to determine reachability of these states. We then adapt an interval constrains propagation based algorithm to explore the parameter space and identify the sets of resulting parameters. We show the applicability of our method by carrying out a thorough case study characterized by highly nonlinear hybrid models. We discriminate two cardiac cell action potential models [15, 16] in terms of cell-type specificity and identify parameter ranges for which a cardiac cell may lose excitability. The results show that our method can obtain biological insights that are consistent with experimental observations, and scales to complex systems. In particular, the analysis we carried out in the cardiac case study is difficult to be performed by — if not out of the scope of — state-of-the-art tools [17–20].

Related Work. A survey of modeling and analysis of biological systems using hybrid models can be found in [21]. Analyzing the properties of biochemical networks using formal verification techniques is being actively pursued by a number of researchers, for which we refer to Brim's *et al.* recent survey [22]. Of particular interest in our context are parameter synthesis methods for qualitative behavior specifications (*e.g.*, temporal logic formulas). The method introduced in [23] can deal with parameter synthesis for piecewise affine linear systems. For ODEs, Donzé *et al.* [24] explore the parameter space using adaptive sampling and simulation, while Palaniappan *et al.* [25] use a statistical model checking approach. Other techniques perform a sweep of the entire (bounded) parameter space, after it has been discretized [26, 27]. Randomized optimization techniques were used for parameter estimation in stochastic hybrid systems [28], while identification techniques for affine systems were used in [29]. The techniques above can handle nonlinear hybrid systems only through sampling and simulation, and so are incomplete. Our approach is instead δ-complete. It is based on verified numerical integration and constraint programming algorithms, which effectively enable an over-approximation of the system dynamics to be computed. Thus, if a model is found to be unfeasible (i.e. an unsat answer is returned, see Section 2 for more details), then this is correct. This behavior better fits with the safety requirements expected by formal verification.

2 δ-Decisions for Hybrid Models

We encode reachability problems of hybrid automata using a first-order language $\mathcal{L}_{\mathbb{R}_\mathcal{F}}$ over the reals, which allows the use of a wide range of real functions including nonlinear ODEs. We then use δ-complete decision procedures to find solutions to these formulas to synthesize parameters.

Definition 1 ($\mathcal{L}_{\mathbb{R}_\mathcal{F}}$-Formulas). *Let $\mathcal{F}$ be a collection of computable real functions. We define:*

$$t := x \mid f(t(\boldsymbol{x})), \text{ where } f \in \mathcal{F} \text{ (constants are 0-ary functions)};$$
$$\varphi := t(\boldsymbol{x}) > 0 \mid t(\boldsymbol{x}) \geq 0 \mid \varphi \wedge \varphi \mid \varphi \vee \varphi \mid \exists x_i \varphi \mid \forall x_i \varphi.$$

By computable real function we mean Type 2 computable, which informally requires that a (real) function can be algorithmically evaluated with arbitrary accuracy. Since in general $\mathcal{L}_{\mathbb{R}_\mathcal{F}}$ formulas are undecidable, the decision problem needs to be relaxed. In particular, for any $\mathcal{L}_{\mathbb{R}_\mathcal{F}}$ formula ϕ and any rational $\delta > 0$ one can obtain a δ-weakening formula ϕ^δ from ϕ by substituting the atoms $t > 0$ with $t > -\delta$ (and similarly for $t \geq 0$). Obviously, ϕ implies ϕ^δ, but not the *vice versa*. Now, the δ-decision problem is deciding correctly whether:

– ϕ is false (unsat);
– ϕ^δ is true (δ-sat).

If both cases are true, then either decision is correct. In previous work [12, 13, 30] we presented algorithms (δ-*complete* decision procedures) for solving δ-decision problems for $\mathcal{L}_{\mathbb{R}_\mathcal{F}}$ and for ODEs. These algorithms have been implemented in the dReal toolset [31]. More details on δ-decision problems are in Appendix.

Now we state the encoding for hybrid models. Recall that hybrid automata generalize finite-state automata by permitting continuous-time evolution (or *flow*) in each discrete state (or *mode*). Also, in each mode an *invariant* must be satisfied by the flow, and mode switches are controlled by *jump* conditions.

Definition 2 ($\mathcal{L}_{\mathbb{R}_\mathcal{F}}$-Representations of Hybrid Automata). *A hybrid automaton in $\mathcal{L}_{\mathbb{R}_\mathcal{F}}$-representation is a tuple*

$$H = \langle X, Q, \{\mathsf{flow}_q(\boldsymbol{x}, \boldsymbol{y}, t) : q \in Q\}, \{\mathsf{inv}_q(\boldsymbol{x}) : q \in Q\},$$
$$\{\mathsf{jump}_{q \to q'}(\boldsymbol{x}, \boldsymbol{y}) : q, q' \in Q\}, \{\mathsf{init}_q(\boldsymbol{x}) : q \in Q\}\rangle$$

where $X \subseteq \mathbb{R}^n$ for some $n \in \mathbb{N}$, $Q = \{q_1, ..., q_m\}$ is a finite set of modes, and the other components are finite sets of quantifier-free $\mathcal{L}_{\mathbb{R}_\mathcal{F}}$-formulas.

Example 1 (Nonlinear Bouncing Ball). The bouncing ball is a standard hybrid system model. It can be $\mathcal{L}_{\mathbb{R}_{\mathcal{F}}}$-represented in the following way:

- $X = \mathbb{R}^2$ and $Q = \{q_u, q_d\}$. We use q_u to represent bounce-back mode and q_d the falling mode.
- flow $= \{\mathsf{flow}_{q_u}(x_0, v_0, x_t, v_t, t), \mathsf{flow}_{q_d}(x_0, v_0, x_t, v_t, t)\}$. We use x to denote the height of the ball and v its velocity. Instead of using time derivatives, we can directly write the flows as integrals over time, using $\mathcal{L}_{\mathbb{R}_{\mathcal{F}}}$-formulas:

  • $\mathsf{flow}_{q_u}(x_0, v_0, x_t, v_t, t)$ defines the dynamics in the bounce-back phase:

  $$(x_t = x_0 + \int_0^t v(s)ds) \wedge (v_t = v_0 + \int_0^t g(1 - \beta v(s)^2)ds)$$

  • $\mathsf{flow}_{q_d}(x_0, v_0, x_t, v_t, t)$ defines the dynamics in the falling phase:

  $$(x_t = x_0 + \int_0^t v(s)ds) \wedge (v_t = v_0 + \int_0^t g(1 + \beta v(s)^2)ds)$$

  where β is a constant. Again, note that the integration terms define Type 2 computable functions.
- jump $= \{\mathsf{jump}_{q_u \to q_d}(x, v, x', v'), \mathsf{jump}_{q_d \to q_u}(x, v, x', v')\}$ where
  • $\mathsf{jump}_{q_u \to q_d}(x, v, x', v')$ is $(v = 0 \wedge x' = x \wedge v' = v)$.
  • $\mathsf{jump}_{q_d \to q_u}(x, v, x', v')$ is $(x = 0 \wedge v' = \alpha v \wedge x' = x)$, for some constant α.
- init_{q_d} is $(x = 10 \wedge v = 0)$ and init_{q_u} is $\perp$.
- inv_{q_d} is $(x >= 0 \wedge v >= 0)$ and inv_{q_u} is $(x >= 0 \wedge v <= 0)$.

We now show the encoding of bounded reachability, which is used for encoding the parameter synthesis problem. We want to decide whether a given hybrid system reaches a particular region of its state space after following a (bounded) number of discrete transitions, *i.e.*, jumps. First, we need to define auxiliary formulas used for ensuring that a particular mode is picked at a certain step.

Definition 3. *Let $Q = \{q_1, ..., q_m\}$ be a set of modes. For any $q \in Q$, and $i \in \mathbb{N}$, use b_q^i to represent a Boolean variable. We now define*

$$\mathsf{enforce}_Q(q, i) = b_q^i \wedge \bigwedge_{p \in Q \backslash \{q\}} \neg b_p^i$$

$$\mathsf{enforce}_Q(q, q', i) = b_q^i \wedge \neg b_{q'}^{i+1} \wedge \bigwedge_{p \in Q \backslash \{q\}} \neg b_p^i \wedge \bigwedge_{p' \in Q \backslash \{q'\}} \neg b_{p'}^{i+1}$$

We omit the subscript Q when the context is clear.

We can now define the following formula that checks whether a *goal* region of the automaton state space is reachable after exactly k discrete transitions. We first state the simpler case of a hybrid system without invariants.

Definition 4 (k-Step Reachability, Invariant-Free Case). *Suppose H is an invariant-free hybrid automaton, U a subset of its state space represented by* goal, *and $M > 0$. The formula* $\text{Reach}_{H,U}(k, M)$ *is defined as:*

$$\exists^X \boldsymbol{x}_0 \exists^X \boldsymbol{x}_0^t \cdots \exists^X \boldsymbol{x}_k \exists^X \boldsymbol{x}_k^t \exists^{[0,M]} t_0 \cdots \exists^{[0,M]} t_k.$$

$$\bigvee_{q \in Q} \left(\text{init}_q(\boldsymbol{x}_0) \wedge \text{flow}_q(\boldsymbol{x}_0, \boldsymbol{x}_0^t, t_0) \wedge \text{enforce}(q, 0) \right)$$

$$\wedge \bigwedge_{i=0}^{k-1} \left(\bigvee_{q,q' \in Q} \left(\text{jump}_{q \to q'}(\boldsymbol{x}_i^t, \boldsymbol{x}_{i+1}) \wedge \text{enforce}(q, q', i) \right. \right.$$

$$\left. \left. \wedge \text{flow}_{q'}(\boldsymbol{x}_{i+1}, \boldsymbol{x}_{i+1}^t, t_{i+1}) \wedge \text{enforce}(q', i+1) \right) \right)$$

$$\wedge \bigvee_{q \in Q} (\text{goal}_q(\boldsymbol{x}_k^t) \wedge \text{enforce}(q, k))$$

where $\exists^X x$ is a shorthand for $\exists x \in X$.

Intuitively, the trajectories start with some initial state satisfying $\text{init}_q(\boldsymbol{x}_0)$ for some q. Then, in each step the trajectory follows $\text{flow}_q(\boldsymbol{x}_i, \boldsymbol{x}_i^t, t)$ and makes a continuous flow from $\boldsymbol{x}_i$ to $\boldsymbol{x}_i^t$ after time t. When the automaton makes a jump from mode q' to q, it resets variables following $\text{jump}_{q' \to q}(\boldsymbol{x}_k^t, \boldsymbol{x}_{k+1})$. The auxiliary enforce formulas ensure that picking $\text{jump}_{q \to q'}$ in the i-the step enforces picking flow_q' in the $(i+1)$-th step.

When the invariants are not trivial, we need to ensure that for all the time points along a continuous flow, the invariant condition holds. We need to universally quantify over time, and the encoding is as follows:

Definition 5 (k-Step Reachability, Nontrivial Invariant). *Suppose H contains invariants, and U is a subset of the state space represented by* goal. *The $\mathcal{L}_{\mathbb{R}_{\mathcal{F}}}$-formula* $\text{Reach}_{H,U}(k, M)$ *is defined as:*

$$\exists^X \boldsymbol{x}_0 \exists^X \boldsymbol{x}_0^t \cdots \exists^X \boldsymbol{x}_k \exists^X \boldsymbol{x}_k^t \exists^{[0,M]} t_0 \cdots \exists^{[0,M]} t_k.$$

$$\bigvee_{q \in Q} \left(\text{init}_q(\boldsymbol{x}_0) \wedge \text{flow}_q(\boldsymbol{x}_0, \boldsymbol{x}_0^t, t_0) \wedge \text{enforce}(q, 0) \right.$$

$$\left. \wedge \forall^{[0,t_0]} t \forall^X \boldsymbol{x} \, (\text{flow}_q(\boldsymbol{x}_0, \boldsymbol{x}, t) \to \text{inv}_q(\boldsymbol{x})) \right)$$

$$\wedge \bigwedge_{i=0}^{k-1} \left(\bigvee_{q,q' \in Q} \left(\text{jump}_{q \to q'}(\boldsymbol{x}_i^t, \boldsymbol{x}_{i+1}) \wedge \text{flow}_{q'}(\boldsymbol{x}_{i+1}, \boldsymbol{x}_{i+1}^t, t_{i+1}) \wedge \text{enforce}(q, q', i) \right. \right.$$

$$\left. \left. \wedge \text{enforce}(q', i+1) \wedge \forall^{[0,t_{i+1}]} t \forall^X \boldsymbol{x} \, (\text{flow}_{q'}(\boldsymbol{x}_{i+1}, \boldsymbol{x}, t) \to \text{inv}_{q'}(\boldsymbol{x}))) \right) \right)$$

$$\wedge \bigvee_{q \in Q} (\text{goal}_q(\boldsymbol{x}_k^t) \wedge \text{enforce}(q, k)).$$

The extra universal quantifier for each continuous flow expresses the requirement that for all the time points between the initial and ending time point ($t \in$

$[0, t_i + 1]$) in a flow, the continuous variables $\boldsymbol{x}$ must take values that satisfy the invariant conditions $\mathrm{inv}_q(\boldsymbol{x})$.

Parameter Identification. The parameter identification problem we tackle is basically a k-step reachability question: Is there a parameter combination for which the model reaches the goal region in k steps? If none exists, then the model is *unfeasible*. Otherwise, a witness (*i.e.*, a value for each parameter) is returned. Note that because we ask for δ-decisions, the returned witness might correspond to a *spurious* behavior of the system. The occurrence of such behaviors can be controlled via the precision δ, but in general cannot be eliminated. We have developed the dReach tool (`http://dreal.cs.cmu.edu/dreach.html`) that automatically builds reachability formulas from a hybrid model and a goal description. Such formulas are then verified by our δ-complete solver dReal [31].

3   Case Study

To exemplify different aspects of our parameter synthesis framework, we carried out a case study on models of cardiac cell electrical dynamics. All experiments reported below were done using a machine with an Intel Core i5 3.4GHz processor and 8GB RAM. The precision δ was set to 10^{-4}. The model files are available at `http://www.cs.cmu.edu/~liubing/cmsb14/`.

3.1   Hybrid Models of Cardiac Cells

The heart rhythm is enabled by the electrical activity of cardiac muscle cells, which make up the atria and ventricles. The electrical dynamics of cardiac cells is governed by the organized opening and closing of ion channel gates on the cell membrane. Improper functioning of the cardiac cell ionic channels can cause the cells to lose excitability, which disorders electric wave propagation and leads to cardiac abnormalities such as ventricular *tachycardia* or *fibrillation*. In order to understand the mechanisms of cardiac disorders, hybrid automata models have been recently developed, including the Fenton-Karma (FK) model [15] and the Bueno-Cherry-Fenton (BCF) model [16].

BCF Model. In this model, the change of cells transmembrane potential u, in response to an external stimulus ϵ from neighboring cells, is regulated by a fast ion channel gate v and two slow gates w and s. Figure 1(a) shows the four modes associated with the BCF model. In Mode 1, gates v and w are open and gate s is closed. The transmembrane potassium current causes the decay of u. The cell is resting and waiting for stimulation. We assume an external stimulus ϵ equal to 1 that lasts for 1 millisecond. The stimulation causes u to increase, which may trigger $\mathrm{jump}_{1\to 2} : u \geq \theta_o$. When this jump takes place, the system switches to Mode 2 and v starts closing, and the decay rate of u changes. The system will jump to Mode 3 if $u \geq \theta_w$. In Mode 3, w is also closing; u is governed by the potassium current and the calcium current. When $u \geq \theta_v$, Mode 4 can be

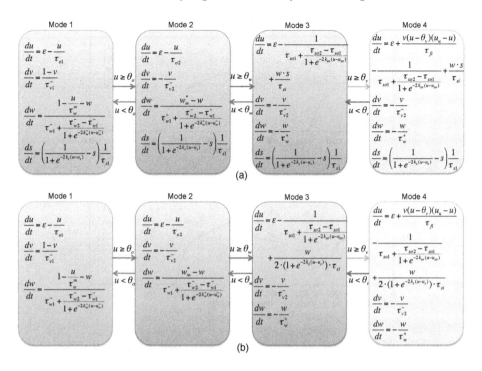

Fig. 1. Hybrid models of cardiac cells. (a) BCF model. (b) FK model.

reached, which signals a successful action potential (AP) initiation. In Mode 4, u reaches its peak due to the fast opening of the sodium channel. The cardiac muscle contracts and u starts decreasing.

FK Model. As shown in Figure 1(b), this model comprises the same four modes and equations of the BCF model, except that the current change induced by gate s is reduced to an explicit term which is integrated in the right-hand side of du/dt. Similarly to the BCF model, an AP can be successfully initiated when Mode 4 is reached.

We specified both the BCF and the FK models using dReach's modeling language. Starting from the state ($u = 0$, $v = 1$, $w = 1$, $s = 0$, $\epsilon \in [0.9, 1.1]$) in Mode 1 (note that the value of s does not matter to FK, which does not contains s), we checked whether Mode 4 is reachable using the parameter values presented in [16]. This was true (*i.e.*, dReach returned δ-sat) for both models (Table 1, Run#1 and Run#2). The simulation of a few witness trajectories are shown in Figure 2 (the stimulus ϵ was reset every 500 milliseconds).

3.2   Model Falsification

Both the BCF and the FK models were able to reproduce essential characteristics (*e.g.*, steady-state action potential duration) observed in human ventricular

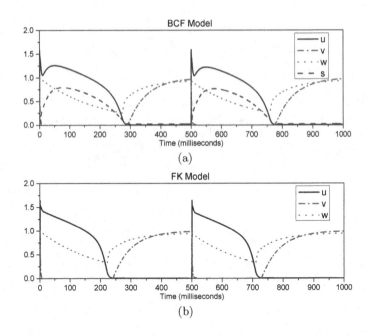

Fig. 2. The simulated witness trajectories of the BCF and the FK models

cells [15, 16]. However, ventricular cells comprise three cell types, which possess different dynamical characteristics. For instance, Figure 3 shows that time courses of APs for epicardial and endocardial human ventricular cells have different morphologies [32]. The important *spike-and-dome* AP morphology can only be observed in epicardial cells but not endocardial cells. Hence, in a model-based study, one needs to identify cell-type-specific parameters to take account into cellular heterogeneity. The feasibility of this task will depend on the model of choice, as for certain models it would be impossible to reproduce a dynamical behavior no matter which parameter values are used. Here we illustrate that such models can be ruled out efficiently using our δ-decision based parameter synthesis framework.

Robustness. We first considered the robustness property of the models. To ensure proper functioning of cardiac cells in noisy environments, an important property of the system is to filter out insignificant stimulation. Thus, we expected to see that AP could not be initiated for small ϵ. Starting from the state ($u = 0$, $v = 1$, $w = 1$, $s = 0$, $\epsilon \in [0.0, 0.25]$) in Mode 1, we checked the reachability of Mode 4. The unsat answer was returned by dReach for both the BCF and FK model (Table 1, Run#3 and Run#4), showing that the models are robust to stimulation amplitude.

AP morphology. Next we tested whether the models could reproduce the spike-and-dome AP morphology of epicardial cells. We introduced three auxiliary

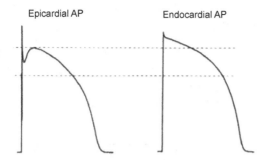

Fig. 3. Different AP morphologies observed in epicardial and endocardial cells [32]

modes (Mode 5, 6 and 7). If $time \geq 1$, the system will jump from Mode 4 to Mode 5, in which ϵ will be reset to 0. The system will jump from Mode 5 to Mode 6 if $time \geq 10$, and will jump from Mode 6 to Mode 7 if $time \geq 30$. In Modes 6 and 7, we enforced invariants $1.0 \leq u \leq 1.15$ and $1.18 \leq u \leq 2.0$, respectively, to depict the spike-and-dome morphology observed experimentally [32]. We then checked reachability of Mode 7, starting from Mode 1 in state $(u = 0, v = 1, w = 1, s = 0, \epsilon \in [0.9, 1.1], \tau_{si} \in [1, 2], u_s \in [0.5, 2])$, where τ_{si} and u_s are two model parameters that govern the dynamics of u and s in Mode 3 and 4 (see Figure 1). The δ-sat answer was returned for BCF (Table 1, Run#5), while unsat was returned for FK (Table 1, Run#6), indicating that the FK model cannot reproduce spike-and-dome shapes using reasonable parameter values. Hence, FK is not suitable to study the dynamics of epicardial cells.

We remark that any unsat answer is guaranteed to be correct. This effectively means that we proved that the FK model cannot reach Mode 7 for *any* starting state in the rectangle $(u = 0, v = 1, w = 1, s = 0, \epsilon \in [0.9, 1.1], \tau_{si} \in [1, 2], u_s \in [0.5, 2])$. Sampling-based approaches cannot have the same level of certainty, while other approaches cannot handle the complexity of the flows in the model.

3.3   Parameter Identification for Cardiac Disorders

When the system cannot reach Mode 4, the cardiac cell loses excitability, which might lead to tachycardia or fibrillation. Starting with Mode 1, our task was to identify parameter ranges for which the system will never go into Mode 4. In what follows, we focused our study on the BCF model. Grosu *et al.* [33] have tackled this parameter identification problem by linearizing the BCF model into a piecewise-multiaffine system (referred as MHA). With this simplification, parameter ranges could be identified using the Rovergene tool [23]. However, the BCF and MHA models have different sets of parameters. Here we identify disease-related ranges of the *original* BCF parameters. It can be derived from the model equations that τ_{o1} and τ_{o2} govern the dynamics of u in Mode 1 and Mode 2 respectively, and hence determine whether $jump_{1 \rightarrow 2}$ and $jump_{2 \rightarrow 3}$ can be triggered. For τ_{o1}, we performed a binary search in value domain $(0, 0.01]$

to obtain a threshold value θ_{o1} such that Mode 4 is unreachable if $\tau_{o1} < \theta_{o1}$ while Mode 4 is reachable if $\tau_{o1} \geq \theta_{to1}$. The search procedure is illustrated in Algorithm 1. Specifically, we set candidate θ_{o1}^i to be the midpoint of the search domain. We then checked the reachability of Mode 4 with the initial state ($u = 0$, $v = 1$, $w = 1$, $s = 0$, $\theta_{o1} = \theta_{o1}^i$). If δ-sat was returned (*e.g.*, Table 1, Run#7), we would recursively check the left-hand half of the search domain; otherwise (*e.g.*, Table 1, Run#8), we would check the other half.

Algorithm 1. Identify parameter threshold value using binary search

---

1 BinarySearch(M, v_{min}, v_{max}, δ)
    input : A dReach model M; lower and upper bounds of parameter v: v_{min},
            v_{max}; precision δ
    output: A threshold value θ_v
2 **initialization**: $\theta_v \leftarrow (v_{min} + v_{max})/2$;
3 **if** $|v_{min} - v_{max}| \leq \delta$ **then**
4 | return θ_v ;
5 **else**
6 | $Res \leftarrow$ dReach(M, θ_v, δ) ;
7 | **if** $Res = \delta$-sat **then**
8 | | return BinarySearch(M, v_{min}, θ_v, δ)
9 | **else**
10 | | return BinarySearch(M, θ_v, v_{max}, δ)
11 | **end**
12 **end**

In this manner, we identified θ_{o1} to be 0.006, which suggest that when $\tau_{o1} \in (0, 0.006)$, the system will always stay in Mode 1 (Table 1, Run#9). Similarly, we also obtained a threshold value of 0.13 for τ_{o2}, such that Mode 3 cannot be reached when $\tau_{o2} \in (0, 0.13)$ (Table 1, Run#10). Furthermore, whether the system can jump from Mode 3 to Mode 4 depends on the interplay between τ_{so1} and τ_{so2}. For each value τ_{so2}^i of τ_{so2} sampled from domain $[0, 100]$, we performed the binary search in $[0, 5]$ to find the threshold value θ_{so1} such that Mode 4 is unreachable when $\tau_{so1} \in [0, \theta_{so1}]$ and $\tau_{so2} = \tau_{so2}^i$. By linear regression of the obtained values of θ_{so1}, we identified one more condition that Mode 4 is unreachable: $6.2 \cdot \tau_{so1} + \tau_{so2} \geq 9.9$ (*e.g.*, $\tau_{so1} \in [10, 40] \wedge \tau_{so1} \in [0.5, 2]$, see Table 1, Run#11). Taken together, we identified the following disease-related parameter ranges:

$$\tau_{o1} \in (0, 0.006) \vee \tau_{o2} \in (0, 0.13) \vee 6.2 \cdot \tau_{so1} + \tau_{so2} \geq 9.9$$

Figure 4 visualizes these results by showing the simulated trajectories using corresponding parameter values.

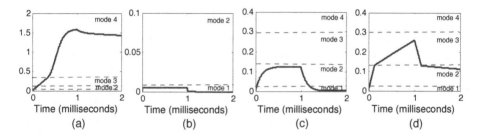

Fig. 4. Simulation results using disease related parameter values. (a) Normal condition (original parameters) (b) $\tau_{o1} = 0.0055$ (c) $\tau_{o2} = 0.125$ (d) $\tau_{so1} = 1.2$, $\tau_{so2} = 1.0$.

Table 1. Experimental results. Var = number of variables in the unrolled formula, Result = bounded model checking result, Time = CPU time (s), $\delta = 10^{-4}$.

Run	Model	Initial State	Var	Result	Time
1	BCF	$(u = 0, v = 1, w = 1, s = 0, \epsilon \in [0.9, 1.1])$	53	δ-sat	303
2	FK	$(u = 0, v = 1, w = 1, \epsilon \in [0.9, 1.1])$	53	δ-sat	216
3	BCF	$(u = 0, v = 1, w = 1, s = 0, \epsilon \in [0, 0.25])$	53	unsat	2.09
4	FK	$(u = 0, v = 1, w = 1, \epsilon \in [0.0, 0.25])$	53	unsat	0.78
5	BCF	$(u = 0, v = 1, w = 1, s = 0, \epsilon \in [0.9, 1.1])$	89	δ-sat	7,904
6	FK	$(u = 0, v = 1, w = 1, \epsilon \in [0.9, 1.1], \tau_{si} \in [1, 2], u_s \in [0.5, 2])$	119	unsat	0.06
7	BCF	$(u = 0, v = 1, w = 1, s = 0, \tau_{o1} = 30.02)$	53	δ-sat	0.89
8	BCF	$(u = 0, v = 1, w = 1, s = 0, \tau_{o1} = 0.0055)$	53	unsat	1.33
9	BCF	$(u = 0, v = 1, w = 1, s = 0, \tau_{o1} \in (0.0, 0.006))$	62	unsat	0.76
10	BCF	$(u = 0, v = 1, w = 1, s = 0, \tau_{o2} \in (0.0, 0.13))$	62	unsat	0.32
11	BCF	$(u = 0, v = 1, w = 1, s = 0, \tau_{so1} \in [10, 40], \tau_{so1} \in [0.5, 2])$	71	unsat	0.11

4 Conclusion

We have presented a framework using δ-complete decision procedures for the parameter identification of hybrid biological systems. We have used δ-satisfiable formulas to describe parameterized hybrid automata and to encode parameter synthesis problems. We have employed δ-decision procedures to perform bounded model checking, and developed an algorithm to obtain the resulting parameters. Our verified numerical integration and constraint programming algorithms effectively compute an over-approximation of the system dynamics. An unsat answer can always be trusted, while a δ-sat answer might be due to the over-approximation (see Section 2 for more details). We chose this behavior as it better fits with the safety requirements expected by formal verification. We have demonstrated the applicability of our method on a highly nonlinear hybrid model of a cardiac cell that are difficult to analyze with other verification tools. We have successfully ruled out a model candidate which did not fit experimental observations, and we have identified critical parameter ranges that can induce cardiac disorders.

It is worth noting that our method can be applied to ODE based models with discrete events, which are special forms of hybrid automata. Such models are often specified using the Systems Biology Markeup Language (SBML) and archived in the BioModels database [34]. Currently, we are currently developing an SBML-to-dReal translator to facilitate the δ-decision based analysis of SBML models. Further, our method also has the potential to be applied to other model formalisms such as hybrid functional Petri nets [35] and the formalisms realized in Ptolemy [36]. We plan to explore this in future work. Another interesting direction is applying our method for parameter estimation from experimental data. By properly encoding the noisy wet-lab experimental data using logic formulas, bounded model checking can be utilized to find the unknown parameter values. In this respect, the specification logic used in [25] promises to offer helpful pointers.

Acknowledgements. This work has been partially supported by award N00014-13-1-0090 of the US Office of Naval Research and award CNS0926181 of the National Science foundation (NSF).

References

1. Liu, B., Thiagarajan, P.: Modeling and analysis of biopathways dynamics. Journal of Bioinformatics and Computational Biology 10(4), 1231001 (2012)
2. Chen, K.C., Calzone, L., Csikasz-Nagy, A., Cross, F.R., Novak, B., Tyson, J.J.: Integrative analysis of cell cycle control in budding yeast. Mol. Biol. Cell. 15(8), 3841–3862 (2004)
3. Ghosh, R., Tomlin, C.: Symbolic reachable set computation of piecewise affine hybrid automata and its application to biological modelling: Delta-notch protein signalling. IET Syst. Biol. 1(1), 170–183 (2004)
4. Hu, J., Wu, W.-C., Sastry, S.S.: Modeling subtilin production in *bacillus subtilis* using stochastic hybrid systems. In: Alur, R., Pappas, G.J. (eds.) HSCC 2004. LNCS, vol. 2993, pp. 417–431. Springer, Heidelberg (2004)
5. Ye, P., Entcheva, E., Smolka, S.A., Grosu, R.: Modelling excitable cells using cycle-linear hybrid automata. IET Syst. Biol. 2(1), 24–32 (2008)
6. Aihara, K., Suzuki, H.: Theory of hybrid dynamical systems and its applications to biological and medical systems. Phil. Trans. R. Soc. A 368(1930), 4893–4914 (2010)
7. Antoniotti, M., Mishra, B., Piazza, C., Policriti, A., Simeoni, M.: Modeling cellular behavior with hybrid automata: Bisimulation and collapsing. In: Priami, C. (ed.) CMSB 2003. LNCS, vol. 2602, pp. 57–74. Springer, Heidelberg (2003)
8. Lincoln, P., Tiwari, A.: Symbolic systems biology: Hybrid modeling and analysis of biological networks. In: Alur, R., Pappas, G.J. (eds.) HSCC 2004. LNCS, vol. 2993, pp. 660–672. Springer, Heidelberg (2004)
9. Baldazzi, V., Monteiro, P.T., Page, M., Ropers, D., Geiselmann, J., Jong, H.D.: Qualitative analysis of genetic regulatory networks in bacteria. In: Understanding the Dynamics of Biological Systems, pp. 111–130. Springer (2011)
10. Tanaka, G., Hirata, Y., Goldenberg, S.L., Bruchovsky, N., Aihara, K.: Mathematical modelling of prostate cancer growth and its application to hormone therapy. Phil. Trans. R. Soc. A 368, 5029–5044 (2010)

11. Henzinger, T.A.: The theory of hybrid automata. In: LICS, pp. 278–292 (1996)
12. Gao, S., Avigad, J., Clarke, E.M.: Delta-complete decision procedures for satisfiability over the reals. In: IJCAR, pp. 286–300 (2012)
13. Gao, S., Avigad, J., Clarke, E.M.: Delta-decidability over the reals. In: LICS, pp. 305–314 (2012)
14. Biere, A., Cimatti, A., Clarke, E., Zhu, Y.: Symbolic model checking without bDDs. In: Cleaveland, W.R. (ed.) TACAS 1999. LNCS, vol. 1579, pp. 193–207. Springer, Heidelberg (1999)
15. Fenton, F., Karma, A.: Vortex dynamics in 3D continuous myocardium with fiber rotation: filament instability and fibrillation. Chaos 8, 20–47 (1998)
16. Bueno-Orovio, A., Cherry, E.M., Fenton, F.H.: Minimal model for human ventricular action potentials in tissue. J. Theor. Biol. 253, 544–560 (2008)
17. Kwiatkowska, M., Norman, G., Parker, D.: PRISM: Probabilistic symbolic model checker. In: Field, T., Harrison, P.G., Bradley, J., Harder, U. (eds.) TOOLS 2002. LNCS, vol. 2324, pp. 200–204. Springer, Heidelberg (2002)
18. Donzé, A.: Breach, a toolbox for verification and parameter synthesis of hybrid systems. In: Touili, T., Cook, B., Jackson, P. (eds.) CAV 2010. LNCS, vol. 6174, pp. 167–170. Springer, Heidelberg (2010)
19. Annpureddy, Y., Liu, C., Fainekos, G., Sankaranarayanan, S.: S-TaLiRo: A tool for temporal logic falsification for hybrid systems. In: Abdulla, P.A., Leino, K.R.M. (eds.) TACAS 2011. LNCS, vol. 6605, pp. 254–257. Springer, Heidelberg (2011)
20. Chabrier-Rivier, N., Fages, F., Soliman, S.: The biochemical abstract machine BIOCHAM. In: Danos, V., Schachter, V. (eds.) CMSB 2004. LNCS (LNBI), vol. 3082, pp. 172–191. Springer, Heidelberg (2005)
21. Bortolussi, L., Policriti, A.: Hybrid systems and biology. In: Bernardo, M., Degano, P., Zavattaro, G. (eds.) SFM 2008. LNCS, vol. 5016, pp. 424–448. Springer, Heidelberg (2008)
22. Brim, L., Češka, M., Šafránek, D.: Model checking of biological systems. In: Bernardo, M., de Vink, E., Di Pierro, A., Wiklicky, H. (eds.) SFM 2013. LNCS, vol. 7938, pp. 63–112. Springer, Heidelberg (2013)
23. Batt, G., Belta, C., Weiss, R.: Temporal logic analysis of gene networks under parameter uncertainty. IEEE T. Automat. Contr. 53, 215–229 (2008)
24. Donzé, A., Clermont, G., Langmead, C.J.: Parameter synthesis in nonlinear dynamical systems: Application to systems biology. J. Comput. Biol. 17(3), 325–336 (2010)
25. Palaniappan, S.K., Gyori, B.M., Liu, B., Hsu, D., Thiagarajan, P.S.: Statistical model checking based calibration and analysis of bio-pathway models. In: Gupta, A., Henzinger, T.A. (eds.) CMSB 2013. LNCS, vol. 8130, pp. 120–134. Springer, Heidelberg (2013)
26. Calzone, L., Chabrier-Rivier, N., Fages, F., Soliman, S.: Machine learning biochemical networks from temporal logic properties. In: Priami, C., Plotkin, G. (eds.) Transactions on Computational Systems Biology VI. LNCS (LNBI), vol. 4220, pp. 68–94. Springer, Heidelberg (2006)
27. Donaldson, R., Gilbert, D.: A model checking approach to the parameter estimation of biochemical pathways. In: Heiner, M., Uhrmacher, A.M. (eds.) CMSB 2008. LNCS (LNBI), vol. 5307, pp. 269–287. Springer, Heidelberg (2008)
28. Koutroumpas, K., Cinquemani, E., Kouretas, P., Lygeros, J.: Parameter identification for stochastic hybrid systems using randomized optimization: A case study on subtilin production by Bacillus subtilis. Nonlinear Anal.-Hybrid Syst. 2, 786–802 (2008)

29. Cinquemani, E., Porreca, R., Ferrari-Trecate, G., Lygeros, J.: Subtilin production by *Bacillus subtilis*: Stochastic hybrid models and parameter identification. IEEE Trans. Automat. Contr. 53, 38–50 (2008)

30. Gao, S., Kong, S., Clarke, E.M.: Satisfiability modulo ODEs. In: FMCAD, pp. 105–112 (2013)

31. Gao, S., Kong, S., Clarke, E.M.: dReal: An SMT solver for nonlinear theories over the reals. In: Bonacina, M.P. (ed.) CADE 2013. LNCS, vol. 7898, pp. 208–214. Springer, Heidelberg (2013)

32. Nabauer, M., Beuckelmann, D.J., Uberfuhr, P., Steinbeck, G.: Regional differences in current density and rate-dependent properties of the transient outward current in subepicardial and subendocardial myocytes of human left ventricle. Circulation 93, 169–177 (1996)

33. Grosu, R., Batt, G., Fenton, F.H., Glimm, J., Le Guernic, C., Smolka, S.A., Bartocci, E.: From cardiac cells to genetic regulatory networks. In: Gopalakrishnan, G., Qadeer, S. (eds.) CAV 2011. LNCS, vol. 6806, pp. 396–411. Springer, Heidelberg (2011)

34. Li, C., Donizelli, M., Rodriguez, N., Dharuri, H., Endler, L., Chelliah, V., Li, L., He, E., Henry, A., Stefan, M.I., Snoep, J.L., Hucka, M., Novere, N.L., Laibe, C.: BioModels Database: An enhanced, curated and annotated resource for published quantitative kinetic models. BMC Sys. Biol. 4, 92 (2010)

35. Matsuno, H., Tanaka, Y., Aoshima, H., Doi, A., Matsui, M., Miyano, S.: Biopathways representation and simulation on hybrid functional petri net. In Silico Biol. 3(3), 389–404 (2003)

36. Ptolemaeus, C. (ed.): System Design, Modeling, and Simulation using Ptolemy II (2014), `Ptolemy.org`

37. Weihrauch, K.: Computable Analysis: An Introduction. Springer (2000)

Appendix: $\mathcal{L}_{\mathbb{R}_\mathcal{F}}$-Formulas and δ-Decidability

We will use a logical language over the real numbers that allows arbitrary *computable real functions* [37]. We write $\mathcal{L}_{\mathbb{R}_\mathcal{F}}$ to represent this language. Intuitively, a real function is computable if it can be numerically simulated up to an arbitrary precision. For the purpose of this paper, it suffices to know that almost all the functions that are needed in describing hybrid systems are Type 2 computable, such as polynomials, exponentiation, logarithm, trigonometric functions, and solution functions of Lipschitz-continuous ordinary differential equations.

More formally, $\mathcal{L}_{\mathbb{R}_\mathcal{F}} = \langle \mathcal{F}, > \rangle$ represents the first-order signature over the reals with the set $\mathcal{F}$ of computable real functions, which contains all the functions mentioned above. Note that constants are included as 0-ary functions. $\mathcal{L}_{\mathbb{R}_\mathcal{F}}$-formulas are evaluated in the standard way over the structure $\mathbb{R}_\mathcal{F} = \langle \mathbb{R}, \mathcal{F}^\mathbb{R}, >^\mathbb{R} \rangle$. It is not hard to see that we can put any $\mathcal{L}_{\mathbb{R}_\mathcal{F}}$-formula in a normal form, such that its atomic formulas are of the form $t(x_1, ..., x_n) > 0$ or $t(x_1, ..., x_n) \geq 0$, with $t(x_1, ..., x_n)$ composed of functions in $\mathcal{F}$. To avoid extra preprocessing of formulas, we can explicitly define $\mathcal{L}_\mathcal{F}$-formulas as follows.

Definition 6 ($\mathcal{L}_{\mathbb{R}_\mathcal{F}}$-Formulas). *Let $\mathcal{F}$ be a collection of computable real functions. We define:*

$$t := x \mid f(t(\boldsymbol{x})), \text{ where } f \in \mathcal{F} \text{ (constants are 0-ary functions)};$$
$$\varphi := t(\boldsymbol{x}) > 0 \mid t(\boldsymbol{x}) \geq 0 \mid \varphi \wedge \varphi \mid \varphi \vee \varphi \mid \exists x_i \varphi \mid \forall x_i \varphi.$$

In this setting $\neg\varphi$ is regarded as an inductively defined operation which replaces atomic formulas $t > 0$ with $-t \geq 0$, atomic formulas $t \geq 0$ with $-t > 0$, switches $\wedge$ and $\vee$, and switches $\forall$ and $\exists$.

Definition 7 (Bounded $\mathcal{L}_{\mathbb{R}_\mathcal{F}}$-Sentences). *We define the bounded quantifiers $\exists^{[u,v]}$ and $\forall^{[u,v]}$ as $\exists^{[u,v]}x.\varphi =_{df} \exists x.(u \leq x \wedge x \leq v \wedge \varphi)$ and $\forall^{[u,v]}x.\varphi =_{df} \forall x.((u \leq x \wedge x \leq v) \to \varphi)$ where u and v denote $\mathcal{L}_{\mathbb{R}_\mathcal{F}}$ terms, whose variables only contain free variables in φ excluding x. A bounded $\mathcal{L}_{\mathbb{R}_\mathcal{F}}$-sentence is*

$$Q_1^{[u_1,v_1]}x_1 \cdots Q_n^{[u_n,v_n]}x_n \; \psi(x_1,...,x_n),$$

where $Q_i^{[u_i,v_i]}$ are bounded quantifiers, and $\psi(x_1,...,x_n)$ is quantifier-free.

Definition 8 (δ-Variants). *Let $\delta \in \mathbb{Q}^+ \cup \{0\}$, and φ an $\mathcal{L}_{\mathbb{R}_\mathcal{F}}$-formula*

$$\varphi: \; Q_1^{I_1}x_1 \cdots Q_n^{I_n}x_n \; \psi[t_i(\boldsymbol{x},\boldsymbol{y}) > 0; t_j(\boldsymbol{x},\boldsymbol{y}) \geq 0],$$

where $i \in \{1,...k\}$ and $j \in \{k+1,...,m\}$. The δ-weakening φ^δ of φ is defined as the result of replacing each atom $t_i > 0$ by $t_i > -\delta$ and $t_j \geq 0$ by $t_j \geq -\delta$:

$$\varphi^\delta: \; Q_1^{I_1}x_1 \cdots Q_n^{I_n}x_n \; \psi[t_i(\boldsymbol{x},\boldsymbol{y}) > -\delta; t_j(\boldsymbol{x},\boldsymbol{y}) \geq -\delta].$$

It is clear that $\varphi \to \varphi^\delta$ (see [13]).

In [12], we have proved that the following δ-decision problem is decidable, which is the basis of our framework.

Theorem 1 (δ-Decidability [12]). *Let $\delta \in \mathbb{Q}^+$ be arbitrary. There is an algorithm which, given any bounded $\mathcal{L}_{\mathbb{R}_\mathcal{F}}$-sentence φ, correctly returns one of the following two answers:*

– *δ-True: φ^δ is true.*
– *False: φ is false.*

When the two cases overlap, either answer is correct.

The following theorem states the (relative) complexity of the δ-decision problem. A bounded Σ_n sentence is a bounded $\mathcal{L}_{\mathbb{R}_\mathcal{F}}$-sentence with n alternating quantifier blocks starting with $\exists$.

Theorem 2 (Complexity [13]). *Let S be a class of $\mathcal{L}_{\mathbb{R}_\mathcal{F}}$-sentences, such that for any φ in S, the terms in φ are in Type 2 complexity class C. Then, for any $\delta \in \mathbb{Q}^+$, the δ-decision problem for bounded Σ_n-sentences in S is in $(\Sigma_n^\mathsf{P})^\mathsf{C}$.*

Basically, the theorem says that increasing the number of quantifier alternations will in general increase the complexity of the problem, unless $\mathsf{P} = \mathsf{NP}$ (recall that $\Sigma_0^\mathsf{P} = \mathsf{P}$ and $\Sigma_1^\mathsf{P} = \mathsf{NP}$). This result can specialized for specific families of functions. For example, with polynomially-computable functions, the δ-decision problem for bounded Σ_n-sentences is (Σ_n^P)-complete. For more details and results we again point the interested reader to [13].

Trace Simplifications Preserving Temporal Logic Formulae with Case Study in a Coupled Model of the Cell Cycle and the Circadian Clock

Pauline Traynard, François Fages, and Sylvain Soliman

Inria Paris-Rocquencourt, Team Lifeware, France

Abstract. Calibrating dynamical models on experimental data time series is a central task in computational systems biology. When numerical values for model parameters can be found to fit the data, the model can be used to make predictions, whereas the absence of any good fit may suggest to revisit the structure of the model and gain new insights in the biology of the system. Temporal logic provides a formal framework to deal with imprecise data and specify a wide variety of dynamical behaviors. It can be used to extract information from numerical traces coming from either experimental data or model simulations, and to specify the expected behaviors for model calibration. The computation time of the different methods depends on the number of points in the trace so the question of trace simplification is important to improve their performance. In this paper we study this problem and provide a series of trace simplifications which are correct to perform for some common temporal logic formulae. We give some general soundness theorems, and apply this approach to period and phase constraints on the circadian clock and the cell cycle. In this application, temporal logic patterns are used to compute the relevant characteristics of the experimental traces, and to measure the adequacy of the model to its specification on simulation traces. Speed-ups by several orders of magnitude are obtained by trace simplification even when produced by smart numerical integration methods.

1 Introduction

Calibrating dynamical models on experimental data time series is a central task in computational systems biology. When numerical values for model parameters can be found to fit the data, the model can be used to make predictions, whereas the absence of any good fit may suggest to revisit the structure of the model and gain new insights in the biology of the system, see for instance [23,15].

Temporal logic provides a formal framework to deal with imprecise data and specify a wide variety of dynamical behaviors. In the early days of systems biology, propositional temporal logic was proposed by computer scientists to formalize the Boolean properties of the behavior of biochemical reaction systems [11,5] or gene regulatory networks [4,3]. Generalizing these techniques to

P. Mendes et al. (Eds.): CMSB 2014, LNBI 8859, pp. 114–128, 2014.

quantitative models can be done in two ways: either by discretizing the different regimes of the dynamics in piece-wise linear or affine models [8,2], or by relying on numerical simulations and taking a first-order version of temporal logic with constraints on concentrations, as query language for the numerical traces [1,13,14]. Such language can be used not only to extract information from numerical traces coming from either experimental data or model simulations, but also to specify the expected behaviors as constraints for model calibration and robustness measure [20,21,9].

The general idea of model-checking a single finite trace has been well known for years, notably in the framework of Runtime Verification [17]. It usually relies on the classical bottom-up algorithm, which is bilinear [22]. This extends even to quantitative model-checking like the continuous interpretation of Signal Temporal Logic [10] since the combination of two booleans or two reals by min/-max is cheap. However, when using the full power of First-Order Linear Time Logic (FO-LTL) to compute validity domains, the dependency of the complexity on the size of the trace is no longer linear but exponential in the number of variables [13], reflecting the computational cost of combining complex domains. The question of trace simplification [14] is therefore important to improve the performance of FO-LTL constraint solving, and with it of the corresponding calibration methods.

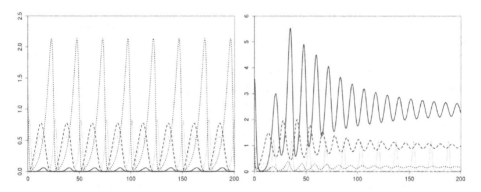

Fig. 1. Traces of some elements of the coupled cell cycle (*MPF* and *Wee1* in grey, respectively solid and dashed lines) and circadian clock (*PerCry*, *Bmal1* and *RevErb*$_\alpha$ in black, respectively solid, dashed and dotted lines) models with different parameter sets

In this paper we provide a series of trace simplifications which are correct to perform for some common temporal logic formulae. We give some general soundness theorems, and apply this approach to period and phase constraints on the circadian clock and the cell cycle. The traces shown in Fig. 1, and detailed in Sect. 6, contain each several thousands of time-points. Computing the domains of the formula describing the period between each pair of successive peaks by polyhedral methods [13] becomes quite computationally expensive. In this application, temporal logic patterns are used to compute the relevant characteristics of the experimental traces, and to measure the adequacy of the model to its specification on simulation traces. Speed-ups by several orders of magnitude are obtained by trace simplification, even when produced by smart numerical integration methods (e.g. Rosenbrock's implicit method), making trace simplification comparable with ad-hoc solvers.

2   Temporal Logic Patterns

The *Linear Time Logic* LTL is a temporal logic [6] which extends classical logic with modal operators for qualifying when a formula is true in a series of timed states. The temporal operators are **X** ("next", for at the next time point), **F** ("finally", for at some time point in the future), **G** ("globally", for at all time points in the future), **U** ("until", for a first formula must be true until a second one becomes true), and **W** (" weak until", a dual operator of **U**). These operators enjoy some simple duality properties, $\neg \mathbf{X}\phi = \mathbf{X}\neg\phi$, $\neg \mathbf{F}\phi = \mathbf{G}\neg\phi$, $\neg\mathbf{G}\phi = \mathbf{F}\neg\phi$, $\neg(\psi \mathbf{U} \phi) = (\neg\phi \mathbf{W} \neg\psi)$, $\neg(\psi \mathbf{W} \phi) = (\neg\psi \mathbf{U} \neg\phi)$, and we have $\mathbf{F}\phi = \text{true } \mathbf{U} \phi$, $\mathbf{G}\phi = \phi \mathbf{W} \text{ false}$.

In this paper we consider a first-order version of LTL, denoted by FO-LTL($\mathbb{R}_{\text{lin}}$), with variables and linear constraints over $\mathbb{R}$, and quantifiers. The grammar of FO-LTL($\mathbb{R}_{\text{lin}}$) formulae is defined as follows: $\phi ::= c \mid \neg\phi \mid \phi \Rightarrow \psi \mid \phi \wedge \phi \mid \phi \vee \phi \mid \exists x \, \phi \mid \forall x \, \phi \mid \mathbf{X}\phi \mid \mathbf{F}\phi \mid \mathbf{G}\phi \mid \phi\mathbf{U}\phi \mid \phi\mathbf{W}\phi$ where c denotes linear constraints between molecular concentrations (written with upper case letters) their first derivative (written dA/dt), free variables (written with lower case letters), real numbers, and the state time variable, denoted by *Time*; e.g., $\mathbf{F}(A < v)$ is an FO-LTL($\mathbb{R}_{\text{lin}}$) formula. To denote the value of state variable A in the state s_i we shall use a subscript notation such as A_{s_i}.

Temporal logic formulae are classically interpreted in a Kripke structure, i.e. a transition relation over a set of states such that each state has at least one successor [6]. In this paper, we consider finite traces obtained either by biological experiments, or by numerical integration. To give meaning to LTL formulae, a finite trace $(s_0, ..., s_n)$ is thus complemented in an infinite trace by adding a loop on the last state, $(s_0, ..., s_n, s_n, ...)$. The practical assumption behind this classical convention for interpreting temporal logic on finite traces [22] is that the time horizon considered is sufficiently long for properly evaluating the formulas of interest. We also replace the computed value of $\frac{dA}{dt}$ by 0 in the last state, in order to maintain the coherence between the concentrations and their derivatives. In this interpretation over finite traces, the formula $\mathbf{G}\phi$ is thus true in the last state if ϕ is true in the last state. The semantics of formulae containing free variables is given by the validity domains of the variables.

Definition 1. *The validity domain* $\mathcal{D}_{(s_0,...,s_n),\phi}$ *of the free variables of an FO-LTL($\mathbb{R}_{\text{lin}}$) formula ϕ on a finite trace $T = (s_0, ..., s_n)$, is a vector of least domains for the variables, noted* $\mathcal{D}_{(s_0,...,s_n),\phi}$, *satisfying the following equations:*

- $\mathcal{D}_{T,\phi} = \mathcal{D}^T_{s_0,\phi}$,
- $\mathcal{D}^T_{s_i,c(\boldsymbol{x})} = \{\boldsymbol{v} \in \mathbb{R}^k \mid s_i \models c[\boldsymbol{v}/\boldsymbol{x}]\}$ *for a constraint $c(\boldsymbol{x})$,*

- $\mathcal{D}^T_{s_i,\phi\wedge\psi} = \mathcal{D}^T_{s_i,\phi} \cap \mathcal{D}^T_{s_i,\psi}$, and $\mathcal{D}^T_{s_i,\phi\vee\psi} = \mathcal{D}^T_{s_i,\phi} \cup \mathcal{D}^T_{s_i,\psi}$,
- $\mathcal{D}^T_{s_i,\neg\phi} = \complement\,\mathcal{D}^T_{s_i,\phi}$,
- $\mathcal{D}^T_{s_i,\exists x\phi} = \Pi_x \mathcal{D}^T_{s_i,\phi}$, and $\mathcal{D}^T_{s_i,\forall x\phi} = \mathcal{D}^T_{s_i,\neg\exists x\neg\phi}$,
- $\mathcal{D}^T_{s_i,\mathbf{X}\phi} = \mathcal{D}^T_{s_{i+1},\phi}$ if $i < n$, and $\mathcal{D}^T_{s_n,\mathbf{X}\phi} = \mathcal{D}^T_{s_n,\phi}$,
- $\mathcal{D}^T_{s_i,\mathbf{F}\phi} = \bigcup_{j=i}^{n} \mathcal{D}^T_{s_j,\phi}$, and $\mathcal{D}^T_{s_i,\mathbf{G}\phi} = \bigcap_{j=i}^{n} \mathcal{D}^T_{s_j,\phi}$,
- $\mathcal{D}^T_{s_i,\phi\mathbf{U}\psi} = \bigcup_{j=i}^{n}(\mathcal{D}^T_{s_j,\psi} \cap \bigcap_{k=i}^{j-1} \mathcal{D}^T_{s_k,\phi})$.

where $\complement$ is the set complement operator over domains, and Π_x is the domain projection operator out of x, restoring domain $\mathbb{R}$ for x, and the other operators are defined by duality.

3   Trace Simplifications

The usual computation of the validity domains involves computing domains for each subformula on each point of the trace s_i. When dealing with temporal data coming from numerical integration, especially of stiff systems, n can be very high, which induces a high computational cost, $\mathcal{O}(n^k)$, where k is the number of variables. As mentionned in [14], and justified in the following sections of this paper, a practical solution to this issue involves simplifying the numerical trace without changing the generic domain solving algorithm. In this section we therefore define more precisely the formal framework for defining such trace simplifications.

Definition 2 (Trace simplification). *Let T be a finite trace $(s_0,\ldots,s_n)$ and ϕ an FO-LTL($\mathbb{R}_{lin}$) formula with constraints over the states of T.*
* T' is a simplification of T for ϕ at i, written $T' \preceq^i_\phi T$ if:*

- *$T' = (s_{j_0},\ldots,s_{j_k})$ for $J = \{j_0,\ldots,j_k\}$ a subset of the indices $\{0,\ldots,n\}$ such that $j_0 < \ldots < j_k$, i.e., T' is a subtrace of T;*
- *$\mathcal{D}^T_{s_i,\phi} = \mathcal{D}^{T'}_{s_{j_i},\phi}$, where j_i is the smallest index in J such that $j_i \geq i$, i.e. the validity domains on T at i and T' at j_i are equal.*

T' is a simplification of T for ϕ, written $T' \preceq_\phi T$ when it is a simplification of T at s_0, i.e., $\mathcal{D}_{T,\phi} = \mathcal{D}_{T',\phi}$.
T' is a strict simplification of T for ϕ, written $T' \prec T$ if $J \subsetneq \{0,\ldots,n\}$.
T' is an optimal simplification of T for ϕ if its cardinal is minimal in the set of the simplifications of T for ϕ.

Property-driven reduction of the system under analysis is a technique that has been addressed many times in the history of computer science. In the framework of abstract interpretation [7], not only the states but also the transitions can be abstracted in a new system for simplifying the analysis of some given properties. The definition above can be seen as a particular instance of this framework where a subset of states on the trace is preserved without abstraction, and the transitions are abstracted accordingly to this subset. This abstraction reflects our motivation of computing exact validity domains for formula variables (no state domain abstraction) more efficiently (transition abstraction).

4   Examples

Most of the equations for $\mathcal{D}^T_{s_i,\phi}$ in Definition 1 are local, in the sense that they only need information about the state at s_i. One obvious case of simplification is when the unions or intersections involved in the domains for **F**, **G** and **U** can be computed on a strict subset of the points, sometimes even a singleton. Since it will come up often in the following examples, let us define a simple subtrace containing all the local extrema and the initial point of the trace.

Definition 3 (Extrema Subtrace). *Let* $T = (s_0, \ldots, s_n)$ *be a trace,* T^e_x *is the subtrace of* T *defined as follows:*

$$T^e_x = \{s_i \in T \mid (dx/dt)_{i-1} > 0 \wedge (dx/dt)_i \le 0\}$$
$$\cup \{s_i \in T \mid (dx/dt)_{i-1} < 0 \wedge (dx/dt)_i \ge 0\} \cup \{s_0\}$$

We shall write $T^e = \bigcup_x T^e_x$

In the following examples, we will use the formulae given in [14] plus a few other ones, and for each, we will compute the corresponding domain and examine possible trace simplifications.

Example 1 (Minimal Amplitude).

Formula: $\phi = \exists v \mid \mathbf{F}(A < v) \wedge \mathbf{F}(A > v + a)$

Validity Domain. Let s_{minA} and s_{maxA} be some points of the trace where A is respectively minimum and maximum.

$$\mathcal{D}_{T,\phi} = \Pi_a(\mathcal{D}^T_{s_0,\mathbf{F}(A<v)} \cap \mathcal{D}^T_{s_0,\mathbf{F}(A>v+a)})$$
$$= \Pi_a((\bigcup_{i=0}^{n} \mathcal{D}^T_{s_i,A<v}) \cap (\bigcup_{i=0}^{n} \mathcal{D}^T_{s_j,A>v+a})) \qquad (*)$$
$$= \Pi_a(\mathcal{D}^T_{s_{minA},A<v} \cap \mathcal{D}^T_{s_{maxA},A>v+a}) \qquad (*)$$

Trace Simplification. From the computation of the domain, equations marked with a ($*$), one can see that both unions are actually equal to a single domain, only dependent on the state but not on T. Therefore any choice of s_{minA}, s_{maxA} leads to an optimal trace simplification T_J where $J = \{minA, maxA\}$.

Note that because of the semantic link between A and $\frac{dA}{dt}$, T^e_A contains s_{minA} and s_{maxA} and therefore will result in the same unions in the computation of the domain, hence T^e_A is a simplification of T for ϕ.

Example 2 (Threshold).

Formula: $\phi = \mathbf{F}(\textit{Time} > 20 \wedge A < v)$

Validity Domain. Let T be a trace $(s_0, \ldots, s_n)$ and $T_{>20}$ its subtrace on the points $J = \{0 \le i \le n \mid Time_{s_i} > 20\}$. As before, we chose some $s_{minA_{>20}}$, a point where A is minimum on $T_{>20}$.

$$\mathcal{D}_{T,\phi} = \mathcal{D}^T_{s_0, \mathbf{F}(Time>20 \wedge A<v)} = \bigcup_{i=0}^{n} \mathcal{D}^T_{s_i, Time>20 \wedge A<v}$$

$$= \bigcup_{i=0}^{n} (\mathcal{D}^T_{s_i, Time>20} \cap \mathcal{D}^T_{s_i, A<v}) \qquad (*)$$

$$= \bigcup_{i \in J} \mathcal{D}^T_{s_i, A<v} = \mathcal{D}^T_{s_{minA_{>20}}, A<v} \qquad (*)$$

Trace Simplification. As shown by the marked equations, the single point $\{s_{minA_{>20}}\}$ is enough to compute the big union of the domain, it defines an optimal trace simplification of T for ϕ.

Notice that T_A^e is not a simplification unless it does contain a local minimum such that $Time > 20$: if that is not the case, e.g. always increasing trace, $s_{minA_{>20}}$ will be the first state after $Time = 20$, which is not a local extremum.

Example 3 (Crossing).

Formula: $\phi = \mathbf{F}(A > B \wedge \mathbf{X}(A \le B \wedge Time = t))$

Validity Domain $\quad \mathcal{D}_{T,\phi} = \bigcup_{i=0}^{n} (\mathcal{D}^T_{s_i, A_{s_i}>B_{s_i}} \cap (\mathcal{D}^T_{s_{i+1}, A_{s_i} \le B_{s_i}} \cap \mathcal{D}^T_{s_{i+1}, Time=t}))$

$$= \bigcup_{i \in \{0,\ldots,n\}] \mid A_{s_i}>B_{s_i} \wedge A_{s_{i+1}} \le B_{s_{i+1}}} \{Time_{s_{i+1}}\}$$

The computation above simply discards from the union the trace points where the intersection is empty because one of the two first members is empty.

Trace Simplification. Once again, for any trace $T = (s_0, \ldots, s_n)$, the validity domain is a big union that can be restricted to the points of $J = \{i, i+1 \in \{0, \ldots, n\}] \mid A_{s_i} > B_{s_i} \wedge A_{s_{i+1}} \le B_{s_{i+1}}\}$, which defines a simplification T_J of T for ϕ. As in Example 2, T_A^e is not a simplification of T for ϕ since it obviously misses the points at which $Time$ has to be computed.

Example 4 (Peak).

Formula: $\phi = \mathbf{F}(\frac{dA}{dt} > 0 \wedge \mathbf{X}(\frac{dA}{dt} \le 0 \wedge Time = t))$

Validity Domain. The reasoning is the same as for Example 3.

$$\mathcal{D}_{T,\phi} = \mathcal{D}^T_{s_0,\phi} = \bigcup_{i=0}^{n} (\mathcal{D}^T_{s_i,\frac{dA}{dt}>0} \cap (\mathcal{D}^T_{s_{i+1},\frac{dA}{dt}\leq 0} \cap \mathcal{D}^T_{s_{i+1},Time=t}))$$

$$= \bigcup_{i\in\{0,\ldots,n\}\mid (\frac{dA}{dt})_{s_i}>0 \wedge (\frac{dA}{dt})_{s_{i+1}}\leq 0} \mathcal{D}^T_{s_{i+1},Time=t}$$

$$= \bigcup_{i\in\{0,\ldots,n\}\mid (\frac{dA}{dt})_{s_i}>0 \wedge (\frac{dA}{dt})_{s_{i+1}}\leq 0} \{Time_{s_{i+1}}\}$$

Trace Simplification. As above, for any trace $T = (s_0,\ldots,s_n)$, $J = \{i, i+1 \in \{0,\ldots,n\}\} \mid \frac{dA}{dt}_{s_i} > 0 \wedge \frac{dA}{dt}_{s_{i+1}} \leq 0\}$ defines a simplification T_J of T for ϕ.

Note that T^e_A is also a simplification of T for ϕ since it contains all $i+1$ at which A_{s_i} is used and a predecessor with the right sign of the derivative, either s_0 or a nadir preceding the peak. Note also that $|T^e_A| \leq |T_J| + 2$ since there can be one nadir more than there are peaks, plus the origin s_0.

Example 5 (Period).

Formula: $\phi = \exists (t1, t2) \mid p = t_2 - t_1 \wedge t_1 < t_2$

$$\wedge \, \mathbf{F}(\frac{dA}{dt} > 0 \wedge \mathbf{X}(\frac{dA}{dt} \leq 0 \wedge Time = t1))$$

$$\wedge \, \mathbf{F}(\frac{dA}{dt} > 0 \wedge \mathbf{X}(\frac{dA}{dt} \leq 0 \wedge Time = t2))$$

$$\wedge \, \neg \exists t3 \mid t_1 < t_3 < t_2 \wedge \mathbf{F}(\frac{dA}{dt} > 0 \wedge \mathbf{X}(\frac{dA}{dt} \leq 0 \wedge Time = t3))$$

ϕ encodes the fact that t_1 and t_2 are peaks, with no peak in between.

Trace Simplification One can notice that the domain is formed of the same kind of union as in Example 4, repeated three times, and under top-level projections/intersections/complementations. Now, remark that a simplification for the formula of Example 4 will, by definition, allow to compute correctly the domains for all three $\mathbf{F}$ formulae, and therefore is a simplification for the compound ϕ. This is a special case of Theorem 1 detailed in the next section.

It follows that T_J of Example 4 and T^e_A are simplifications of T for ϕ.

Equivalent Formula:

$$\phi = \exists (t_1, t_2) \mid p = t_2 - t_1 \wedge \mathbf{F}(\frac{dA}{dt} > 0 \wedge \mathbf{X}(\frac{dA}{dt} \leq 0 \wedge Time = t_1$$

$$\wedge \, (\frac{dA}{dt} \leq 0) \mathbf{U}(\frac{dA}{dt} > 0$$

$$\wedge \, ((\frac{dA}{dt} > 0) \mathbf{U}(\frac{dA}{dt} \leq 0 \wedge Time = t_2)))))$$

Validity Domain. Note first that the validity domain of the subformula $\psi = \frac{dA}{dt} > 0 \wedge ((\frac{dA}{dt} > 0)\mathbf{U}(\frac{dA}{dt} \leq 0 \wedge Time = t_2))$ is computed at each time point s_i like this:

$$\mathcal{D}^T_{s_i,\psi} = \mathcal{D}^T_{s_i,\frac{dA}{dt}>0} \cap \bigcup_{j=i}^{n}(\mathcal{D}^T_{s_j,\frac{dA}{dt}\leq 0 \wedge Time=t_2} \cap (\bigcap_{k=i}^{j-1}(\mathcal{D}^T_{s_k,\frac{dA}{dt}>0})))$$

Since $\mathcal{D}^T_{s_i,\frac{dA}{dt}>0}$ is either empty or equal to the whole space when $\frac{dA}{dt}_{s_i}$ is respectively negative or strictly positive, it holds that $D^T_{s_i,\psi}$ is empty if $\frac{dA}{dt}_{s_i} \leq 0$, otherwise:

$$\mathcal{D}^T_{s_i,\psi} = \bigcup_{j=i}^{n}(\mathcal{D}^T_{s_j,\frac{dA}{dt}\leq 0 \wedge Time=t_2} \cap (\bigcap_{k=i}^{j-1}(\mathcal{D}^T_{s_k,\frac{dA}{dt}>0})))$$

$$= \bigcup_{j=i}^{n}(\mathcal{D}^T_{s_j,\frac{dA}{dt}\leq 0} \cap \mathcal{D}^T_{s_j,Time=t_2} \cap (\bigcap_{k=i}^{j-1}(\mathcal{D}^T_{s_k,\frac{dA}{dt}>0})))$$

$$= \bigcup_{j\in\{i,\dots,n\}|(\frac{dA}{dt})_{s_j}\leq 0 \wedge \forall k\in\{i,\dots,j-1\},(\frac{dA}{dt})_{s_k}>0} \mathcal{D}^T_{s_j,Time=t_2}$$

$$= \bigcup_{j\in\{i,\dots,n\}|(\frac{dA}{dt})_{s_j}\leq 0 \wedge \forall k\in\{i,\dots,j-1\},(\frac{dA}{dt})_{s_k}>0} \{Time_{s_j}\}$$

This union is in fact restricted to the first point s_j after s_i where $\frac{dA}{dt}$ is no longer strictly positive.

With the same reasoning, the validity domain for the whole formula becomes:

$$\mathcal{D}_{T,\phi} = \bigcup_{(i,j)\in P} \{Time_{s_{j+1}} - Time_{s_{t+1}}\}$$

where P is the set of pairs of successive peaks:

$$P = \{(i,j) \,|\, (\frac{dA}{dt})_{s_i} > 0 \wedge (\frac{dA}{dt})_{s_{i+1}} \leq 0 \wedge (\frac{dA}{dt})_{s_j} > 0 \wedge (\frac{dA}{dt})_{s_{j+1}} \leq 0$$

$$\wedge \,\neg\exists i < k < j \,|\, (\frac{dA}{dt})_{s_k} > 0 \wedge (\frac{dA}{dt})_{s_{k+1}} \leq 0\}$$

T^e_A is a simplification of T for ϕ since it contains all the peaks of the trace.

5   General Simplification Results

Example 5 shows that if one can simplify subformulae, one might obtain a simplification for the whole formula. Indeed, with some hypotheses, the patterns described in the previous section can actually be composed.

The first theorem simply notices that if the highest-level temporal subformulae have a simplification, it also holds for the compound formula.

Theorem 1. *Let T be a trace containing a state s_i, ϕ and ψ two formulae and T' such that $T' \preceq^i_\phi T$ and $T' \preceq^i_\psi T$. Then $T' \preceq^i_\mu T$ for μ equal to*

$$\phi \wedge \psi \text{ or } \phi \vee \psi \text{ or } \neg\phi \text{ or } \exists x\phi \text{ or } \forall x\phi$$

Proof. We have $\mathcal{D}^T_{s_i,\phi} = \mathcal{D}^{T'}_{s_{j_i},\phi}$ and the same for ψ, therefore $\mathcal{D}^T_{s_i,\phi\wedge\psi} = \mathcal{D}^T_{s_i,\phi} \cap \mathcal{D}^T_{s_i,\psi} = \mathcal{D}^{T'}_{s_{j_i},\phi} \cap \mathcal{D}^{T'}_{s_{j_i},\psi} = \mathcal{D}^{T'}_{s_{j_i},\phi\wedge\psi}$ and the same for the other operators.   □

Note that it is not true that if T' is a simplification for ϕ and T'' a simplification for ψ, then the union of the points in T' and T'' defines a simplification for $\phi \vee \psi$: indeed, adding points to a simplification can invalidate it, for instance if the formula contains **X**. Now, remark that if a subtrace contains extreme domains, it is a simplification for **F** and **G**:

Theorem 2. *Let $T = (s_0, \dots, s_n)$ be a trace, ϕ a formula and $T' = T_J$ a substrace of T such that:*
$\forall j \in J, T' \preceq^j_\phi T$ *and* $\forall 0 \le i \le n, \exists j \in J, \mathcal{D}^T_{s_i,\phi} \subset \mathcal{D}^{T'}_{s_j,\phi}$ *(resp. $\mathcal{D}^T_{s_i,\phi} \supset \mathcal{D}^{T'}_{s_j,\phi}$)*
then: $T' \preceq_{\mathbf{F}\phi} T$ (resp. $T' \preceq_{\mathbf{G}\phi} T$)

Proof. We have, $\forall 0 \le i \le n, \mathcal{D}^T_{s_i,\phi} \subset \mathcal{D}^{T'}_{s_j,\phi}$ it follows that $\bigcup_{i=0}^n \mathcal{D}^T_{s_i,\phi} \subset \bigcup_{j\in J} \mathcal{D}^{T'}_{s_j,\phi}$. The other inclusion is immediate since J is a subset of the indices $\{0, \dots, n\}$ and we have simplification for ϕ at those indices. The result for **G** is obtained similarly.   □

Consider now the case of formulae without free variables, their domain is either empty or full, which can be taken advantage of:

Corollary 1. *Let $T = (s_0, \dots, s_n)$ be a trace, ϕ a formula, c a constraint without free variables and J_c be the subset of indices defined by $J_c = \{0 \le i \le n \mid s_i \models c\}$ If $\forall i \in J_c, T_{J_c} \preceq^i_\phi T$ then $T_{J_c} \preceq_{\mathbf{F}(c\wedge\phi)} T$ and $T_{J_c} \preceq_{\mathbf{G}(\neg c\vee\phi)} T$*

Proof. Let us prove the result for **F**, then Thm. 1 can give it for **G**. We will simply apply the above theorem to $c \wedge \phi$. The first hypothesis of Thm. 2 is satisfied by T_{J_c} since $T_{J_c} \preceq^i_\phi T \Rightarrow T_{J_c} \preceq^i_{c\wedge\phi} T$. For the second hypothesis, it is enough to notice that if $i \notin J_c$ then $\mathcal{D}^T_{s_i,c\wedge\phi} = \mathcal{D}^T_{s_i,c} \cap \mathcal{D}^T_{s_i,\phi} = \emptyset$.   □

Note that in general $\mathbf{F}\phi\wedge\psi$ is not easy to simplify. On the contrary $\mathcal{D}_{T,\mathbf{F}(\phi\vee\psi)} = \mathcal{D}_{T,\mathbf{F}(\phi)\vee\mathbf{F}(\psi)}$ which can benefit from Theorem 1.

In many cases it is worth noticing that T^e_A satisfies the hypothesis of Thm. 2 for any formula $\mathbf{F}(\frac{dA}{dt} > 0 \wedge \mathbf{X}(\frac{dA}{dt} \le 0 \wedge c))$.

Proposition 1. *Let $\phi = \mathbf{F}(\frac{dA}{dt} > 0 \wedge \mathbf{X}(\frac{dA}{dt} \le 0 \wedge c))$ be a formula, $T^e_A \preceq_\phi T$*

Proof. We will apply Thm. 2. First note that for any extremum j in T^e_A we have $T^e_A \preceq^j_\phi T$. Indeed, s_0 is in T^e_A but will not be used to compute $\mathcal{D}_c$, on the other hand it ensures that even the first extremum does have a predecessor of the correct sign for the derivative. Now, notice that $\mathcal{D}^T_{s_i,\phi}$ will be empty at each point not a predecessor of a state of T^e_A. At those points the domain on T is the same as that at the preceding extremum (or s_0 for the first) on T^e_A. This enforces the inclusion needed for the second hypothesis of Thm. 2.   □

Taken together, these results prove all the simplifications of the previous examples except the second formula of Example 5, which is a deeply nested formula with **U** that relies on the semantics of the *Time* variable.

6   Evaluation on Oscillation Constraints between the Cell Cycle and Circadian Clock

Cellular rhythms represent an interesting field of research for systems biology, where models should satisfy qualitative properties like oscillations, synchronization among elements, and stability, as well as quantitative properties on the lengths of the oscillations and phases. FO-LTL($\mathbb{R}_{lin}$) formulae are particularly adequate to constraint biological oscillators models after these considerations.

We illustrate the use of FO-LTL($\mathbb{R}_{lin}$) constraints on a coupled model of the cell cycle and the circadian clock, which are two such biological oscillators also inter-regulated through clock-controlled cell cycle components. This gives rise to complex behaviors as suggested in a detailed study by Nagoshi et al. [12].

We use a reference model of the mammalian circadian clock [16] and a model of a generic cell cycle oscillator focusing on the G2/M transition [19]. A molecular link between the two systems is introduced with the regulation of the cell cycle kinase *Wee1* by the clock gene *bmal1* [18].

Figure 1 shows two examples of traces obtained with different sets of parameters values, simulated over a time horizon of 200 hours. They give different dynamical behaviors with correct oscillations of the components on the first one, and damped oscillations on the other. By applying specifications expressed with the temporal logic formalism on these traces, we investigate the behavior of the system, or evaluate how far each set of parameter values is from reproducing desired properties in a calibrating process.

The chosen FO-LTL($\mathbb{R}_{lin}$) formulae express constraints on the periods of each module, phases between the components, as well as stability constraints. Each formula accept T_M^e as a simplification of T, where M is the set of molecules appearing in the formula. They correspond to patterns associated to dedicated solvers defined in [14] and listed below with the corresponding properties. Detailed formulae are given in Appendix B with justifications for the simplifications.

- Constraints on the amplitude: *MinAmpl(A,min)*. This constraints the molecule A to an amplitude of at least *min*.
- Constraints on the period: *DistanceSuccPeaks(A,d)* specifies that there should be two successive peaks of the molecule A distant by *d*. The results for the evaluation on the first trace, computed either with the FO-LTL($\mathbb{R}_{lin}$) formula and the generic solver or with the *ad hoc* pattern and dedicated solver are the same and shown in Appendix A. This gives an example of information extraction from a trace with a FO-LTL($\mathbb{R}_{lin}$) formula.
- Constraints on the phases: *DistancePeaks(A,B,d)*. Here *d* take as values the possible distances between a peak of A and the following peak of B.
- Stability constraints on the oscillations: the specification *MaxDiffDistance-Peaks(A,d)* ensures that two successive peak-to-peak distances are not too

different, with a maximum difference of d, so that the oscillations of the molecule have a relative regularity over time. A second stability constraint, *MaxDiffDistancePeaks(A,d)*, constraints the differences between the peak amplitudes, and is thus useful to filter out damped oscillations. The evaluation of *MaxDiffAmplPeaks(PerCry,d)* on the trace gives $[d > 8.48801e - 05]$ as the validity domain for the first trace and $[d > 1.90466]$ for the second trace. Thus the evaluation of the constraint extracts the maximum difference in amplitudes between two successive peaks, and this result can be used as a penalty for the set of parameter values that result in damped oscillations.

We apply these constraints to the traces presented above, before and after performing the generic trace simplification where the trace T is replaced by the trace T_M^e, that is T_{PerCry}^e for all constrains in Table 1, except for *Distance-Peaks(MPF,PerCry)* where the simplified trace is $T_{MPF,PerCry}^e$.

The initial traces are obtained with two different integration methods:

- In Biocham the default simulation method is the Rosenbrock's numerical integration method. This implicit method with variable step-size avoids generating too many points and does an impressively good job in producing relatively sparse traces. With this method the first trace counts 971 point, 18 of which are kept in the simplified trace T_{PerCry}^e and 34 in $T_{MPF,PerCry}^e$. The second trace T counts 1047 points, T_{PerCry}^e counts 35 points and $T_{MPF,PerCry}^e$ counts 58 points. Since the initial traces have reasonable sizes the computing times for the simplifications are short: between 8 and 16ms.
- However in some cases, the Rosenbrock method is less adequate than other non-adaptive methods. For example, this is the case when the model involves events, since the approximation done for numerical integration with big steps, may not be valid for determining when an event becomes true. Therefore we also consider the fourth order Runge-Kutta method with a fixed step size. With this method, the trace optimisation is all the more beneficial since the traces originally count more points: 20002 points here for a time horizon of 200 hours. However the same trace simplifications take longer: around 160ms for T_{PerCry}^e and 250ms for $T_{MPF,PerCry}^e$.

The execution times are compared in Table 1 where each constraint is identified by the equivalent pattern. We compare the evaluation of the constraints on a trace with a high number of points (fixed Runge-Kutta method) or a reduced size (adaptive Rosenbrock method), and either complete or simplified. Furthermore the generic solver is compared to the dedicated solvers defined in [14].

Table 1 clearly shows that trace simplification provides a faster evaluation for all constraints on all traces, with a speed-up up to 100 fold for the more complex ones. The dedicated solvers benefit as well from this speed-up, however it has to be noted that applying the dedicated solver on the full trace is faster than the time needed for the trace simplification in this example. Although the simplification can be done just once on a trace that can be then evaluated repeatedly for different patterns, the number of evaluations would have to be unlikely high for any real benefit. In contrast, the time gain obtained with the combined use

Table 1. Computing time (in ms) for the validity domain of different formula patterns. Comparison between the first and second parameter sets, with variable or fixed step-size over 200h, before (**Bef.**) and after (**Aft.**) simplification.

Formula	Solver	First trace				Second trace			
		variable		fixed		variable		fixed	
		Bef.	Aft.	Bef.	Aft.	Bef.	Aft.	Bef.	Aft.
Reached(PerCry)	generic	12	0	260	4	12	0	204	0
	dedicated	0	0	16	0	4	0	16	0
MinAmpl(PerCry)	generic	132	0	2728	0	132	4	2516	4
	dedicated	0	0	16	0	4	0	16	0
LocalMax(PerCry)	generic	64	0	1308	4	72	4	1316	4
	dedicated	0	0	36	8	4	0	44	4
DistancePeaks(PerCry)	generic	512	12	9584	12	708	80	12373	104
	dedicated	4	4	40	8	32	28	80	48
DistanceSuccPeaks(PerCry)	generic	532	12	10980	12	1188	36	23101	156
	dedicated	4	0	40	8	4	0	28	4
MaxDiffDistancePeaks(PerCry)	generic	1700	32	34818	32	3056	96	60776	108
	dedicated	0	0	36	0	4	0	52	20
DistancePeaks(MPF,PerCry)	generic	456	16	9332	16	496	32	9365	32
	dedicated	4	4	68	12	4	0	76	20

of the trace simplification and the generic solver is clear. This suggests that the trace simplification is a good strategy when the desired constraint is not covered by the patterns with dedicated solvers, provided that the FO-LTL($\mathbb{R}_{lin}$) formula accepts a good trace simplification accordingly with the theorems presented in Sect. 3.

7   Conclusion

We have shown that trace simplifications can result in speed-ups by several orders of magnitude for the evaluation of temporal logic constraints. In particular we have given some general conditions on the syntax of the formulae under which it is correct to keep in the trace only the time points corresponding to the local extrema of the molecules, or the crossing points between molecular concentrations.

On an application concerning the modeling of the coupling between the circadian clock and the cell cycle, we have shown that temporal logic patterns provide an elegant way to extract information on the periods and phases from numerical traces, and to use these formulae as constraints for parameter search. On simulation traces, the speedup obtained in computation time was by several orders of magnitude, even on relatively sparse simulation traces obtained by Rosenbrock's implicit method for numerical integration.

The trace simplifications described in this paper are implemented in Biocham release 3.6.

Acknowledgements. This work has been supported by the French OSEO Biointelligence project. We acknowledge discussions with our partners at Dassault-Systèmes and the CMSB reviewers for their remarks.

References

1. Antoniotti, M., Policriti, A., Ugel, N., Mishra, B.: Model building and model checking for biochemical processes. Cell Biochemistry and Biophysics 38, 271–286 (2003)
2. Batt, G., Page, M., Cantone, I., Goessler, G., Monteiro, P., de Jong, H.: Efficient parameter search for qualitative models of regulatory networks using symbolic model checking. Bioinformatics 26(18), i603–i610 (2010)
3. Batt, G., Ropers, D., de Jong, H., Geiselmann, J., Mateescu, R., Page, M., Schneider, D.: Validation of qualitative models of genetic regulatory networks by model checking: Analysis of the nutritional stress response in Escherichia coli. Bioinformatics 21(suppl.1), i19–i28 (2005)
4. Bernot, G., Comet, J.-P., Richard, A., Guespin, J.: A fruitful application of formal methods to biological regulatory networks: Extending Thomas' asynchronous logical approach with temporal logic. Journal of Theoretical Biology 229(3), 339–347 (2004)
5. Chabrier, N., Fages, F.: Symbolic model checking of biochemical networks. In: Priami, C. (ed.) CMSB 2003. LNCS, vol. 2602, pp. 149–162. Springer, Heidelberg (2003)
6. Clarke, E.M., Grumberg, O., Peled, D.A.: Model Checking. MIT Press (1999)
7. Cousot, P., Cousot, R.: Abstract interpretation: A unified lattice model for static analysis of programs by construction or approximation of fixpoints. In: POPL 1977: Proceedings of the 6th ACM Symposium on Principles of Programming Languages, pp. 238–252. ACM Press, New York (1977)
8. de Jong, H., Gouzé, J.-L., Hernandez, C., Page, M., Sari, T., Geiselmann, J.: Qualitative simulation of genetic regulatory networks using piecewise-linear models. Bulletin of Mathematical Biology 66(2), 301–340 (2004)
9. Donzé, A., Ferrère, T., Maler, O.: Efficient robust monitoring for STL. In: Sharygina, N., Veith, H. (eds.) CAV 2013. LNCS, vol. 8044, pp. 264–279. Springer, Heidelberg (2013)
10. Donzé, A., Maler, O.: Robust satisfaction of temporal logic over real-valued signals. In: Chatterjee, K., Henzinger, T.A. (eds.) FORMATS 2010. LNCS, vol. 6246, pp. 92–106. Springer, Heidelberg (2010)
11. Eker, S., Knapp, M., Laderoute, K., Lincoln, P., Meseguer, J., Sönmez, M.K.: Pathway logic: Symbolic analysis of biological signaling. In: Proceedings of the seventh Pacific Symposium on Biocomputing, pp. 400–412 (January 2002)
12. Emi, N., Camille, S., Christoph, B., Thierry, L., Felix, N., Schibler, U.: Circadian gene expression in individual fibroblasts: cell-autonomous and self-sustained oscillators pass time to daughter cells. Cell 119, 693–705 (2004)
13. Fages, F., Rizk, A.: On temporal logic constraint solving for the analysis of numerical data time series. Theoretical Computer Science 408(1), 55–65 (2008)
14. Fages, F., Traynard, P.: Temporal logic modeling of dynamical behaviors: First-order patterns and solvers. In: del Cerro, L.F., Inoue, K. (eds.) Logical Modeling of Biological Systems, ch.8, pp. 307–338. ISTE Ltd. (2014)
15. Heitzler, D., Durand, G., Gallay, N., Rizk, A., Ahn, S., Kim, J., Violin, J.D., Dupuy, L., Gauthier, C., Piketty, V., Crépieux, P., Poupon, A., Clément, F., Fages, F., Lefkowitz, R.J., Reiter, E.: Competing G protein-coupled receptor kinases balance G protein and β-arrestin signaling. Molecular Systems Biology 8(590) (June 2012)
16. Leloup, J.-C., Goldbeter, A.: Toward a detailed computational model for the mammalian circadian clock. Proceedings of the National Academy of Sciences 100, 7051–7056 (2003)

17. Markey, N., Schnoebelen, P.: Model checking a path. In: Amadio, R.M., Lugiez, D. (eds.) CONCUR 2003. LNCS, vol. 2761, pp. 251–265. Springer, Heidelberg (2003)
18. Matsuo, T., Yamaguchi, S., Mitsui, S., Emi, A., Shimoda, F., Okamura, H.: Control mechanism of the circadian clock for timing of cell division in vivo. Science 302(5643), 255–259 (2003)
19. Qu, Z., MacLellan, W.R., Weiss, J.N.: Dynamics of the cell cycle: checkpoints, sizers, and timers. Biophysics Journal 85(6), 3600–3611 (2003)
20. Rizk, A., Batt, G., Fages, F., Soliman, S.: A general computational method for robustness analysis with applications to synthetic gene networks. Bioinformatics 12(25), il69–il78 (2009)
21. Rizk, A., Batt, G., Fages, F., Soliman, S.: Continuous valuations of temporal logic specifications with applications to parameter optimization and robustness measures. Theoretical Computer Science 412(26), 2827–2839 (2011)
22. Roşu, G., Havelund, K.: Rewriting-based techniques for runtime verification. Automated Software Engineering 12(2), 151–197 (2005)
23. Stoma, S., Donzé, A., Bertaux, F., Maler, O., Batt, G.: STL-based analysis of TRAIL-induced apoptosis challenges the notion of type I/type II cell line classification. PLoS Computational Biology 9(5), e1003056 (2013)

A   Example of Computation with Both the Generic Solver and a Dedicated One

```
domains(t2-t1=d & F(d([CRY_nucl-PER_nucl])/dt>0 & X(Time=t1 & d([CRY_nucl-PER_nucl
    ])/dt=<0 & (d([CRY_nucl-PER_nucl])/dt=<0) U (d([CRY_nucl-PER_nucl])/dt>0 & ((
    d([CRY_nucl-PER_nucl])/dt>0) U (d([CRY_nucl-PER_nucl])/dt=<0 & Time=t2)))))).
Domain computed in 532 ms
d = 24.6095, t1 = 15.2848, t2 = 39.8944
| d = 24.7193, t1 = 39.8944, t2 = 64.6137
| d = 25.1225, t1 = 64.6137, t2 = 89.7362
| d = 24.7623, t1 = 89.7362, t2 = 114.499
| d = 24.7984, t1 = 114.499, t2 = 139.297
| d = 24.8047, t1 = 139.297, t2 = 164.102
| d = 24.7704, t1 = 164.102, t2 = 188.872
```

```
domains(distanceSuccPeaks([CRY_nucl-PER_nucl],[d])).
Domain computed in 4 ms
d = 24.6095
| d = 24.7193
| d = 25.1225
| d = 24.7623
| d = 24.7984
| d = 24.8047
| d = 24.7704
```

B   Oscillation Constraints

Constraints on the Amplitude. As shown in Ex. 1, the following formula, accepting T_A^e as a simplification of T, ensures that a molecule A has an amplitude of at least min: $\phi = \exists v \mid \mathbf{F}(A < v) \wedge \mathbf{F}(A > v + min)$.

It is equivalent to the pattern $MinAmpl(A,min)$ described in [14] and associated to a specific solver which computes the amplitude of A directly from the trace.

Constraints on the Period. This formula extracts the distances between successive peaks:

$$\phi = \exists(t_1, t_2) \mid d = t_2 - t_1 \wedge \mathbf{F}(\frac{dA}{dt} > 0 \wedge \mathbf{X}(\frac{dA}{dt} \leq 0 \wedge Time = t_1$$

$$\wedge \, (\frac{dA}{dt} \leq 0)\mathbf{U}(\frac{dA}{dt} > 0$$

$$\wedge \, ((\frac{dA}{dt} > 0)\mathbf{U}(\frac{dA}{dt} \leq 0 \wedge Time = t_2))))))$$

This formula accepts T_A^e as a simplification of T, as shown in Ex. 4 and with Thm. 1. It is equivalent to the pattern *DistanceSuccPeaks(A,d)*. The specific solver associated to this pattern computes the list of peaks of A directly from the trace and exhibits the possible distances between two successive peaks. Computing the validity domain of this formula enables to extract each peak-to-peak distance from the trace, giving an estimation of the period of the oscillations.

Constraints on the Phases

$$\phi = \exists(t_1, t_2) \mid t2 - t1 = d \wedge \mathbf{F}(\frac{dA}{dt} \geq 0 \wedge \mathbf{X}(\frac{dA}{dt} < 0 \wedge Time = t1))$$

$$\wedge \, \mathbf{F}(\frac{dB}{dt} \geq 0 \wedge \mathbf{X}(\frac{dB}{dt} < 0 \wedge Time = t2))$$

corresponds to *DistancePeaks([A,B],d)*. $T_A^e \cup T_B^e$ is a simplification of T for ϕ.

Stability Constraints. The following formula constraints two successive peak-to-peak distances to be similar by setting a maximum for the difference between the two distances.

$$\phi = \exists(t_1, t_2, t_3) \mid t2 - t1 = d1 \wedge t3 - t2 = d2 \wedge d2 - d1 \leq d \wedge d1 - d2 \leq d$$

$$\wedge \, \mathbf{F}(\frac{dA}{dt} > 0 \wedge \mathbf{X}(\frac{dA}{dt} \leq 0 \wedge Time = t_1$$

$$\wedge \, (\frac{dA}{dt} \leq 0)\mathbf{U}(\frac{dA}{dt} > 0$$

$$\wedge \, ((\frac{dA}{dt} > 0)\mathbf{U}(\frac{dA}{dt} \leq 0 \wedge Time = t_2$$

$$\wedge \, (\frac{dA}{dt} \leq 0)\mathbf{U}(\frac{dA}{dt} > 0$$

$$\wedge \, ((\frac{dA}{dt} > 0)\mathbf{U}(\frac{dA}{dt} \leq 0 \wedge Time = t_3))))))))$$

This formula accepts T_A^e as a simplification of T and the equivalent pattern is *MaxDiffDistancePeaks(A,d)*. A similar formula, useful to filter out damped oscillations, constraints the differences between the peak amplitudes, and is equivalent to the pattern *MaxDiffAmplPeaks(A,d)*.

Characterization of Reachable Attractors Using Petri Net Unfoldings

Thomas Chatain[1], Stefan Haar[1], Loïg Jezequel[1], Loïc Paulevé[2,3], and Stefan Schwoon[1]

[1] LSV, ENS Cachan, INRIA, CNRS, France
[2] CNRS & Laboratoire de Recherche en Informatique UMR CNRS 8623
Université Paris-Sud, 91405 Orsay Cedex, France
[3] Inria Saclay - Île de France, team AMIB, Palaiseau, France

Abstract. Attractors of network dynamics represent the long-term behaviours of the modelled system. Their characterization is therefore crucial for understanding the response and differentiation capabilities of a dynamical system. In the scope of qualitative models of interaction networks, the computation of attractors reachable from a given state of the network faces combinatorial issues due to the state space explosion.

In this paper, we present a new algorithm that exploits the concurrency between transitions of parallel acting components in order to reduce the search space. The algorithm relies on Petri net unfoldings that can be used to compute a compact representation of the dynamics. We illustrate the applicability of the algorithm with Petri net models of cell signalling and regulation networks, Boolean and multi-valued. The proposed approach aims at being complementary to existing methods for deriving the attractors of Boolean models, while being generic since it applies to any safe Petri net.

Keywords: dynamical systems, attractors, concurrency, qualitative models, biological networks.

1 Introduction

Living cells embed multiple regulation processes that lead to several emerging phenotypes such as cell differentiation, division, or response to environmental stress or signals. A large part of these processes are often represented as interaction networks (e.g., signalling networks, gene regulation networks) that describe the influences between numerous entities (genes, RNA, proteins). The global dynamics of such networks can then be captured using qualitative modelling frameworks, such as Boolean or discrete networks, that describe the possible transitions between the qualitative states of the system.

In the landscape of dynamics of a network model, one can distinguish between the transient and long-run dynamics, the latter being our focus in this article. In qualitative models, the long-run dynamics are referred to as *attractors*, and are formally defined as the Bottom Strongly Connected Components (BSCCs) of the

P. Mendes et al. (Eds.): CMSB 2014, LNBI 8859, pp. 129–142, 2014.

transition graph whose nodes are the global states of the network, and directed edges are the possible direct transitions between those states. One typically distinguishes between two kind of attractors: the fixed points, that are the states from which no further transition is possible; and the cyclic attractors, that are a set of states that can be visited infinitely often.

Characterizing the attractors of network dynamics is key for capturing the potential adaptation and differentiation processes the cell can undergo. In particular, one could verify, from a given state of the network, if the dynamics always converges toward a unique attractor or may diverge towards different attractors. The former indicates a deterministic long-term behaviour, whereas the latter suggests an indeterministic differentiation, potentially controlled by additional mechanisms not captured by the level of abstraction of the model.

In practice, given a qualitative model of a network, the computation of the attractors reachable from one (set of) states can become very expensive as the size of the network grows. The naive approach consisting in generating the transition graph and computing the BSSCs suffers from the combinatorial explosion of the state space (exponential with the number of components of the network) and the explosion of the number of transitions.

A part of the combinatorial explosion of dynamics is due to the concurrency between the asynchronous transitions: in a given state, several transitions may be independently fired, which results in numerous redundant interleavings of transitions in the concrete transition space.

Contribution. In this paper, we propose a new algorithm for characterizing all the attractors that are reachable from a given initial state in a qualitative model expressed with *safe* Petri nets [18], a broad class of nets which encompasses asynchronous Boolean or multi-valued networks. Our algorithm exploits the *unfoldings* of safe Petri net in order to reduce the size of the state space to explore.

Petri net unfoldings [7,8] aim at representing the state space rechable from an initial state by exploiting concurrency between transitions to prune redundant interleavings of these transitions. To our knowledge this is the first algorithm for computing all the reachable attractors which relies on unfolding structures. Our algorithm is applicable to any safe Petri net.

Whereas experiments on particular cases of biological networks show room for improvement, such a technique exploiting concurrency is foremost complementary to existing algorithms (which ignore this dynamical feature) and therefore appeals for designing combinations of techniques to make tractable the analysis of very large networks.

Related work. In the scope of Boolean networks, there has been numerous work to link the topology of the network (the interaction graph, giving signed relations between the components) with the fixed points - resulting in bounds or characterization of a subset of fixed points (e.g., [1,23,22]); and with cyclic attractors, e.g., [24,17]. While the full characterization of the fixed points of Boolean/multivalued networks can be quite efficient for large networks [20,21,12], the complete characterization of cyclic attractors is still a challenging task due to the combinatoric explosion of the state space to explore. Symbolic representation of the state space using binary decision diagrams has been used by [11] to characterize attractors in synchronous and asynchronous Boolean networks. The relationships between attractors in synchronous and asynchronous settings has then been exploited in [3] to speed the exploration of all possible attractors in Boolean networks, as well as Boolean network reduction techniques in [29]. Approximate methods are also largely considered, such as in [30], which selectively explore appropriate regions of the state space to derive a subset of cyclic attractors of the global network dynamics.

In this paper, we will focus on the use of unfolding to compute *finite complete prefixes* [14] of safe Petri nets. Finite complete prefixes contain all the reachable markings in a compact representation (the prefix is always smaller than the reachability graph). Unfoldings are very well suited to capture concurrent system dynamics, and can be efficient for reachability analysis [9], for instance.

Outline. In Sect. 2, we give a formal definition of (safe) Petri nets and their attractors, and introduce a running example. In Sect. 3, we present the unfolding of safe Petri nets. In Sect. 4, we detail our new algorithm to derive the attractors of a safe Petri net using its unfolding. Finally, implementation and experimations on Petri net models of biological networks are discussed in Sect. 5.

2 Petri Nets and Attractors

A Petri net is a bipartite graph where nodes are either places or transitions. In this paper, we consider only *safe* Petri nets where a place is either active or inactive (in oppositon to general Petri nets where each place can receive an arbitrary number of tokens, safe Petri nets allow at most one token per place). The set of active places form the state, or marking, of the net. A transition is said enabled if all the places that are parents of the transition are active. In the semantics, the *firing* of a transition makes inactive the parent places and then makes active the children places, modifying the current marking of the net.

Formally, a *(safe) Petri net* is a tuple $\mathcal{N} = \langle P, T, F, M_0 \rangle$ where P and T are sets of *nodes* (called *places* and *transitions* respectively), and $F \subseteq (P \times T) \cup (T \times P)$ is a *flow relation* (whose elements are called *arcs*). A subset $M \subseteq P$ of the places is called a *marking*, and $M_0 = \{p_0^1, \ldots, p_0^n\}$ is a distinguished *initial marking*. For any node $x \in P \cup T$, we call *pre-set* of x the set ${}^\bullet x = \{y \in P \cup T \mid (y, x) \in F\}$ and *post-set* of x the set $x^\bullet = \{y \in P \cup T \mid (x, y) \in F\}$.

A transition $t \in T$ is *enabled* at a marking M if and only if ${}^\bullet t \subseteq M$. Then t can *fire*, leading to the new marking $M' = (M \setminus {}^\bullet t) \cup t^\bullet$. We write $M \xrightarrow{t} M'$. A *firing sequence* is a (finite or infinite) word $w = t_1 t_2 \ldots$ over T such that there exist markings $M_1, M_2, \ldots$ such that $M_0 \xrightarrow{t_1} M_1 \xrightarrow{t_2} M_2 \ldots$ For any such firing sequence w, the markings $M_1, M_2, \ldots$ are called *reachable markings*.

The Petri nets we consider are said to be *safe* because we will assume that any reachable marking M is such that for any $t \in T$ that can fire from M leading to M', the following property holds: $\forall p \in M \cap M', p \in {}^\bullet t \cap t^\bullet \lor p \notin {}^\bullet t \cup t^\bullet$.

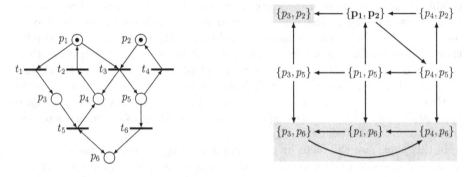

Fig. 1. A safe Petri net (left) and the corresponding marking graph (right) - the initial marking is in bold

Figure 1 (left) shows an example of a safe Petri net. The places are represented by circles and the transitions by horizontal lines (each one with a label identifying it). The arrows represent the arcs. The initial marking is represented by dots (or tokens) in the marked places.

From an initial marking of the net, one can recursively derive all possible transitions and reachable markings, resulting in the *marking graph* (Def. 1). The marking graph is always finite in the case of safe Petri nets. The attractors reachable from the initial marking of the net can then be fully characterized by the bottom strongly connected components of the marking graph (Def. 2).

Definition 1 (Marking Graph). *The marking graph of a Petri Net $\mathcal{N}$ is a directed graph $\mathcal{G} = (V, A)$ such that V is the set of all reachable markings (obtained from all the possible firing sequences) and $A \subseteq V \times V$ is such that $(M, M') \in A$ if and only if $M \xrightarrow{t} M'$ for some $t \in T$.*

Definition 2 (Attractors). *An attractor is a bottom strongly connected component of $\mathcal{G}$, that is a set $\mathcal{A}$ of markings such that either $\mathcal{A} = \{M\}$ and no transition is enabled from M; or for every $M \in \mathcal{A}$, the set of markings reachable from M is precisely $\mathcal{A}$.*

Figure 1 (right) represents the marking graph of the Petri net of Figure 1 (left). The two attractors of the Petri net of Figure 1 (left) are evidenced by the grey parts of its marking graph in Figure 1 (right).

3 Unfoldings

In this section, we explain the basics of Petri net unfoldings. A more extensive treatment of the theory explained here can be found, e.g., in [7]. Roughly speaking, the unfolding of a Petri net $\mathcal{N}$ is an "acyclic" Petri net $\mathcal{U}$ that has the same behaviours as $\mathcal{N}$ (modulo homomorphism). In general, $\mathcal{U}$ is an infinite net, but if $\mathcal{N}$ is safe, then it is possible [16] to compute a finite prefix $\mathcal{P}$

of $\mathcal{U}$ that is "complete" in the sense that every reachable marking of $\mathcal{N}$ has a reachable counterpart in $\mathcal{P}$. Thus, $\mathcal{P}$ represents the set of reachable markings of $\mathcal{N}$. Figure 2 shows a finite complete prefix of the unfolding of the Petri net of Figure 1.

In principle, the set of reachable markings can also be computed by constructing the marking graph (see Definition 1). However, the marking graph suffers from combinatorial explosion due to concurrency. For instance, suppose that $\mathcal{N}$ simply contains n independent concurrent actions. Then the only attractor of the net is reached by executing all n actions in any arbitrary order. However, the marking graph will uselessly explore all $n!$ different schedules for executing them, and all 2^n intermediate markings.

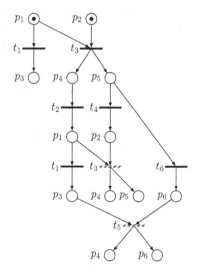

Research into concurrent systems has produced a number of solutions to alleviate the problem of combinatorial explosion due to concurrency (and eventually other sources). In [16], McMillan first proposed the use of finite unfolding prefixes. Esparza et al [8] later improved this solution. For instance, the unfolding of the previous example with n independent actions is simply of size $\mathcal{O}(n)$. With respect to the marking graph, an unfolding represents a time-space tradeoff: in general, a complete unfolding prefix $\mathcal{P}$ is much smaller than the marking graph of $\mathcal{N}$, but the problem whether a marking M of $\mathcal{N}$ is reachable, given $\mathcal{P}$, is NP-complete. However, this tradeoff is usually favourable [9].

Fig. 2. A finite complete prefix of the unfolding of the Petri net of Figure 1. Dashed events are flagged as *cut-offs*: the unfolding procedure does not continue beyond them.

We now give some technical definitions to introduce unfoldings formally.

Definition 3 (Causality, Conflict, Concurrency). *Let* $\mathcal{N} = \langle P, T, F, M_0 \rangle$ *be a net and* $x, y \in P \cup T$ *two nodes of* $\mathcal{N}$. *We say that* x *is a* causal predecessor *of* y, *noted* $x < y$, *if there exists a non-empty path of arcs from* x *to* y. *We note* $x \leq y$ *if* $x < y$ *or* $x = y$. *If* $x \leq y$ *or* $y \leq x$, *then* x *and* y *are said to be causally related.* x *and* y *are in* conflict, *noted* $x \# y$, *if there exist* $u, v \in T$ *such that* $u \leq x$, $v \leq y$, *and* ${}^\bullet u \cap {}^\bullet v \neq \emptyset$. *We call* x *and* y concurrent, *noted* x co y, *if they are neither causally related nor in conflict.*

As we said before, an unfolding is an "acyclic" version of a net $\mathcal{N}$. This notion of acyclicity is captured by Definition 4.

Definition 4 (Occurrence net). *Let* $\mathcal{N} = \langle P, T, F, M_0 \rangle$ *be a Petri net. We say that* $\mathcal{N}$ *is an* occurrence net *if it satisfies the following properties:*

1. The causality relation $<$ is acyclic;

2. $|{}^\bullet p| \leq 1$ *for all places* p, *and* $M(p) = 1$ *iff* $|{}^\bullet p| = 0$;
3. *for every transition* t, $t \# t$ *does not hold, and* $\{x \mid x \leq t\}$ *is finite.*

As is convention in the unfolding literature, we shall refer to the places of an occurrence net as *conditions* and to its transitions as *events*. Due to the structural constraints, the firing sequences of occurrence nets have special properties: if some condition c is marked during a run, then the token on c was either present initially or produced by one particular event (the single event in ${}^\bullet c$); moreover, once the token on c is consumed, it can never be replaced by another token, due to the acyclicity constraint on $<$.

Definition 5 (Configuration, Cut). *Let* $\mathcal{N} = \langle C, E, F, M \rangle$ *be an occurrence net. A set* $\mathcal{C} \subseteq E$ *is called* configuration *of* $\mathcal{N}$ *if (i)* $\mathcal{C}$ *is causally closed, i.e. for all* $e, e' \in E$ *with* $e' < e$, *if* $e \in \mathcal{C}$ *then* $e' \in \mathcal{C}$; *and (ii)* $\mathcal{C}$ *is conflict-free, i.e. if* $e, e' \in \mathcal{C}$, *then* $\neg(e \# e')$. *The* cut *of* $\mathcal{C}$, *denoted* $Cut(\mathcal{C})$, *is the set of conditions* $(M \cup \mathcal{C}^\bullet) \setminus {}^\bullet \mathcal{C}$.

Intuitively, a configuration is a set of events that can fire during a firing sequence of $\mathcal{N}$, and its cut is the set of conditions marked after that firing sequence.

We can now define the notion of unfoldings. Let $\mathcal{N} = \langle P, T, F, M_0 \rangle$ be a safe Petri net. The unfolding $\mathcal{U} = \langle C, E, G, M_0' \rangle$ of $\mathcal{N}$ is an (infinite) occurrence net (equipped with a homomorphism h) such that the firing sequences and reachable markings of $\mathcal{U}$ are exactly the firing sequences and reachable markings of $\mathcal{N}$ (modulo h). $\mathcal{U}$ can be inductively constructed as follows:

1. The conditions C are a subset of $(E \cup \{\bot\}) \times P$. For a condition $c = \langle x, p \rangle$, we will have $x = \bot$ iff $x \in M_0'$; otherwise x is the singleton event in ${}^\bullet c$. Moreover, $h(c) = p$. The initial marking M_0' contains exactly one condition $\langle \bot, p \rangle$ for each initially marked place p of $\mathcal{N}$.
2. The events of E are a subset of $2^C \times T$. More precisely, we have an event $e = \langle C', t \rangle$ for every set $C' \subseteq C$ such that c co c' holds for all $c, c' \in C'$ and $\{ h(c) \mid c \in C' \} = {}^\bullet t$. In this case, we add edges $\langle c, e \rangle$ for each $c \in C'$ (i.e. ${}^\bullet e = C'$), we set $h(e) = t$, and for each $p \in t^\bullet$, we add to C a condition $c = \langle e, p \rangle$, connected by an edge $\langle e, c \rangle$.

Intuitively, a condition $\langle x, p \rangle$ represents the possibility of putting a token onto place p through a particular firing sequence, while an event $\langle C', t \rangle$ represents a possibility of firing transition t in a particular context.

Recall that a configuration $\mathcal{C}$ of $\mathcal{U}$ represents a possible firing sequence, whose resulting marking corresponds, due to the construction of $\mathcal{U}$, to a reachable marking of $\mathcal{N}$. This marking is defined as $Mark(\mathcal{C}) := \{ h(c) \mid c \in Cut(\mathcal{C}) \}$. Since $\mathcal{U}$ is infinite in general, we are interested in computing an initial portion of it (a *prefix*) that completely characterizes the behaviour of $\mathcal{N}$.

Definition 6 (Complete Prefix). *Let* $\mathcal{N} = \langle P, T, F, M_0 \rangle$ *be a safe Petri net and* $\mathcal{U} = \langle C, E, G, M_0' \rangle$ *its unfolding. A finite occurrence net* $\mathcal{P} = \langle C', E', G', M_0' \rangle$

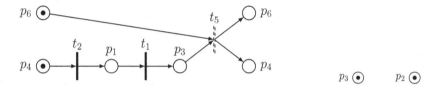

Fig. 3. Two attractors of the Petri net of Figure 1 represented as finite complete prefixes $\mathcal{U}_{\{p_4,p_6\}}$ (on the left) and $\mathcal{U}_{\{p_3,p_2\}}$ (on the right).

is said to be a prefix *of* $\mathcal{U}$ *if* $E' \subseteq E$ *is causally closed,* $C' = M_0' \cup E'^{\bullet}$, *and* G' *is the restriction of* G *to* C' *and* E'. *A prefix* $\mathcal{P}$ *is said to be* complete *if for every reachable marking* M *of* $\mathcal{N}$ *there exists a configuration* $\mathcal{C}$ *of* $\mathcal{P}$ *such that (i) Mark$(\mathcal{C}) = M$, and (ii) for each transition* $t \in T$ *enabled in* M, *there is an event* $\langle C'', t \rangle \in E'$ *enabled in Cut$(\mathcal{C})$.*

It is known [16,8] that the construction of such a complete prefix is indeed possible, and efficient tools [26,13] exist for this purpose. The precise details of this construction are out of scope for this paper; for what follows it suffices to know that it essentially follows the construction of $\mathcal{U}$ outlined above but that certain events are flagged as *cut-offs* when they do not "contribute any new reachable markings" (these events are represented by dashed lines in Figure 2). The construction then does not continue "beyond" such a cut-off event.

4   Extracting Attractors from Unfoldings

Our method for finding the attractors of a Petri net $\mathcal{N}$ uses unfoldings in two steps: first to find a set of markings which intersects all the attractors, second to output the attractors as a set of finite complete prefixes.

4.1   Representation of Attractors as Finite Complete Prefixes

We first remark that finite complete prefixes of unfoldings are particularly well suited for the representation of attractors. In fact, every attractor $\mathcal{A}$ is a set of states which form a maximal strongly connected component of the marking graph of $\mathcal{N}$. For every marking M in $\mathcal{A}$, the attractor $\mathcal{A}$ is precisely the set of markings reachable from M. Hence it can be compactly represented as a finite complete prefix of the unfolding of the Petri net $\mathcal{N}$ initialized at M. Denote this prefix $\mathcal{U}_M$: the markings associated to the configurations of $\mathcal{U}_M$ are precisely those of the attractor, moreover the prefix shows the dynamics of the net while in the attractor. Last, the size of $\mathcal{U}_M$ (as number of non cut-off events) can be up to exponentially smaller (in case of highly concurrent behaviour) than the number of markings in the attractor, and never exceeds it.

Figure 3 shows two attractors of the Petri net $\mathcal{N}$ of Figure 1 represented as finite complete prefixes. The one on the left, $\mathcal{U}_{\{p_4,p_6\}}$, represents the attractor containing the marking $\{p_4, p_6\}$. The one on the right represents the attractor made of the single marking $\{p_3, p_2\}$ which is a deadlock.

4.2   Maximal Configurations and Attractors

We have shown the interest of prefixes for representing attractors. What we need is a way to find a set $\mathcal{M}$ of markings of the Petri net $\mathcal{N}$ which contains one marking per attractor. Given such $\mathcal{M}$, the set $\{\mathcal{U}_M \mid M \in \mathcal{M}\}$ gives a complete characterization of the attractors of $\mathcal{N}$.

Now we show that the desired set $\mathcal{M}$ of markings can be obtained from the maximal configurations of a finite complete prefix of the unfolding of $\mathcal{N}$.

Definition 7 (Maximal Configuration). *A configuration of a prefix is called maximal if no other event of the prefix can be added to the configuration. Equivalently, the configuration is a deadlock of the prefix viewed as a Petri net.*

For example, in the prefix shown on Figure 2, the configuration corresponding to the firing sequence $t_3 t_2$ is not maximal because it can be extended, for instance by t_4 and the other event labeled t_3, yielding this time a maximal configuration, which reaches the marking $\{p_4, p_5\}$. Notice that this marking is not a deadlock in the original Petri net, yet the configuration is maximal in the prefix.

Property 1. Let $\mathcal{N}$ be a Petri net and $\mathcal{U}$ a finite complete prefix of its unfolding. For every attractor $\mathcal{A}$ of $\mathcal{N}$, there exists a maximal configuration of $\mathcal{U}$ whose associated marking belongs to $\mathcal{A}$.

Proof. Choose a marking M in $\mathcal{A}$. Because $\mathcal{U}$ is complete, it has a configuration $\mathcal{C}$ whose associated marking is M. Now because $\mathcal{U}$ is finite, it has a maximal configuration $\mathcal{C}'$ which extends $\mathcal{C}$. The marking M' associated to $\mathcal{C}'$ is reachable from M, therefore it is also in $\mathcal{A}$.

The prefix shown on Figure 2 has four maximal configurations. One is obtained after firing only t_1; its associated marking is the deadlock $\{p_3, p_2\}$. This marking is associated to another maximal configuration: the one obtained by firing t_3 and then $t_2 t_1$ concurrently with t_4. The third maximal configurations is obtained by firing t_3, then t_2 and t_4 concurrently, and then t_3 again; it reaches the marking $\{p_4, p_5\}$. Finally, one can fire t_3, then $t_2 t_1$ and t_6 concurrently, and then t_5; it reaches the marking $\{p_4, p_6\}$.

One can check that every attractor has a marking in this set: the deadlock $\{p_3, p_2\}$ is represented twice; the marking $\{p_4, p_6\}$ represents the attractor $\{\{p_3, p_6\}, \{p_1, p_6\}, \{p_4, p_6\}\}$. The marking $\{p_4, p_5\}$, also associated to a maximal configuration of $\mathcal{U}$, is not in an attractor.

4.3   Algorithm

Property 1 allows one to use finite complete prefixes to identify attractors: the set $\mathcal{M}_{\max}$ of markings corresponding to maximal configurations intersects all the attractors. But not all the markings in $\mathcal{M}_{\mathrm{max}}$ belong to an attractor. Also an attractor may contain several markings in $\mathcal{M}_{\mathrm{max}}$.

In order to characterize the attractors of a safe Petri net $\mathcal{N}$, we filter the set $\mathcal{M}_{\mathrm{max}}$ and keep only one marking per attractor. The idea is to remove iteratively the markings from which another marking of the set is reachable. The reachability checking is done again using unfoldings.

The algorithm is the following:

1. Compute a finite complete prefix $\mathcal{U}$ of the unfolding of $\mathcal{N}$.
2. Initialize $\mathcal{M}$ to the set $\mathcal{M}_{\mathrm{max}}$ of markings corresponding to maximal configurations of $\mathcal{U}$.
3. Initialize the set of attractors to $\emptyset$.
4. Loop for M in $\mathcal{M}$
   - Compute a finite complete prefix $\mathcal{U}_M$ of the net $\mathcal{N}$ initialized at M.
   - If a marking $M' \in \mathcal{M}$ other than M is reachable from M (the reachability check is done using $\mathcal{U}_M$),
     Then remove M from $\mathcal{M}$,
     Else add $\mathcal{U}_M$ to the set of attractors.
5. Output the set of attractors.

Termination of the algorithm is straightforward. We prove that at every step of the algorithm, the set $\mathcal{M}$ intersects all the attractors. This property is preserved when we remove a marking M from $\mathcal{M}$ because, if M is in an attractor $\mathcal{A}$, then the marking $M' \in \mathcal{M}$ reachable from M is also in $\mathcal{A}$. Notice also that, if M is not in an attractor, then at least one attractor $\mathcal{A}$ is reachable from M; and because $\mathcal{M} \cap \mathcal{A} \neq \emptyset$, $\mathcal{M}$ contains a marking $M' \in \mathcal{A}$ which is reachable from M. This ensures that $\mathcal{U}_M$ is added to the set of attractors iff M is in an attractor $\mathcal{A}$ and $\mathcal{M} \cap \mathcal{A} = \{M\}$.

4.4   Illustration on the Running Example

For our running example the set $\mathcal{M}$ is initialized to $\{\{p_3, p_2\}, \{p_4, p_5\}, \{p_4, p_6\}\}$. The algorithm computes the prefix $\mathcal{U}_M$ for every $M \in \mathcal{M}$, but outputs only $\mathcal{U}_{\{p_3,p_2\}}$ and $\mathcal{U}_{\{p_4,p_6\}}$, pictured in Figure 3. $\mathcal{U}_{\{p_4,p_5\}}$ is dropped because $\{p_4, p_6\}$ is reachable from $\{p_4, p_5\}$.

5   Implementation and Experimental Results

In order to test the applicability of our approach, we implemented a prototype of the algorithm described above using Mole[26] for computing the complete prefixes, and Minisat[6] for extracting the maximal configurations[1].

We applied our algorithm for the identification of attractors of three qualitative models of biological networks taken from the literature. In all cases, we

---

[1] Executables, scripts, and Petri net models are available for Linux 64bits at http://loicpauleve.name/cmsb2014.tbz2

applied a transformation to safe Petri net (Appendix A) that consists in having one place for each qualitative level of each component of the network, and transitions corresponding to the asynchronous semantics, i.e., each transition will actually change the level of only one component. By construction, the places corresponding to the levels of each components are mutually exclusive. The initial marking may then correspond either to a single state of the qualitative model, or to several possible initial states by adding transitions that non-deterministically select the initial state for some components (Sect. A.2). In this latter encoding, the returned attractors are the attractors reachable from at least one of the possible initial state of the qualitative model.

Table 1. Results of the attractors characterization using Petri net unfoldings. For each model (which includes the initial state), we give the number of maximal configurations and the number of attractors reachable from the initial state.

Model	Nb. nodes	Nb. max. conf.	Nb. attractors
Lambda switch	4	15	2
Cell cycle	10	12	1
ERBB (1)	20	301	1
ERBB (2)	20	302	2
VPC C. elegans	88	1240	1

Tab. 1 sums up the results of computing the finite complete prefixes on the following regulatory networks: a multi-valued model of the lambda switch [27], a Boolean model of he mammalian cell cycle [10], a Boolean model of the ERBB receptor [25], and a multi-valued model of fate determination in the Vulval Precursor Cells (VPC) in C. elegans [28]. For the ERBB model, two different initial settings have been tested: (1) when EGF is active; (2) when either EGF is active or inactive. For the other models, the initial state is the level 0 for all the components. The execution times are in the order of a fraction of a second for the two first models; in the order of a few seconds for ERBB; and around 15 minutes for the VPC model. For the latter, we note that starting from a different initial state leads to a combinatoric explosion of the complete prefix, showing room for improvement to handle large model in general. It is difficult to compare with other existing tools as most of them handle only Boolean networks and do not support the search from a given initial state. GINsim [19] also provides attractors computations from a given initial state, but it relies on explicit state transition graph computation, which is always larger than a complete prefix (Section 3). For instance on the VPC example, GINsim has been stopped after one hour.

6 Discussion

We presented a new algorithm for computing all the attractors reachable from a given state in the general class of safe Petri nets, i.e., Petri nets having at most one token in each place. This class includes Boolean and multi-valued networks that are typically used to model the qualitative dynamics of biological networks. Our approach relies on Petri net unfoldings that natively take into account the concurrency between transitions to produce a compact representation of the reachable states. Then, we use the notion of *maximal configuration* to derive, from the computed unfolding, a set of states that includes at least one state of each reachable attractor. This set of maximal configuration is then filtered to output exactly one state per reachable attractors. The identification of attractors is complete in the sense that all the attractors reachable from the supplied markings are detected, including fixed points and cyclic and complex attractors.

By the use of Petri unfoldings, we aim at reducing the complexity of exploring the full reachable space by inherently pruning redundant transitions due to some interleaving of concurrent transitions. We applied a prototype implementation of the algorithm to four biological networks ranging from four to eighty interacting components, either multi-valued or Boolean.

The unfolding technique mentioned in this paper is generic to any safe Petri net. This indicates several directions for improving its computation (including the extraction of maximal configurations) in the particular case of biological interaction networks, such as the use of contextual Petri nets [2], merge processes [15], unravelings [4], and the use of the network topology and static analysis to prune non-necessary transitions and decompose the detection of attractors.

References

1. Aracena, J.: Maximum number of fixed points in regulatory boolean networks. Bull. Math. Biol. 70(5), 1398–1409 (2008)
2. Baldan, P., Bruni, A., Corradini, A., König, B., Rodríguez, C., Schwoon, S.: Efficient unfolding of contextual Petri nets. TCS 449, 2–22 (2012)
3. Berntenis, N., Ebeling, M.: Detection of attractors of large boolean networks via exhaustive enumeration of appropriate subspaces of the state space. BMC Bioinformatics 14(1), 361 (2013)
4. Casu, G., Pinna, G.M.: Flow unfolding of safe nets. In: Petri Nets (2014)
5. Chaouiya, C., Naldi, A., Remy, E., Thieffry, D.: Petri net representation of multi-valued logical regulatory graphs. Natural Computing 10(2), 727–750 (2011)
6. Eén, N., Sörensson, N.: An extensible SAT-solver. In: Giunchiglia, E., Tacchella, A. (eds.) SAT 2003. LNCS, vol. 2919, pp. 502–518. Springer, Heidelberg (2004)
7. Esparza, J., Heljanko, K.: Unfoldings – A Partial-Order Approach to Model Checking. Springer (2008)
8. Esparza, J., Römer, S., Vogler, W.: An improvement of McMillan's unfolding algorithm. FMSD 20, 285–310 (2002)
9. Esparza, J., Schröter, C.: Unfolding based algorithms for the reachability problem. Fund. Inf. 47(3-4), 231–245 (2001)
10. Faure, A., Naldi, A., Chaouiya, C., Thieffry, D.: Dynamical analysis of a generic Boolean model for the control of the mammalian cell cycle. Bioinformatics 22(14), 124–131 (2006)
11. Garg, A., Di Cara, A., Xenarios, I., Mendoza, L., De Micheli, G.: Synchronous versus asynchronous modeling of gene regulatory networks. Bioinformatics 24(17), 1917–1925 (2008)

12. Hinkelmann, F., Brandon, M., Guang, B., McNeill, R., Blekherman, G., Veliz-Cuba, A., Laubenbacher, R.: ADAM: Analysis of discrete models of biological systems using computer algebra. BMC Bioinformatics 12(1), 295 (2011)

13. Khomenko, V.: Punf,
    http://homepages.cs.ncl.ac.uk/victor.khomenko/tools/punf/

14. Khomenko, V., Koutny, M., Vogler, W.: Canonical prefixes of Petri net unfoldings. Acta Inf. 40(2), 95–118 (2003)

15. Khomenko, V., Mokhov, A.: An algorithm for direct construction of complete merged processes. In: Kristensen, L.M., Petrucci, L. (eds.) PETRI NETS 2011. LNCS, vol. 6709, pp. 89–108. Springer, Heidelberg (2011)

16. McMillan, K.L.: Using unfoldings to avoid the state explosion problem inthe verification of asynchronous circuits. In: Probst, D.K., von Bochmann, G. (eds.) CAV 1992. LNCS, vol. 663, pp. 164–177. Springer, Heidelberg (1993)

17. Melliti, T., Noual, M., Regnault, D., Sené, S., Sobieraj, J.: Full characterization of attractors for two intersected asynchronous boolean automata cycles. CoRR, abs/1310.5747 (2013)

18. Murata, T.: Petri nets: Properties, analysis and applications. Proc. of the IEEE 77(4), 541–580 (1989)

19. Naldi, A., Berenguier, D.: Logical modelling of regulatory networks with GINsim. Biosystems 97(2), 134–139 (2009)

20. Naldi, A., Thieffry, D., Chaouiya, C.: Decision diagrams for the representation and analysis of logical models of genetic networks. In: Calder, M., Gilmore, S. (eds.) CMSB 2007. LNCS (LNBI), vol. 4695, pp. 233–247. Springer, Heidelberg (2007)

21. Paulevé, L., Magnin, M., Roux, O.: Refining dynamics of gene regulatory networks in a stochastic π-calculus framework. In: Priami, C., Back, R.-J., Petre, I., de Vink, E. (eds.) Transactions on Computational Systems Biology XIII. LNCS, vol. 6575, pp. 171–191. Springer, Heidelberg (2011)

22. Paulevé, L., Richard, A.: Topological Fixed Points in Boolean Networks. C. R. Acad. Sci. - Series I - Mathematics 348(15-16), 825–828 (2010)

23. Richard, A.: Positive circuits and maximal number of fixed points in discrete dynamical systems. Discrete Appl. Math. 157(15), 3281–3288 (2009)

24. Richard, A.: Negative circuits and sustained oscillations in asynchronous automata networks. Adv. in Appl. Math. 44(4), 378–392 (2010)

25. Sahin, O., Frohlich, H., Lobke, C., Korf, U., Burmester, S., Majety, M., Mattern, J., Schupp, I., Chaouiya, C., Thieffry, D., Poustka, A., Wiemann, S., Beissbarth, T., Arlt, D.: Modeling ERBB receptor-regulated G1/S transition to find novel targets for de novo trastuzumab resistance. BMC Systems Biology 3(1) (2009)

26. Schwoon, S.: Mole, http://www.lsv.ens-cachan.fr/~schwoon/tools/mole/

27. Thieffry, D., Thomas, R.: Dynamical behaviour of biological regulatory networks – II. Immunity control in bacteriophage lambda. Bull. Math. Biol. 57, 277–297 (1995)

28. Weinstein, N., Mendoza, L.: A network model for the specification of vulval precursor cells and cell fusion control in caenorhabditis elegans. Frontiers in Genetics 4(112) (2013)

29. Zañudo, J.G.T., Albert, R.: An effective network reduction approach to find the dynamical repertoire of discrete dynamic networks. Chaos 23, 025111 (2013)

30. Zheng, D., Yang, G., Li, X., Wang, Z., Liu, F., He, L.: An efficient algorithm for computing attractors of synchronous and asynchronous boolean networks. PLoS ONE 8(4), e60593 (2013)

A   Encoding Asynchronous Discrete Networks with Safe Petri Nets

A.1   Encoding with One Initial State

In literature, Boolean and multi-valued networks modelling dynamics of biological influence networks are typically represented by functions associating for each component the levels towards which it evolves with respect to each possible level of its regulators. In order to encode their asynchronous dynamics in Petri nets, one need to have a transition-centered representation, instead of a function-centered. Informally, this can be achieved by having one place per possible level of each component (we note i_u the place corresponding to the level u of component i), and listing the conditions for moving a token from i_u to i_{u+1} and i_{u-1}. Such conditions can typically be built from the expression of the discrete functions of the network. Our encoding always results in safe Petri nets, which makes it different from [5] which relies on more advanced Petri nets semantics (multiple tokens on places and weighted arcs).

A Discrete Network gathers a finite number of components $i \in \{1, \cdots, n\}$ having a discrete finite domain $\mathbb{F}^i$ that we note $\mathbb{F}^i = \{0, \cdots, l_i\}$, l_i being the maximum level for the component i. For each component $i \in \{1, \cdots, n\}$, a map $f^i : \mathbb{F} \to \mathbb{F}^i$ is defined, where $\mathbb{F} = \mathbb{F}^1 \times \cdots \times \mathbb{F}^n$, giving the next value of the component with respect to the global state of the network. Typically f^i depends on a subset of components (its regulators) that we denote $\mathsf{dep}(f^i)$. In the case of Asynchronous Discrete Networks (ADN), a transition relation $\to_{ADN} \subseteq \mathbb{F} \times \mathbb{F}$ is defined such that $x \to_{ADN} x'$ if and only if there exists a unique $i \in \{1, \cdots, n\}$ such that $x'[i] = f^i(x)$ and $\forall j \in \{1, \cdots, n\}, j \neq i, x'[j] = x[j]$, i.e. one and only one component has been updated. This is formalised in Def. 8.

Definition 8 (Asynchronous Discrete Network (ADN)). *An ADN is defined by a couple* $(\mathbb{F}, \langle f^1, \ldots, f^n \rangle)$ *where* $\mathbb{F} = \mathbb{F}^1 \times \cdots \times \mathbb{F}^n$, *and* $\forall i \in \{1, \cdots, n\}$, $f^i : \mathbb{F} \to \mathbb{F}^i$ *with* $\mathbb{F}^i = \{0, \cdots, l_i\}$. *Given two states* $x, x' \in \mathbb{F}$, *the transition relation* $\to_{ADN}$ *is given by*

$$x \to_{ADN} x' \iff \exists i \in \{1, \cdots, n\}, f^i(x) = x'[i] \land \forall j \in \{1, \cdots, n\}, j \neq i, x[j] = x'[j] ,$$

where $x[i]$ *is the i-th component of* x. *We note* $\mathsf{dep}(f^i) \subseteq \{1, \cdots, n\}$ *the set of components on which the value of* f^i *depends:* $\forall x, x' \in \mathbb{F}$ *such that* $\forall j \in \mathsf{dep}(f^i), x[j] = x'[j]$, *necessarily* $f^i(x) = f^i(x')$.

In the scope of an ADN $(\mathbb{F}, \langle f^1, \ldots, f^n \rangle)$, we use $\mathsf{cond}(x)$ to map a state to the set of literals for the presence of the components at the corresponding state, e.g., $\mathsf{cond}(\langle 1, 0, 1 \rangle) = \{1_1, 2_0, 3_1\}$: $\mathsf{cond}(x) \stackrel{\Delta}{=} \{i_u \mid i \in \{1, \cdots, n\}, x[i] = u\}$. Given a component i at a state u, we note $\mathbf{conds}^i_{u+}$ and $\mathbf{conds}^i_{u-}$ the set of conditions where i can respectively increase or decrease. This set of conditions can be read as a disjunctive normal form expressing the possibility of the transition: $\mathbf{conds}^i_{u+} \stackrel{\Delta}{=} \mathsf{simplify}(\{\mathsf{cond}(x)|_{\mathsf{dep}(i)} \mid x \in \mathbb{F}, x[i] = u, f^i(x) > u\})$; $\mathbf{conds}^i_{u-} \stackrel{\Delta}{=} \mathsf{simplify}(\{\mathsf{cond}(x)|_{\mathsf{dep}(i)} \mid x \in \mathbb{F}, x[i] = u, f^i(x) < u\})$; where simplify is an operator to reduce the number of conditions, and $\mathsf{cond}(x)|_{\mathsf{dep}(i)}$ restricts the literals to those corresponding to components influencing i.

Finally, given an ADN $(\mathbb{F}, \langle f^1, \ldots, f^n \rangle)$ and an initial state $x_0 \in \mathbb{F}$, the corresponding safe Petri net is defined by (P, T, F, M_0) where $P = \{i_u \mid i \in \{1, \cdots, n\}, u \in \{0, \cdots, l_i\}\}$, $M_0 = \mathsf{cond}(x_0)$, and T and F are the smallest sets (w.r.t. inclusion) such that $\forall i \in \{1, \cdots, n\}, \forall u \in \{0, \cdots, l_i\}, \forall \Phi \in \mathbf{conds}^i_{u+} \cup \mathbf{conds}^i_{u-}, \exists t \in T : {}^\bullet t = \Phi \cup \{a_i\} \wedge t^\bullet = (\Phi \setminus \{a_i\}) \cup \{a_j\}$. By construction, the Petri net is safe and for each $i \in \{1, \cdots, n\}$, the places $i_0, \cdots, i_{l_i}$ are mutually exclusive.

A.2 Encoding with Multiple Initial States

When studying the dynamics of a qualitative network, one may want to consider several initial states, chosen nondeterministically. Using the construction depicted in the previous section, one can encode this indeterministic choice by adding a place, initially marked, per (independent) indeterministic choice, and a transition per corresponding (local) state. Fig. 4 illustrate this construction with either an indeterministic initial state for one component, or for several components.

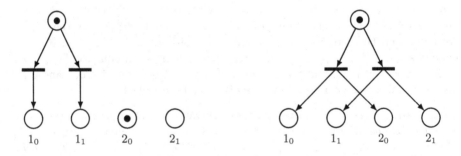

Fig. 4. (left) Encoding of the indeterministic choice between the initial state of 1; (right) indeterministic choice between the initial state of the couple 1 and 2: either $\langle 0, 0 \rangle$ or $\langle 1, 1 \rangle$

Dynamic Modeling and Simulation of Leukocyte Integrin Activation through an Electronic Design Automation Framework

Nicola Bombieri[1], Rosario Distefano[1], Giovanni Scardoni[2],
Franco Fummi[1], Carlo Laudanna[2], and Rosalba Giugno[3]

[1] Dept. Computer Science, University of Verona, Italy
[2] Dept. Patology and Diagnostics, University of Verona, Italy
[3] Dept. Clinical and Molecular Biomedicine, University of Catania, Italy
{nicola.bombieri,rosario.distefano,giovanni.scardoni,franco.fummi,
carlo.laudanna}@univr.it, giugno@dmi.unict.it

Abstract. Model development and analysis of biological systems is recognized as a key requirement for integrating in-vitro and in-vivo experimental data. In-silico simulations of a biochemical model allows one to test different experimental conditions, helping in the discovery of the dynamics that regulate the system. Several characteristics and issues of biological system modeling are common to the electronics system modeling, such as concurrency, reactivity, abstraction levels, as well as state space explosion during verification. This paper proposes a modeling and simulation framework for discrete event-based execution of biochemical systems based on SystemC. SystemC is the reference language in the electronic design automation (EDA) field for modeling and verifying complex systems at different abstraction levels. SystemC-based verification is the de-facto an alternative to model checking when such a formal verification technique cannot deal with the state space complexity of the model. The paper presents how the framework has been applied to model the intracellular signalling network controlling integrin activation mediating leukocyte recruitment from the blood into the tissues, by handling the solution space complexity through different levels of simulation accuracy.

Keywords: Biochemical networks, Dynamic modeling and simulation, SystemC.

1 Introduction

Cells are the fundamental units of the living organisms. They interact with the environment and with other cells by processing and exchanging environmental informations. Each different input coming from the environment produces a set of chemical reactions, which are the *answer* of the cell to the input. Those reactions depend on some parameters, such as the concentration of the reactants and the chemical properties regulating the reaction speed, and generate linear reaction pathways in turn organized in concurrent non-linear complex networks [16].

P. Mendes et al. (Eds.): CMSB 2014, LNBI 8859, pp. 143–154, 2014.
© Springer International Publishing Switzerland 2014

Dynamic network modeling in systems biology aims at describing how such interactions among defined elements determine the time course of the state of the elements, and of the whole system, under different conditions. A validated dynamic model that correctly captures experimentally observed normal behavior allows researchers to track the changes in the system due to perturbations, to discover possible covariation between coupled variables, and to identify conditions in which the dynamics of variables are qualitatively similar [19].

Mathematical models, such as those based on differential equations [7], have definitely gained consensus in the network modeling community as they have the highest potential to accurately describe the system. Nevertheless, since they have the highest requirement for input information, they are difficult to obtain and analyse if the number of independent variables grows and if the relationships depend on quantitative events, such as concentration reaching a threshold value.

Computational models, such as Boolean networks [25], Petri nets [10], interactive state machines [24], and Process Calculi [23], offer an effective alternative if precise quantitative relationships are unknown, if they involve many different variables, or if they change over time [14]. A common way to explain a certain class of complex dynamical systems is to view them as highly *concurrent reactive systems*. Hand-in-hand with the central notion of reactivity go (i) the discrete event-based execution and simulation of dynamical systems, which requires a fundamental understanding of parallelism, interaction, and causality; (ii) the design of complex systems from building blocks, requiring means for composition and encapsulation; and (iii) the description of systems at different levels of granularity, requiring methods for abstraction and refinement [13].

All these issues related to concurrent reactive systems have been largely addressed in the past decades in the electronic design automation (EDA) field and a large body of methodologies and tools are at the state of the art. In this context, SystemC [4] has become the de-facto reference standard language for system-level modelling and simulation of Hardware/Software/Network electronic systems at different abstraction levels [8].

In this paper, we propose a framework for modeling and simulation of biochemical networks based on SystemC. The framework relies on a state machine-based computational model to model the behavior of each network element. The element models are implemented and connected to realize a system-level network in SystemC. Finally, the network is connected to a stimuli generator and monitor of results to run a discrete and deterministic network simulation. To handle the complexity of exploring the solution space, the proposed framework allows us to discretize the range of the variable values with different levels of accuracy. In addition, the framework allows us to reuse existing EDA techniques and tools to parallelize the SystemC simulation, both on GPUs [22] and on clusters [12].

The paper presents how the framework has been applied to model the signaling network controlling LFA-1 beta2 integrin activation mediating leukocyte recruitment from the blood into the tissues, a central event during the immune response. Such a case study has been chosen for the large number of independent variables, for the lack of quantitative information such as molecular concentrations, activation and inhibition delays and lifetimes, and for the relationships strongly depending on qualitative events. The dynamic simulation of the model has been conducted with the aim of exploring the occurrence of emergent properties in signaling events controlling leukocyte recruitment, such as oscillating behaviors and, more in general, to help in better understanding the overall dynamics of leukocyte recruitment.

The paper is organized as follows. Section 2 summarizes the most important concepts and constructs of SystemC for modeling protein networks. Section 3 presents the leukocyte integrin activation case study. Section 4 presents the proposed framework, while Section 5 reports the obtained experimental results. Section 6 is devoted to concluding remarks.

2 Background on SystemC

SystemC [4] is a set of C++ classes and macros that provide an *event-driven* simulation interface in C++. These facilities enable a designer to simulate *concurrent processes*, each described using plain C++ syntax. SystemC processes can communicate in a simulated real-time environment, using signals of all the datatypes provided by C++, some additional ones provided by the SystemC library, as well as user defined.

SystemC has been applied to system-level modeling, architectural exploration, performance modeling, software development, functional verification, and high-level synthesis of digital circuits since 2000. Nowadays, SystemC is the de-facto reference standard in the EDA community. SystemC is defined and promoted by the Open SystemC Initiative (OSCI) - Accellera Systems Initiative, and has been approved by the IEEE Standards Association as IEEE 1666-2005. The SystemC Language Reference Manual (LRM) [5] provides the definitive statement of the semantics of SystemC. OSCI also provides an open-source proof-of-concept simulator, which can be downloaded from the SystemC website [4]. Several optimized simulators are also available in the commerce [1,3,2].

SystemC offers a greater range of expression, similar to object-oriented design partitioning and template classes. Although strictly a C++ class library, SystemC is sometimes viewed as being a language in its own right. Source code can be compiled with the SystemC library (which includes a simulation kernel) to give an executable. SystemC allows designers to model systems at different abstraction levels (i.e., with different levels of details) by providing modeling features such as structural hierarchy and connectivity, communication abstraction, dynamic processes, timed event notifications, transaction-level modeling [9].

The most important language features, which have been used for modeling and simulating the signaling network presented in this paper are the following:

- *Modules.* They are the basic building blocks of a SystemC design hierarchy. A SystemC model usually consists of several modules that communicate via ports. As explained in the following sections, each network element (i.e., protein and cofactor) has been modelled as a module, and all the elements have been hierarchically organized into a module representing the whole network.

- *Ports.* They allow communication from inside a module to the outside (usually to other modules) via signals.
- *Signals.* They are the communication elements of SystemC models. They have been used to model the activation/inhibition activity between elements.
- *Processes.* They are the main computation elements and they are concurrent. Each protein behaviour has been modelled through a process, which reacts to any activation or inhibition by an upstream protein and, in turn, activates or inhibits a downstream protein.
- *Events.* They allow for synchronization between processes. Events are the key objects in SystemC models to provide event-driven simulation.

3   The Case Study

In order to better explain how the proposed framework can be applied for modelling and simulation of signaling networks, we first present the case study, which will be used as a model system in the subsequent sections.

As a model system, we analysed the signaling mechanism controlling beta2 integrin LFA-1 affinity regulation by chemokines, a crucial event mandatory to the fulfilment of the leukocyte recruitment process from the blood into the tissues. This process is critical to immune system function and is modeled as a concurrent ensemble of leukocyte behaviors under flow, including tethering, rolling, firm adhesion, crawling, and transmigration [18]. A central step is the integrin-mediated arrest, comprising a series of adhesive events including increase of integrin affinity, valency and binding stabilization altogether controlling cell avidity. In this context, modulation of integrin affinity is widely recognized as the prominent event in rapid leukocyte arrest induced by chemokines [6,11,15,17]. Regulation of integrin activation depends on a plethora of signaling proteins [6]. At least 67 signaling molecules modulate integrin activity by chemokines [21,20]. In this context, we have previously described an integrated group of signaling proteins including RhoA, Rac1 and CDC42 small GTPases, along with the two effectors PLD1 and PIP5K1C, modulating conformer-selective LFA-1 affinity triggering and homing to secondary lymphoid organs by chemokines of human primary lymphocytes [6]. To date, signaling by rho- and rap-small GTPases are the best-studied mechanisms of integrin activation by chemokines.

Furthermore, and more recently, we have demonstrated that, in human primary T lymphocytes, chemokines control integrin affinity triggering and in vivo homing by means of tyrosine kinases of the JAK family acting as upstream transducer linking chemokine receptors to the activation of the rho and rap module of integrin [20]. Overall, an integrated macro module comprising JAKs, rho and rap small GTPases and a variety of upstream regulators and downstream effectors finely control integrin triggering and mediated lymphocyte recruitment by chemokines. Beside arrest under flow, integrin activation is also critical to support leukocyte crawling and transmigration (diapedesis) along with directional

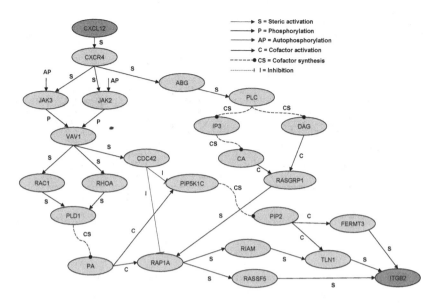

Fig. 1. The protein-protein interaction network of the leukocite integrin activation

movement toward a gradient of chemotactic factors that is chemotaxis. Figure 1 depicts the protein network and each different interaction between proteins or cofactors of the case study.

Notably, cell motility needs an on-off kinetic of integrin activation, allowing cycling between adhesion and de-adhesion event thus ensuring cell movement. Thus, control of the duration of cell adhesion is critical to control cell migration. This on-off, oscillatory, kinetics of integrin triggering likely depends on on-off kinetics of the signaling transduction machinery triggered by chemokines and controlling integrin-mediated cell adhesion. This suggests an equal relevance for both activators as well as inhibitors on integrin triggering. Although negative regulators of cell adhesion have been described, a comprehensive dynamic model of signaling events controlling on-off cycling of integrin activation is still lacking. Such a modeling is an important approach to explore the occurrence of emergent properties in signaling events controlling leukocyte recruitment, such as oscillating behaviors characterized by frequency and amplitude of agonist triggering. In turn, identification of these properties could help to better understand the overall dynamics of leukocyte recruitment.

4   The SystemC Framework for Modelling and Simulation of the Protein Network

The framework relies on three main steps. First, the behavior of each network element (i.e., protein and cofactor) is modeled through the Finite State Machine (FSM) formal model. The element models are then implemented in SystemC

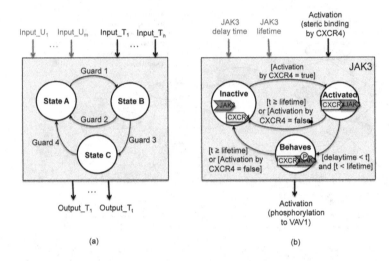

Fig. 2. The protein behavior representation through Finite State Machines. The protein template (a), and the JAK3 example.

modules and connected through SystemC *signals* to realize a system-level network. Finally, the network is connected to a stimuli generator and monitor of results to run a reactive, event-driven network simulation.

4.1 Modelling Proteins through Finite State Machines

The finite state machine model allows us to formally model each protein behavior and, similarly, each cofactor behaviour, in terms of states (e.g., inactive, activated/inhibited, activating/inhibiting, etc.), transitions between states, and guard conditions (i.e., boolean conditions).

Figure 2(a) depicts the proposed FSM template, while Figure 2(b) shows a modelling example of the JAK3 protein of the case study in Figure 1. Each protein changes state (i.e., a transition occurs) when the guard condition is evaluated to be true. The condition may be set on a particular reaction *event* (e.g., activation via phosphorylation, steric, auto-phosphorylation, cofactor or inhibition via phosphatase) generated by any upstream protein or on any environment status. As an example, JAK3 moves from the inactive state to the activated state (which represents the steric binding with CXCR4) as soon as CXCR4 activates JAK3. Once activated, JAK3 seeks for the phosphorylation of its own protein target (VAV1), which occurs after a *delay time* (i.e., the time spent to encounter a molecule of VAV1, to pick up an atom of phosphorus from an ATP molecule, and to add it to VAV1). *t* represents the time elapsed, which is constantly updated during simulation, while *lifetime* represents the maximum lifetime from the activation instant in which the protein carries out its biological function. JAK3 continues to phosphorylate new VAV1 molecules (*Behaves* state) as long as it is bounded with CXCR4 and the lifetime has not expired.

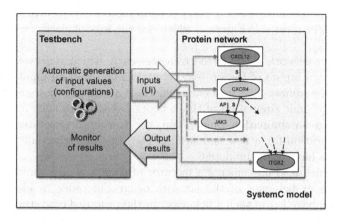

Fig. 3. The SystemC framework

The template distinguishes two sets of input data that can affect the model behavior and one set of generated output:

- *Topological inputs (Input_Ti):* They are inputs whose values are calculated at simulation time and depend on the topological interaction of the modelled protein with upstream proteins. Some examples are the activation via phosphorylation, steric, cofactor, or inhibition.
- *Unknown inputs (Input_Ui):* They are inputs whose values depends on the environment characteristics and status, which are unknown at modeling time. Some examples are the *delay time* (i.e., time spent by the protein to encounter a protein target), the molecular concentrations of the downstream proteins (which affect the delay time), the protein lifetime, etc. For each unknown input, the framework generates different values with the aim of observing, via simulation, how such values affect the system dynamics.
- *Topological outputs (Output_Ti):* They are outputs whose values are calculated at simulation time and depend on the role of the protein towards downstream proteins (e.g., the ouput of the JAK3 module is set to true when JAK3 encounters and activates VAV1 via phosphorylation) .

4.2 Implementation of the Protein Models through SystemC

Each protein is implemented through a SystemC *module*, with both the topological and unknown inputs and outputs as SystemC *ports* (see Section 2). The protein behavior represented by FSM in Figure 2 is implemented through a SystemC *process*, which is sensitive to any *event* on the input signals. An activation/inhibition from an upstream protein is represented by an input (boolean) signal set to *true*. Being event-driven, the process *wakes up* and updates both the internal state and the output signals whenever a new event on inputs occurs.

The network model consists of every protein modules connected via SystemC signals (see right-most side Figure 3, which, for the sake of clarity, reports a part of the network).

The protein network is connected to a *testbench*, which generates the values for the unknown inputs of the protein network. The set of all input values represents a *configuration*. The testbench generates a configuration and runs (i.e., executes) a dynamic simulation of the network behavior for such a set of input values for a given simulation time. Then, the testbench generates a new different configuration for a new simulation. The run ends when all the possible configurations have been simulated.

The testbench also implements a monitor of results, which controls whether any condition or behavior of the network occurs, in order to identify which configurations have led to such a behavior. In the proposed case study, the monitored condition consists of the on-off, oscillatory, kinetics of integrin triggering represented, in the model, by the oscillatory state of ITGB2 between inactive and activate affinity state. Particularly, the monitoring activity of the testbench aims at identifying which configurations, in terms of protein lifetime, activation delays, and protein concentrations lead to a given number of oscillations, with a given oscillation period.

4.3   Simulation of the System-Level Network

The main problem in exploring the dynamics of protein networks is the complexity of the solution space. The solution space, that is, the number of configurations to simulate, grows exponentially over the number of unknown inputs. In addition, several inputs are continuous magnitudes (e.g., delay and lifetime), which would lead to an intractable problem if not properly discretized.

To handle such a complexity, the proposed framework allows us to discretize the range of the input values with different levels of accuracy. As an example, the lifetime of CDC42 in Figure 1 is an unknown input, whose value has to be generated by the testbench. Different values have been simulated, starting form a minimum to a maximum value, by steps of a given time period. The finer the step, the more accurate the space solution exploration, and, on the other hand, the higher the configuration number and the consequent overall simulation time. EDA techniques and tools at the state of the art can be applied to parallelize the SystemC simulation, both on GPUs [22] and on clusters [12] in order to improve the accuracy over the simulation time ratio.

In general, given a number of network elements, n, the total number of input configurations to be generated by the testbench is the following:

$$\prod_{i=1}^{n} \left(\frac{MConcentration_i}{MCStep_i} \right) (Targets_i) \left(\frac{MaxDelayT_i - MinDelayT_i}{DelayStep_i} \right) \left(\frac{MaxLifeT_i - MinLifeT_i}{LifetimeStep_i} \right)$$

where *MConcentration* represents the molecular concentration of the protein (or cofactor), *Targets* represents the number of the downstream targets, *Max* and *MinDelayT* represent the observed range of delay time, *Max* and *MinLifeT* represent the range of the lifetime, while *MCStep*, *DelayStep*, and *LifetimeStep* represent the chosen periods of discretization.

The overall simulation time is linear over the number of configurations. It is possible to associate, before simulation, the required time for simulating the network dynamics with a chosen space exploration accuracy. In addition, parameters *MaxDelayT*, *MinDelayT*, *MaxLifeT*, *MinLifeT*, *MCStep*, *DelayStep*, and *LifetimeStep* can be tuned for each single element of the network. This allows us to explore, with different levels of detail, the behavior and the influence of each protein in the overall network dynamics.

The modular structure of the framework allows us to adopt different simulation models (e.g., stochastic simulations), by modifying the testbench module. The development of a testbench for stochastic simulations with the aim of relaxing the constraints on the input values is part of our current and future work.

5   Experimental Results

The case study presented in Section 3 has been implemented in SystemC with the aim of exploring pro-adhesive signaling events and to better understand the overall dynamics of leukocyte recruitment.

The main goal of the model simulation was identifying the system properties that lead to oscillating behaviors, which are characterized by frequency and amplitude of integrin triggering. In particular, the testbench has been implemented to monitor which configurations of input values lead to oscillations of ITGB2 with a period of 30-40 ms (15-20 ms in active state, 15-20 ms inactive state), which represents the average stopping time of a cell when it interacts with the blood vessel epithelium. Notably, although accurate experimental measurement of on-off dynamics of integrin triggering is, at the present, unavailable, the extremely rapid kinetics of leukocyte arrest under-flow conditions, occurring in the experimentally-determined range of few milliseconds clearly suggest that it is reasonable to consider this rapid time-frame as a correct reference time to simulate on-off dynamics of integrin triggering. Furthermore, since directional leukocyte motility (chemotaixs) appears to maintain constant speed, at least in the context of a chemotactic gradient, it is reasonable to expect the emergence of regular oscillatory dynamics of signaling mechanisms controlling integrin triggering.

In order to reduce the explosion of the exploration space, we assumed the following characteristics of the system, which are summarized in Table 1. Each protein and cofactor (listed in Table 1 with (P) and (C), respectively) have been simulated with three different molecular concentrations (1, half, and maximum molecular number). The delay time of each element has been fixed as a function of the molecular concentration of the target element, with minimum value equal to 2 ms.

The lifetime of each single protein (cofactor) has been explored by discretizing the time intervals, which have been fixed for each element as shown in the table. To better explore the behavior of the most interesting proteins of the network

Table 1. The protein network characteristics

	Unknown inputs				Topological signals	
	MConcentration (# molecules)	downstream targets (#)	delay time (ms)	lifetime (ms)	inputs	outputs
CXCL12 (P)	[1,400]	[1,1]	-	[250,250]	-	sig_CXCR4
CXCR4 (P)	[1,325]	[1,3]	[2,3]	[250,250]	sig_CXCL12	sig_JAK3 sig_JAK2 sig_ABG
JAK3 (P)	[1,300]	[1,1]	[2,5]	[250,250]	sig_CXCR4	pho_VAV1
JAK2 (P)	[1,175]	[1,1]	[2,5]	[42,42]	sig_CXCR4	pho_VAV1
ABG (P)	[1,200]	[1,1]	[2,5]	[31,37]	sig_CXCR4	sig_PLC
VAV1 (P)	[1,168]	[1,3]	[2,2]	[45,51]	pho_JAK3 pho_JAK2	sig_RAC1 sig_RHOA sig_CDC42
RAC1 (P)	[1,235]	[1,1]	[2,6]	[34,40]	sig_VAV1	sig_PLD1
RHOA (P)	[1,146]	[1,1]	[2,6]	[29,35]	sig_VAV1	sig_PLD1
CDC42 (P)	[1,256]	[1,2]	[2,2]	[35,41]	sig_VAV1	sig_PIP5K1C sig_RAP1A
PLC (P)	[1,210]	[1,2]	[2,4]	[33,33]	sig_ABG	syn_IP3 syn_DAG
IP3 (C)	[1,115]	[1,1]	[2,5]	[51,57]	syn_PLC	syn_CA
CA (C)	[1,140]	[1,1]	[2,5]	[44,50]	syn_IP3	sig_RASGRP1
DAG (C)	[1,123]	[1,1]	[2,5]	[56,62]	syn_PLC	sig_RASGRP1
RASGRP1 (P)	[1,127]	[1,1]	[2,4]	[32,38]	sig_CA sig_DAG	sig_RAP1A
PLD1 (P)	[1,67]	[1,1]	[2,4]	[28,28]	sig_RAC1 sig_RHOA	sig_PA
PIP5K1C (P)	[1,234]	[1,1]	[2,4]	[27,33]	sig_CDC42 sig_PA	sys_PIP2
PA (C)	[1,322]	[1,2]	[2,2]	[63,69]	sys_PLD1	sig_RAP1A sig_PIP5K1C
RAP1A (P)	[1,364]	[1,2]	[2,2]	[34,40]	sig_PA sig_RASGRP1 sig_CDC42	sig_RASSF5 sig_RIAM
PIP2 (C)	[1,243]	[1,2]	[2,3]	[55,61]	sys_PIP5K1C	sig_FERMT3 sig_TLN1
RIAM (P)	[1,435]	[1,1]	[2,4]	[39,39]	sig_RAP1A	sig_TLN1
RASSF5 (P)	[1,134]	[1,1]	[2,5]	[32,38]	sig_RAP1A	sig_ITGB2
FERMT3 (P)	[1,123]	[1,1]	[2,5]	[31,31]	sig_PIP2	sig_ITGB2
TLN1 (P)	[1,364]	[1,1]	[2,5]	[36,36]	sig_PIP2	sig_ITGB2
ITGB2 (P)	[1,125]	-	-	[43,49]	sig_FERMT3 sig_TLN1 sig_RASSF5	-

(e.g., CDC42, RAP1A, and PIP5K1C that can lead to oscillations upon inhibition), the lifetime ranges explored in simulation for such elements have been extended. JAK3 and JAK2 have a fixed lifetime (40 ms and 42 ms, respectively) since it has been accepted that, at present, there is not a known phosphatase process that can influence their behavior.

Each protein or cofactor can activate (inhibit) one target at a time. Activation (inhibition) of different targets are explored through different configurations. As an example, VAV1 activates either RAC1 or RHOA or CDC42 (see Figure 1) in a configuration run. Activation of all the targets is guaranteed and covered in different configuration runs.

For each configuration, the network dynamics has been simulated and monitored for a total time of 250 ms. For each configuration run, CXCL12 is always active. CXCL12 and ITGB2 have not delay time. In total, we run around seven billion configurations on a cluster of 16 dual-core CPUs for a total of 278 hours run time. As a result, we filtered the configurations that lead to periodic oscillations (0.06% of the total) from the configurations that lead to aperiodic oscillations or no oscillation (41.7% and 58.33%, respectively). Among the periodic oscillations, the majority of configurations (57.75%) lead to three oscillations in the overall simulated time (250 ms), 21% oscillations lead to two oscillations, while 11.14% and 9.45% lead to five and four oscillations, respectively. Such configurations represent different settings of the *unknown inputs* (see Section 4.1) that lead the model behavior close enough to what experimentally observed. These results encourage us for further model refinements and deeper investigations of the case study.

6   Concluding Remarks

The paper presented a SystemC-based framework for modeling and simulation of the signaling network controlling LFA-1 beta2 integrin activation mediating leukocyte recruitment from the blood into the tissues. The framework relies on the FSM model to formally model the behavior of each network element and on the SystemC EDA language, which allows us to implement the network elements as concurrent and reactive processes. The framework also consists of a testbench, which generates *configurations* of values for each unknown parameters (e.g., molecular concentrations, activation delays, etc.). The framework simulates the system for each configuration to identify the system properties that lead to any experimentally observed behavior, such as the periodic oscillations of ITGB2 in the leukocyte integrin activation case study. The proposed approach allows us to handle the solution space complexity through different levels of simulation accuracy and to apply EDA techniques and tools at the state of the art to parallelize the SystemC simulation, both on GPUs and on clusters, to improve the accuracy over the simulation time ratio.

References

1. Cadence Palladium - System Design and Verification,
   http://www.cadence.com/products/sd/Pages/default.aspx
2. Mentor Graphics SystemVisio, http://www.mentor.com/products/sm/
3. Synopsys System Studio,
   http://www.synopsys.com/Systems/Pages/default.aspx
4. SystemC - Accellera Systems Initiative, http://www.systemc.org
5. IEEE 1666 Standard: SystemC Language Reference Manual (2011),
   http://ieeexplore.ieee.org
6. Bolomini-Vittori, M., Montresor, A., Giagulli, C., Staunton, D., Rossi, B., Martinello, M., Constantin, G., Laudanna, C.: Regulation of conformer-specific activation of the integrin lfa-1 by a chemokine-triggered rho signaling module. Nat. Immunol. 10, 185–194 (2009)
7. Butcher, J.C.: Numerical Methods for Ordinary Differential Equations. Wiley, Chichester (2003)

8. Cai, L., Gajski, D.: Transaction level modeling: An overview. In: ACM/IEEE CODES+ISSS, pp. 19–24 (2003)
9. Cai, L., Gajski, D.: Transaction level modeling: An overview. In: Proceedings of the 1st IEEE/ACM/IFIP International Conference on Hardware/Software Codesign and System Synthesis, pp. 19–24. CODES+ISSS (2003)
10. Chaouiya, C.: Petri net modelling of biological networks. Briefings in Bioinformatics 8(4), 210–219 (2007)
11. Constantin, G., Majeed, M., Giagulli, C., Piccio, L., Kim, J., Butcher, E., Laudanna, C.: Chemokines trigger immediate beta2 integrin affinity and mobility changes: differential regulation and roles in lymphocyte arrest under flow. Immunity 13, 759–769 (2000)
12. Ezudheen, P., Chandran, P., Chandra, J., Simon, B.P., Ravi, D.: Parallelizing SystemC kernel for fast hardware simulation on SMP machines. In: Proc. of ACM/IEEE PADS, pp. 80–87 (2009)
13. Fisher, J., Harel, D., Henzinger, T.A.: Biology as reactivity. Commun. ACM 54(10), 72–82 (2011)
14. Fisher, J., Henzinger, T.A.: Executable cell biology. Nature Biotechnology 25, 1239–1249 (2007)
15. Giagulli, C., Ottoboni, L., Caveggion, E., Rossi, B., Lowell, C., Constantin, G., Laudanna, C., Berton, G.: The src family kinases hck and fgr are dispensable for inside-out, chemoattractant-induced signaling regulating beta 2 integrin affinity and valency in neutrophils, but are required for beta 2 integrin-mediated outside-in signaling involved in sustained adhesion. J. Immunol. 177, 604–611 (2006)
16. Gilbert, D., Fuss, H., Gu, X., Orton, R., Robinson, S., Vyshemirsky, V., Kurth, M.J., Downes, C.S., Dubitzky, W.: Computational methodologies for modelling, analysis and simulation of signalling networks. Briefings in Bioinformatics 7(4), 339–353 (2006)
17. Kim, M., Carman, C., Yang, W., Salas, A., Springer, T.: The primacy of affinity over clustering in regulation of adhesiveness of the integrin $\alpha l\beta 2$. J. Cell. Biol. 167, 1241–1253 (2004)
18. Ley, K., Laudanna, C., Cybulsky, M., Nourshargh, S.: Getting to the site of inflammation: the leukocyte adhesion cascade updated. Nat. Rev. Immunol. 7, 678–689 (2007)
19. Melham, T.: Modelling, abstraction, and computation in systems biology: A view from computer science. Progress in Biophysics and Molecular Biology 111, 129–136 (2013)
20. Montresor, A., Bolomini-Vittori, M., Toffali, L., Rossi, B., Constantin, G., Laudanna, C.: Jak tyrosine kinases promote hierarchical activation of rho and rap modules of integrin activation. J. Cell. Biol. 203(6), 1003–1019 (2013)
21. Montresor, A., Toffali, L., Constantin, G., Laudanna, C.: Chemokines and the signaling modules regulating integrin affinity. Front Immunol. 3, 127 (2012)
22. Nanjundappa, M., Patel, H.D., Jose, B.A., Shukla, S.K.: Scgpsim: a fast systemc simulator on gpus. In: Proceedings of the 2010 Asia and South Pacific Design Automation Conference, ASPDAC 2010, pp. 149–154 (2010)
23. Priami, C.: Stochastic pi-calculus. The Computer Journal 38(7), 578–589 (1995)
24. Sadot, A., Fisher, J., Barak, D., Admanit, Y., Stern, M.J., Hubbard, E.J., Harel, D.: Toward verified biological models. IEEE/ACM Transactions on Computational Biology and Bioinformatics 5(2), 223–234 (2008)
25. Srihari, S., Raman, V., Leong, H.W., Ragan, M.A.: Evolution and controllability of cancer networks: A boolean perspective. IEEE/ACM Transactions on Computational Biology and Bioinformatics 11(1), 83–94 (2013)

Towards Real-Time Control of Gene Expression at the Single Cell Level: A Stochastic Control Approach

Lakshmeesh R.M. Maruthi[1], Ilya Tkachev[1], Alfonso Carta[2], Eugenio Cinquemani[3], Pascal Hersen[4], Gregory Batt[5], and Alessandro Abate[1,6]

[1] Delft Center for Systems and Control, TU Delft, The Netherlands
[2] INRIA Sophia-Antipolis - Méditerranée, France
[3] INRIA Grenoble - Rhône-Alpes, France
[4] Laboratoire Matière et Systèmes Complexes, UMR 7057, Paris, France
[5] INRIA Paris-Rocquencourt, France
[6] Department of Computer Science, University of Oxford, UK
gregory.batt@inria.fr, alessandro.abate@cs.ox.ac.uk

Abstract. Recent works have demonstrated the experimental feasibility of real-time gene expression control based on deterministic controllers. By taking control of the level of intracellular proteins, one can probe single-cell dynamics with unprecedented flexibility. However, single-cell dynamics are stochastic in nature, and a control framework explicitly accounting for this variability is presently lacking. Here we devise a stochastic control framework, based on Model Predictive Control, which fills this gap. Based on a stochastic modelling of the gene response dynamics, our approach combines a full state-feedback receding-horizon controller with a real-time estimation method that compensates for unobserved state variables. Using previously developed models of osmostress-inducible gene expression in yeast, we show *in silico* that our stochastic control approach outperforms deterministic control design in the regulation of single cells. The present new contribution leads to envision the application of the proposed framework to *wetlab* experiments on yeast.

1 Introduction

Gene expression plays a central role in the orchestration of cellular processes. The use of inducible promoters to change the expression level of a gene from its physiological level has significantly contributed to the understanding of the functioning of regulatory networks. Whereas the precise time-varying perturbation of the level of a target protein has the potential to be highly informative on the functioning of cellular processes, so far inducible promoters have been used for either static perturbations or simple dynamic perturbations with limited accuracy (see [14] for a notable exception). Alternative solutions, based on real-time control, have recently been proposed [11,12,16,18]. In real-time, the level of the protein is observed and gene induction is modulated based on the distance to the objective. Thanks to the implementation of such external feedback

P. Mendes et al. (Eds.): CMSB 2014, LNBI 8859, pp. 155–172, 2014.

loops, one can maintain the mean level of a fluorescent protein at some target value over extended time durations (set point experiments) and even follow time-varying profiles with good quantitative accuracy (tracking experiments). However, because of the significant cell-to-cell variability and the stochasticity of gene expression, even if the mean level of the protein follows precisely the objective, the performance of the controller is significantly worse when applied and measured at the single cell level. Yet if one wants to understand the effect of a perturbation of the level of a protein on a given process, one needs to control the level of this protein at the single cell level, that is, one needs to perform single cell control.

In [18] we have shown that single cell control is indeed effective: we have obtained better control performances when controlling single cells individually than when controlling the mean of the cell population. This slightly improved performance has been obtained by controlling the level of a particular, randomly-chosen cell using a deterministic model of gene expression. Given the stochasticity of cellular processes, one might wonder whether better control performances can be obtained by using a more appropriate stochastic model of gene expression. This question is actually not trivial: while the stochastic model is supposed to be closer to reality, it requires the use of complex controller architectures and the solution of computationally challenging optimization problems under tight time constraints.

In this work we investigate to what extent stochastic control techniques outperform more traditional deterministic control approaches. To do so, we consider a stochastic model of gene expression at the single cell level, alongside its deterministic counterpart, and develop state estimators and controllers for deterministic and stochastic control. We then compare the efficiency of the two approaches for set point regulation and tracking control in *in silico* experiments. Methodologically, in this work we introduce a stochastic receding horizon design approach of broad applicability, and a generalizable hybrid approach to state estimation. To our knowledge this is the first work on single cell control that accounts for gene expression noise.

The paper is structured as follows. In Section 2, we present the biological system, alongside the control platform used in [18] that has motivated this work, as well as the models used, inspired from [8,21]. In Section 3 we present control algorithms for deterministic and stochastic control assuming full state observability, whereas in Section 4 we present a state estimation approach for stochastic models. The performances of deterministic and stochastic controllers are compared in Section 5 on two *in silico* control experiments.

2   Osmostress-Induced Gene Expression in Yeast

2.1   Hyper-Osmotic Stress Response in Yeast

In the budding yeast *S. cerevisiae*, an increase of the environmental osmolarity creates a water outflow and a cell shrinkage. The adaptation response to such an osmotic shock is mainly mediated by the high osmolarity glycerol (HOG) signal transduction pathway, leading to an increase of the cellular glycerol level via various mechanisms, one of which is the upregulation of genes involved in glycerol production. In [18], we have used the promoter of the osmoresponsive gene *STL1* to drive the expression of a yellow fluorescent protein, yECitrine, so as to monitor the gene expression response of the cells to repeated osmotic stresses (Fig. 1(a)).

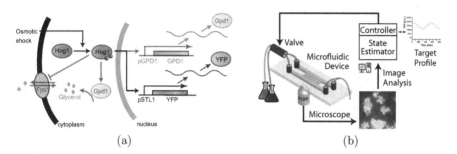

(a)   (b)

Fig. 1. The experimental setup. **(a)** Hyperosmotic shocks trigger the activation of the Hog1 protein and the intracellular accumulation of glycerol via short- and long-term adaptation responses (grayed). This system can be used to induce the production of a protein of interest, here a yellow fluorescent protein (YFP), by repeatedly applying hyperosmotic stresses. **(b)** Real-time control platform: single-cell and population control problems are defined respectively as controlling the fluorescence of a single randomly-chosen cell and the mean fluorescence of all the cells.

2.2 Platform for Control of Osmostress-Induced Gene Expression

Using microfluidic devices one can grow yeast cells in monolayers over extended time durations. Because cells can be trapped in imaging chambers, their response can be tracked by fluorescence microscopy and their environment can be rapidly changed, thus enabling the repeated application of osmotic shocks (Fig. 1(b)). The addition of software for image analysis and for state estimation, and the computation of a control strategy closes the feedback loop. Experiments typically last 10-15 hours, with fluorescence measurements every 5-10 minutes.

2.3 Modeling Osmostress-Induced Gene Expression

We describe the osmostress induced gene expression by the reactions [21]

$$
\begin{aligned}
pSTL1^{off} &\underset{c_2}{\overset{c_1 u}{\rightleftharpoons}} pSTL1^{on} \\
pSTL1^{on} + CR &\underset{c_4}{\overset{c_3}{\rightleftharpoons}} CR.pSTL1^{on} \\
CR.pSTL1^{on} &\overset{c_5}{\rightarrow} CR.pSTL1^{on} + mRNA \\
mRNA &\overset{c_6}{\rightarrow} mRNA + YFP \\
YFP &\overset{c_7}{\rightarrow} \phi \\
mRNA &\overset{c_8}{\rightarrow} \phi
\end{aligned}
\tag{1}
$$

Here $pSTL1^{off}$ and $pSTL1^{on}$ represent the inactive and the active states of the pSTL1 promoter, respectively. Furthermore, the interaction of $pSTL1^{on}$ with chromatin remodeling complexes (CR) enables the formation of the $CR.pSTL1^{on}$ complex and the effective transcription of mRNA, and the subsequent production of the fluorescent protein YFP. The degradations of the mRNA and the YFP protein follow first order kinetics. A change in the valve status from OFF to ON leads to an increase in the osmolarity of the cells environment, in the activation of the Hog1 protein, and in the increase of the effective input function u affecting promoter transition rates. The modeling of these processes is detailed in Appendix A.1, whereas the initial concentrations and the rate coefficients are listed in Table 2 in Appendix A.2.

A stochastic interpretation of the above reactions leads to a Chemical Master Equation (CME) model [6], characterized by a distribution accounting for the probability that the state of the system (represented by variables denoting molecular count) at time instant $t \in \mathbb{R}^+$ is $x(t)$, given its initial state $x(0)$ and an input signal $u(s), s \in [0, t]$. These stochastic semantics will be employed for testing the behavior of the model in *in silico* control experiments: in particular, we will use (a discrete-time version of) the Stochastic Simulation Algorithm (SSA) [6] to simulate the model. The dynamics can be approximated by a system of coupled deterministic dynamical equations, known as the Reaction Rate Equations (RRE) [6], operating over the concentrations x of the species as:

$$\dot{x}[i](t) = \sum_{j=1}^{M} v_{ij} a_j(x(t), u(t)), \qquad i = 1, \dots, N. \tag{2}$$

Here the quantity M is the total number of reactions and N is the total number of species ($x[i]$ being the ith). The vector $v_j := (v_{ij})_{i=1}^{N}$ is the state change vector for each reaction R_j: in particular v_{ij} represents the stoichiometry coefficients, defined as the change in the molecular population of a species S_i caused by the reaction R_j. Finally, the coefficients $a_j(\cdot)$ are the reaction rates, derived from the law of mass-action applied to (1): the control input in particular directly affects the affinity term a_1. The model in (2) is employed to synthetise a deterministic controller that will be used as a reference to assess the performances of the stochastic controller newly developed in this work. For the latter objective, a second approximation of the CME dynamics is introduced in Section 4, in order to derive an efficient state estimation scheme developed in the context of noisy partial observations, which combines the original CME semantics with a Chemical Langevin Equation (CLE) approximation.

3   Single-Cell Control with Full State Information

The control of gene expression is treated as a model-based optimal control problem. The goal of the control synthesis problem is to track a given profile of protein concentration over a finite time horizon T. As in [18], we require that the controller complies with particular timing constraints: the valve should remain ON at least 5 minutes and at most 8 minutes, and two stress inputs must be separated by at least 20 minutes (see Appendix A.1 for more details). These constraints are imposed in order to prevent cell adaptation to hyperosmotic environments.

In this section, the availability of full-state information (namely, knowledge of the values of *all the variables*) is assumed. Above we have formulated two models: a stochastic discrete-state one and a deterministic continuous-state one. For both cases, a control synthesis architecture based on the classical dynamic programming (DP) paradigm is proposed. As the classical DP suffers from the curse of dimensionality, we employ an approximate DP method called Fitted Q-Iteration (FQI) [4,9], tailored here to the finite-horizon setting. The FQI algorithm applies the idea of value iteration to the so-called Q-functions: a Q-function approximation is used in place of a value function approximation, and it allows for an immediate computation of the optimal actions at each optimisation stage. The FQI algorithm offers the possibility to employ powerful regression algorithms from supervised learning to interpolate the Q-function computed over a finite set of states to cover the entire state space [9].

Optimal Controller Synthesis via DP. For the controller synthesis problem, we will adopt a discrete time simulation framework. Let us denote the state space by X, the action space by U, and the space supporting the noise term by W. For each $x \in X$ we denote by $U(x) \subseteq U$ the set of actions enabled at x. A stochastic discrete-time dynamical system is described by the following difference equation:

$$x_{k+1} = f(x_k, u_k, w_k), \qquad k = 1, \ldots, T-1, \tag{3}$$

where $x_k \in X$ is the state of the system at time k, $u_k \in U(x_k)$ is the action taken at time k, and $w_k \in W$ is the noise variable with a specified distribution: let us remark that the recursive dynamics in (3) can be equivalently expressed by a conditional distribution $x_{k+1} \sim P(\cdot|x_k, u_k)$ [10], which in our instance can be derived from a discrete-time version of the CME that we have discussed in the previous section.

A control policy is a sequential decision rule $\pi = (\pi_k)_{k=0}^{T-1}$, where $\pi_k : X \to U$ has to be chosen over admissible controls only: $\pi_k(x) \in U(x)$ for all $x \in X$. The instantaneous cost $c_k(x_k, u_k)$ is comprised within an (expected) additive performance criterion over a finite time horizon, which for a fixed policy π is given by

$$Q_0^\pi(x_0, u_0) := \mathbb{E}\left[c_T(x_T) + \sum_{k=0}^{T-1} c_k(x_k, \pi_k(x_k))\right]. \tag{4}$$

Notice that the terminal cost, c_T, depends only on the state variable. In the following, we shall employ the cost function $c_k(x_k, u_k) = |YFP_k - YFP_{ref,k}|$, which penalises deviations from the reference profile $YFP_{ref,k}$, and a null terminal cost c_T. We are interested in the policy π^* that minimizes the cost:

$$Q_0^*(x, u) := \inf_\pi Q_0^\pi(x, u) = Q_0^{\pi^*}(x, u).$$

This cost can be obtained via DP by the backward recursion, initialised at the value c_T and propagated as:

$$Q_k^*(x, u) = \mathcal{T} Q_{k+1}^*(x, u), \tag{5}$$

where $\mathcal{T}$ is an operator acting on functions $H : X \times U \to \mathbb{R}$ as follows:

$$\mathcal{T}H(x, u) := c(x, u) + \inf_{u' \in U} \mathbb{E} H(f(x, u, w), u'). \tag{6}$$

An optimal policy can be computed as

$$\pi_k^*(x) \in \arg \min_{u \in U} Q_{k+1}^*(x, u), \qquad k = 0, \ldots, T - 1. \tag{7}$$

The Q-iteration in (5)-(7) is computationally unfeasible for problems with extended state spaces, and in particular with the single-cell control problem we are dealing with: we approximate its solution by means of a stochastic FQI [9].

FQI for the Stochastic Model. The FQI is a batch-mode algorithm computed offline, which fits an approximation architecture to the Q-function defined over $X \times U$ using a set of tuples

$$\mathcal{F} = (x^i, u^i, c^{ij}, z^{ij}), \qquad i = \{1, \ldots, m_x\}, \ j = \{1, \ldots, m_z\}, \tag{8}$$

where $x^i \in X$ is the instance of the current (or reference) state, $u^i \in U(x^i)$ is the corresponding action, $z^{ij} \in X$ is a possible successor state under the action u^i, c^{ij} is the cost associated with a transition of the state from x^i to z^{ij}, m_x is the number of current states, m_z is the number of successor states that are needed for the evaluation of the expectation operator in (6) using Monte-Carlo integration.

We adopt an offline approach, owing to the computational complexity of the optimisation problem and to the stringent online time requirements. Using the batch of samples in (8), Algorithm 1 (in the Appendix) computes an approximation of the Q-function through a backward recursion from time instant T to 1. Each iteration of the algorithm consists of the following two steps:

- In the first step, the backward recursion for the Q-function at time $k + 1$ is evaluated using a Monte-Carlo integration. The operator $\mathcal{T}$ is approximated by an empirical operation $\hat{\mathcal{T}}_\mathcal{F}$ as shortly defined in (9): namely the value of $\mathcal{T}\hat{Q}_{k+1}$ is estimated as $\hat{\mathcal{T}}_\mathcal{F}\hat{Q}_{k+1}$, for all x^i, $i = 1, \ldots, m_x$.
- The second step involves fitting the approximation function $\hat{Q}_k$ to $\hat{\mathcal{T}}_\mathcal{F}\hat{Q}_{k+1}$: the optimal fit $\hat{Q}_k$ is achieved by means of a regression algorithm.

The overall performance and computational complexity of the FQI method heavily hinges on the choice of the regression algorithm. The supervised learning paradigm offers a wide range of algorithms that can be used for regression [3]. We have made use of the Fixed-Size Least-Squares Support Vector Machine (LS-SVM) [5], due to its computational efficiency and its powerful capability of generalisation. The LS-SVM model provides two parameters for tuning: the squared bandwidth σ^2 and the regularization parameter γ, which have here been tuned manually through trial and error (but could be as well be optimised over with a more sophisticated alternative). These parameters are crucial to determine the trade-off between the training error minimization, the smoothness and the generalization. Algorithm 1 is detailed in the Appendix: there, we assume that the regression algorithm is fixed, and denote by $\mathcal{G}$ the corresponding space of test functions $G : X \times U \to \mathbb{R}$. For a given tuple $\mathcal{F}$ we denote

$$\hat{\mathcal{T}}_{\mathcal{F}}H(x^i, u^i) := \inf_{u' \in U(x^i)} \frac{1}{m_z} \sum_{j=1}^{m_z} \left[c^{ij} + H(z^{ij}, u') \right] \tag{9}$$

and the corresponding 2-norm as $\| H' - H'' \|_{\mathcal{F}} := \sum_{i=1}^{m_x} \left| H'(x^i, u^i) - H''(x^i, u^i) \right|^2$.

FQI for the Deterministic Model. A discrete-time deterministic model is a special case of (3) where the update law f does not depend on the noise variable w. In our work, we refer to the deterministic dynamics discussed in (2), after time discretization. For this simpler setup, the DP operator takes the form

$$\mathcal{T}H(x, u) = c(x, u) + \inf_{u' \in U} H(f(x, u), u'),$$

and no expectation evaluations are needed. Thus, we have $m_z = 1$, so that only one successor state is needed for each instance of the state. As a result,

$$\hat{\mathcal{T}}_{\mathcal{F}}H(x^i, u^i) = \inf_{u' \in U(x^i)} \left[c^i + H(z^i, u') \right].$$

One can therefore directly tailor Algorithm 1 to the deterministic case.

Practical Implementation of the Stochastic FQI via Receding Horizon Strategy. Although the FQI for the deterministic model works well within our setup, the FQI algorithm for the stochastic model over the entire experimental duration (denoted by the time horizon T) has been found to be computationally infeasible, since parameters achieving a good generalisation for the regression algorithm over the complete time horizon T are not easily found, and because of the Monte-Carlo computations that are instead absent in the deterministic case. In order to overcome this issue, we have embedded the FQI algorithm into a receding horizon strategy, resulting in a stochastic receding horizon scheme (see Algorithm 2 in the Appendix) [1]. In short, over a finite prediction horizon $T_p \ll T$, the Q-functions are approximated offline using Algorithm 1. After the computation of the optimal control sequence and the application of the current control action, the horizon is shifted by one sample and the optimisation is performed again, until the whole horizon T is covered.

4 Partial Information Case: Estimation of System States

Typically not all state variables of a biological model are observed directly. This is in particular the case for the yeast osmotic shock response system, where only protein levels are observable via noisy measurements:

$$y_k = YFP_k + e_k, \quad e_k = (e_a + e_b \cdot YFP_k)\eta_k, \tag{10}$$

where for $k = 1, \ldots, T$, y_k is the measurement at time k for a given cell, and η_k are i.i.d. standard normal random variables, whereas e_a and e_b are the intensity of the additive and multiplicative parts of the measurement noise.

In practice, the state-feedback control must rely on estimates of the state that are generated online from the available measurements. Here we develop a strategy for real-time state estimation with reference to yeast osmotic shock response. We observe that the strategy can be applied to other biological scenarios.

We start from the continuous-time stochastic Markov model of the CME, which is expressed in terms of discrete-valued state variables x. One possible approach for estimating state x from measurements y_k is particle filtering [2]. In particle filtering, N hypothetical evolutions of the system state are randomly simulated up to the next measurement. When the latter becomes available, state estimates are produced by weighting the simulated trajectories, where the weights quantify the relevance of every simulated trajectory to the new (partial) state measurement. Since particles have to explore a large (possibly infinite) state space, in practice particle filtering requires many (e.g. $N > 1000$) simulations of the system, which makes it poorly suited for online applications. In [2], we have proposed an alternative approach using Unscented Kalman Filtering (UKF) [19] and based on the CLE, a continuous-valued approximation of the CME model [7]. In the current context, this approach is partly inappropriate, since the promoter state variables are inherently discrete (they take values 0 or 1 only). In order to combine the flexibility of particle filtering with the computational advantages of UKF, we propose to limit the Langevin approximation to the mRNA and protein dynamics.

We first note that promoter dynamics do not depend on mRNA and protein abundance. Let us partition the state variables as $x = (x^d, x^c)$, where

$$x^d = (pSTL1^{off}, pSTL1^{on}, CR \cdot pSTL1^{on}), \quad x^c = (mRNA, YFP).$$

Consider a model where the dynamics of x^d (not depending on x^c) are left unchanged (i.e., follow the original CME), while for any given trajectory of x^d, the dynamics of x^c are approximated by the Langevin equation

$$dx^c[i] = \sum_{j=1}^{M} v_{ij}^c a_j(x^c, x^d)dt + \sum_{j=1}^{M} v_{ij}^c \sqrt{a_j(x^c, x^d)}dW_j, \quad i = 1, 2. \tag{11}$$

Here, for $j = 1, \ldots, M$, W_j are independent Wiener processes and $v_{\cdot j}^c$ is the subvector of v_j corresponding to x^c. The relevance of the Langevin approximation to mRNA and protein dynamics has been discussed in [7] and, for filtering applications, it has been assessed on a different but relevant system in [2]. Note that, while x^d remains discrete-valued, x^c may now take continuous values.

Based on this hybrid model, a filtering procedure iterating over subsequent measurement indices k combining importance (particle) filtering with UKF is obtained as follows. At time t_{k-1}, let $\hat{x}_{k-1|k-1}^c$ be the estimate of the current state

x^c based on measurements $y_0, \ldots, y_{k-1}$, and let $\hat{x}^{d,i}_{k-1|k-1}$, with $i = 1, \ldots, N$, be N putative values of the current state x_d (with N small, see below). For every i, a hypothetical discrete-state trajectory $\hat{x}^{d,i}_{k-1}(t)$, with $t \in [t_{k-1}, t_k)$, is generated by stochastic simulation of the discrete-state dynamics starting from $\hat{x}^{d,i}_{k-1|k-1}$. Over the same time horizon, for every i, mRNA and protein state predictions $\hat{x}^{c,i}_{k-1}(t)$ are computed along trajectory $\hat{x}^{d,i}_{k-1}(t)$ via UKF. When the next protein measurement y_k becomes available, based on measurement model (10), an importance weight w_i, proportional to the likelihood of y_k given the hypothetical state value $\hat{x}^{c,i}_{k-1}(t_k)$, is computed for every particle i. Note that weights w_i play the role of a-posteriori probabilities of the different particles. Also, continuous-state predictions $\hat{x}^{c,i}_{k-1}(t_k)$ are updated to estimates $\hat{x}^{c,i}_{k|k}$ of the current state x^c by integrating the new piece of information provided by $y(t_k)$, in accordance with the so-called measurement-update step of UKF. At this stage, the ensemble (Conditional Expectation) estimate $\hat{x}^c_{k|k}$ as well as an ensemble (Maximum-A-Posteriori) estimate $\hat{x}^d_{k|k}$ for the discrete state are computed as

$$\hat{x}^d_{k|k} = \arg \max_{z \in \{0,1\}^3} \sum_i 1_z\big(\hat{x}^{d,i}_{k-1}(t_k)\big) \cdot w_i, \qquad \hat{x}^c_{k|k} = \sum_i \hat{x}^{c,i}_{k|k} \cdot w_i, \qquad (12)$$

where $1_z(\cdot)$ is the indicator function. For control purposes, these are the estimates that are passed to the controller with entries of $\hat{x}^c_{k|k}$ rounded to the nearest integers. To proceed for the next iteration of the algorithm, the new putative values of the discrete state $\hat{x}^{d,j}_{k|k}$, with $j = 1, \ldots, N$, are set equal to the result of N independent random extractions from the pool of particles $\{\hat{x}^{d,i}_{k-1}(t_k)\}_{i=1,\ldots,N}$, with sampling probabilities equal to w_i (resampling step of particle filtering). The whole procedure is summarized in Algorithm 3 in the Appendix.

The initialization of the procedure at the starting time $k = 0$ is performed based on the a priori statistics of x^d and x^c. Given the small (finite) discrete state space of x^d, a number of particles N much smaller than traditional particle filter implementations is expected to suffice. Empirical evaluation (not reported here) has led to select $N = 50$, a value above which no significant improvement of filtering performance has been observed. The implementation of the UKF procedure is analogous to that of [2] and is omitted for brevity. We just note that, at every step k and for every particle i, UKF requires the numerical solution of $2n^c + 1$ ODEs over the time span $[t_{k-1}, t_k)$, with $n^c = 2$ being the number of continuous states. The solution of these ODEs can be carried on in parallel with the simulation of $\hat{x}^{d,i}_{k-1}$. Contrary to the control module, resorting to time discretization is not needed, although it can be considered towards higher computational efficiency.

5   Results

5.1   Deterministic and Stochastic Control in the Full Information Case

In this section we present the results of the control of gene expression to track time-homogeneous and time-varying target profiles, using the deterministic and stochastic controllers detailed in Section 3. In order to test the effectiveness of the proposed algorithms, the controller trained using the deterministic FQI was first tested over the deterministic RRE model. As expected in this case, the controller has successfully been able to track the signals (see Appendix A.4 for implementation details). To test the control performance in a realistic biological context, this controller has then been used over the stochastic CME model. At the maximum, the controller is able to track the reference signal to within a deviation of 10% as shown in Fig. 2(a)(b). The deterministic controller has then been replaced with the stochastic controller (see Appendix A.3 for implementation details) and it has been found that the stochastic controller is able to track the reference signal to within a deviation of 5% from the reference trajectory (Fig. 2(d)(e)).

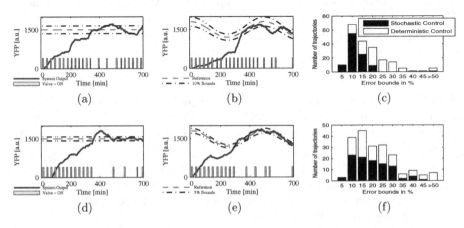

Fig. 2. Comparison of stochastic and deterministic control schemes in the full information case, run over the probabilistic model. (a)(b) Deterministic controller tracking the desired profiles with a shown deviation of 10% from reference trajectories. (d)(e) Stochastic controller showing improved performance with a deviation of 5%. (c)(f) Monte-Carlo simulations validating the superior performance of the stochastic controller over its deterministic counterpart: the histogram plots the number of closed-loop trajectories falling within specific error bounds from the reference trajectory.

In order to get a quantitative comparison of the performance of the stochastic controller over the deterministic controller, 100 runs of each algorithm have been performed using Monte-Carlo simulations. To measure the quality of the control, we have used $\epsilon := \frac{1}{T-T_0} \sum_{k=T_0}^{T} |YFP_k - YFP_{ref,k}| / YFP_{ref,k}$, where T_0 is the time it takes the system to reach the desired trajectory. In practice, we have chosen $T_0 = 400$ and $T_0 = 300$ minutes for the set point and signal tracking experiments, respectively. These results are presented in Fig. 2(c)(f). It is evident from the figure that the controller developed considering the stochastic nature of the gene expression yields superior performance than the controller developed ignoring it.

5.2   Stochastic Control with Partial Information

The control laws obtained in the full information case are functions of the current state x_k: at each time k it is supposed that the controller observes the exact value of the full current state x_k and that it applies the appropriate action. In reality the measurements y_k are limited to the fluorescent protein. The hybrid filter detailed in Section 4 has been used to extract information about the states of the gene expression network using 50 particles. The filter does not succeed to accurately track the switching of the discrete states of the promoter but is able to track the $mRNA$ and YFP protein concentrations fairly accurately (Fig. 3(a)). The filter has then been used in conjunction with the stochastic controller: the simulation results presented in Figure 3(b)(c) show that the controller is robust to state estimation errors and is able to successfully track the reference profiles.

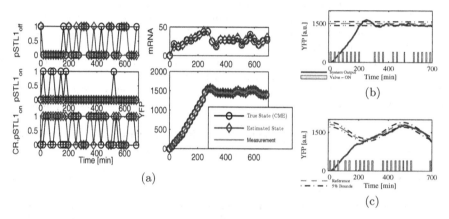

Fig. 3. Results of the stochastic control scheme run with the hybrid filter in the partial information case. (a) State estimation shows accurate results for mRNA and YFP, whereas the filter faces difficulties estimating the switching action of the promoter. (b)(c) Controller robustness over state estimation errors and ability to track reference signals to within a deviation of 5%.

6   Discussion and Conclusions

The main contribution of this paper is the development of a complete model-based control framework adapted to stochastic models of gene expression. Although the identification of stochastic models of gene expression has recently been extensively studied, the control of gene expression using stochastic models has been barely addressed so far. This goal requires the non-trivial development of integrated stochastic state estimators and controllers. We have demonstrated *in silico* that stochastic control has the potential to deliver superior performances in comparison to a deterministic counterpart explored in earlier literature. This work paves the way for the development of an experimental platform for single-cell control based on optogenetics solutions, which enable the independent stimulation of live single cells in real-time [16,20].

Acknowledgments. The first two authors are students and the main contributors to this work. This research has been supported by a van Gogh grant from the FR-NL Academie van Wetenschappen, by the European Commission IAPP project AMBI 324432, by the GeMCo (ANR-10-BLAN-0201) and ICEBERG (ANR-10-BINF-06-01) grants from the Agence Nationale de la Recherche, and by the Action d'Envergure ColAge from INRIA/INSERM.

References

1. Bemporad, A., Morari, M.: Control of systems integrating logic, dynamics, and constraints. Automatica 35(3), 407–427 (1999)
2. Carta, A., Cinquemani, E.: State estimation for gene networks with intrinsic and extrinsic noise: a case study on E.coli arabinose uptake dynamics. In: European Control Conference, ECC 2013, Zurich, Suisse, pp. 3658–3663 (2013)
3. Caruana, R., Niculescu-Mizil, A.: An empirical comparison of supervised learning algorithms. In: Proc. of the 23rd International Conference on Machine Learning, pp. 161–168. ACM (2006)
4. Ernst, D., Geurts, P., Wehenkel, L.: Tree-based batch mode reinforcement learning. Journal of Machine Learning Research, 503–556 (2005)
5. Espinoza, M., Suykens, J.A.K., De Moor, B.: Fixed-size least squares support vector machines: A large scale application in electrical load forecasting. Computational Management Science 3(2), 113–129 (2006)
6. Gillespie, D.T.: A general method for numerically simulating the stochastic time evolution of coupled chemical reactions. Journal of Computational Physics 22(4), 403–434 (1976)
7. Gillespie, D.T.: The chemical Langevin equation. Journal of Chemical Physics 113(1), 297–306 (2000)
8. Gonzalez, A.M., Uhlendorf, J., Cinquemani, E., Batt, G., Ferrari-Trecate, G.: Identification of biological models from single-cell data: A comparison between mixed-effects and moment-based inference. In: European Control Conference, ECC 2013, pp. 3652–3657 (2013)
9. Haesaert, S., Babuska, R., Abate, A.: Sampling-based approximations with quantitative performance for the probabilistic reach-avoid problem over general Markov processes. arXiv preprint, arXiv:1409.0553 (2014)
10. Kallenberg, O.: Foundations of modern probability. Probability and its Applications. Springer, New York (2002)
11. Menolascina, F., Fiore, G., Orabona, E., De Stefano, L., Ferry, M., Hasty, J., di Bernardo, M., di Bernardo, D.: In-vivo real-time control of protein expression from endogenous and synthetic gene networks. PLoS Computational Biology 10(5), e1003625 (2014)
12. Milias-Argeitis, A., Summers, S., Stewart-Ornstein, J., Zuleta, I., Pincus, D., El-Samad, H., Khammash, M., Lygeros, J.: In silico feedback for in vivo regulation of a gene expression circuit. Nature Biotechnology 29, 1114–1116 (2011)
13. Muzzey, D., Gómez-Uribe, C.A., Mettetal, J.T., van Oudenaarden, A.: A systems-level analysis of perfect adaptation in yeast osmoregulation. Cell 138(1), 160–171 (2009)
14. Olson, E.J., Hartsough, L.L., Landry, B.P., Shroff, R., Tabor, J.J.: Characterizing bacterial gene circuit dynamics with optically programmed gene expression signals. Nature Methods 11, 449–455 (2014)

15. Pelckmans, K., Suykens, J.A.K., Van Gestel, T., De Brabanter, J., Lukas, L., Hamers, B., De Moor, B., Vandewalle, J.: LS-SVMlab: a matlab/c toolbox for least squares support vector machines. Tutorial, Leuven, Belgium (2002)
16. Toettcher, J.E., Gong, D., Lim, W.A., Weiner, O.D.: Light-based feedback for controlling intracellular signaling dynamics. Nature Methods 8, 837–839 (2011)
17. Uhlendorf, J., Bottani, S., Fages, F., Hersen, P., Batt, G.: Towards real-time control of gene expression: controlling the HOG signaling cascade. In: 16th Pacific Symposium of Biocomputing, pp. 338–349 (2011)
18. Uhlendorf, J., Miermont, A., Delaveau, T., Charvin, G., Fages, F., Bottani, S., Batt, G., Hersen, P.: Long-term model predictive control of gene expression at the population and single-cell levels. PNAS 109(35), 14271–14276 (2012)
19. Wan, E.A., Van Der Merwe, R.: The unscented kalman filter for nonlinear estimation. In: Adaptive Systems for Signal Processing, Communications, and Control Symposium, AS-SPCC 2000, pp. 153–158. IEEE (2000)
20. Yang, X., Payne-Tobin Jost, A., Weiner, O.D., Tang, C.: A light-inducible organelle-targeting system for dynamically activating and inactivating signaling in budding yeast. Molecular Biology of the Cell 24(15), 2419–2430 (2013)
21. Zechner, C., Ruess, J., Krenn, P., Pelet, S., Peter, M., Lygeros, J., Koeppl, H.: Moment-based inference predicts bimodality in transient gene expression. PNAS 109(21), 8340–8345 (2012)

A   Appendix

A.1   Implementation Constraints on the Control of Gene Expression

In order to limit cell adaptation to hyperosmotic environments, we delimit the duration of hyperosmotic shocks to 8 minutes and impose at least a 20 minute time lag between two successive shocks. We also require that shocks last at minimum 5 minutes.

As shown in [18], there is a known lag between the valve actuation and the actual change of osmolarity of the cellular environment in the imaging chamber. Formally, for a given osmotic shock, we denote by t_{on} and t_{off} the times at which the valve switches to ON and to OFF positions, respectively, and represent the osmolarity h in the imaging chamber as follows (see Figure 4).

$$h(t) = \begin{cases} 0 & \text{if } t < t_{on} + 2, \\ t - (t_{on} + 2) & \text{if } t_{on} + 2 < t < t_{on} + 3, \\ 1 & \text{if } t_{on} + 3 < t < t_{off} + 2, \\ 1 - (t - (t_{off} + 2))/4 & \text{if } t_{off} + 2 < t < t_{off} + 6, \\ 0 & \text{otherwise.} \end{cases} \tag{13}$$

As in [8,13,17], we assume that the activity s of the Hog1 protein depends on the osmolarity of the environment h as follows.

$$\dot{s}(t) = \kappa h(t) - \Gamma s(t), \tag{14}$$

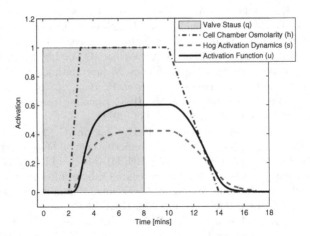

Fig. 4. Temporal evolution of the osmolarity of the cellular environment h, of the Hog1 activity s, and of the promoter activation stochastic rate u, as a function of the position of the microfluidic valve (0/1: normal/hyper-osmotic medium)

with $s(0) = 0$; we further assume that the *pSTL1* promoter activation stochastic rate u is a function of the Hog1 activity s, following Hill-type kinetics as

$$u(t) = \frac{(s(t) + a_0)^{n_H}}{K_d{}^{n_H} + (s(t) + a_0)^{n_H}}. \tag{15}$$

Note that we assume here that there is no significant stochasticity in signal transduction. The rate parameters used for model simulation are listed in the table below. Also in practice, because the controller uses a discrete time representation, we refer to the input at instant k as $u_k = u(t_k)$.

Table 1. Rates of the activation function

Parameter	Value	Parameter	Value
κ	0.3968 (a.u.)	Γ	0.9225 (a.u.)
K_d	0.34906 (a.u.)	a_0	0.0027998 (a.u.)
n_H	2.1199 (a.u.)		

A.2   Parameters Employed in the Simulation and Analysis of the Model

The rate parameters and initial concentrations used for the simulation of the model are listed in the two tables below.

Table 2. Initial concentrations and rates of the stochastic gene expression network

Parameter	Value	Parameter	Value
$(pSTL11^{off})_0$	1 (a.u.)	$(pSTL11^{on})_0$	0 (a.u.)
CR_0	102.51 (a.u.)	$(CR \cdot pSTL11^{on})_0$	0 (a.u.)
$mRNA_0$	0 (a.u.)	YFP_0	0 (a.u.)
c_1	23.604 $(min)^{-1}$	c_5	12.256 $(min)^{-1}$
c_2	180.03 $(min)^{-1}$	c_6	0.36113 $(min)^{-1}$
c_3	0.024559 $(min)^{-1}$	c_7	0.025091 $(min)^{-1}$
c_4	0.9384 $(min)^{-1}$	c_8	0.003354 $(min)^{-1}$

Table 3. Parameters of the measurement model

Parameter	Value	Parameter	Value
e_a	1.0115 $(min)^{-1}$	e_b	0.0037 $(min)^{-1}$

A.3 Implementation Details of the FQI Algorithm Over the Stochastic CME Model

For the stochastic receding horizon control approach, the samples x^i have been drawn corresponding to a single system trajectory. The trajectory has been generated by simulating the system using a discrete time version of the stochastic simulation algorithm. The intrinsic variability results from the stochasticity of the CME.

For each x^i, 250 tuples ($m_x = 250$) of the form (x^i, u^i) have been generated. For each tuple, the system has been simulated 100 times ($m_z = 100$) to obtain the next state z^{ij} to evaluate the Monte-Carlo integration. The cost c^{ij} has been computed as explained in main text and a single batch of 25000 tuples ($\mathcal{F}_s = 25000$) has been obtained. The optimization has been performed for a prediction horizon T_p of 8 minutes and for a time horizon T of 700 minutes. The discretization interval Δt has been set to 0.008 min. The squared bandwidth σ^2 and the regularization parameter γ of the regression algorithm have been tuned by a trial and error method and the final parameters have been reported below.

Table 4. Tuned LS-SVM parameters to track time varying and time constant profiles using the controller trained on the stochastic CME model

Squared Bandwidth (σ^2)	Regularization Parameter (γ)
600000	500

The stochastic FQI and receding horizon algorithms respectively are detailed below.

Algorithm 1. Stochastic Finite Horizon FQI $(T, \mathcal{F})$

1: Initialize the parameters of the regression algorithm σ^2 and γ and set $\hat{Q}_T$ to 0
2: **for** $k := T - 1$ **to** 0 **do**
3:    Estimate $\hat{\mathcal{T}}_{\mathcal{F}}\hat{Q}_{k+1}$.
4:    Find the fit that minimizes the 2-norm loss by means of a regression algorithm

$$\hat{Q}_k(x, u) = \arg\min_{G \in \mathcal{G}} \|G - \hat{\mathcal{T}}_{\mathcal{F}}\hat{Q}_{k+1}\|_{\mathcal{F}}.$$

5: **end for**

Algorithm 2. Stochastic Receding Horizon Control $(T, T_p, \mathcal{F})$

1: **for** k:=1 **to** T **do**
2:    Initialize the parameters of the regression algorithm σ^2 and γ and set Q_T to 0
3:    **for** $l := k + T_p$ **to** k **do**
4:       Estimate $\hat{\mathcal{T}}_{\mathcal{F}}\hat{Q}_{l+1}$.
5:       Find the fit minimizing the 2-norm loss by means of a regression algorithm

$$\hat{Q}_l(x, u) = \arg\min_{G \in \mathcal{G}} \|G - \hat{\mathcal{T}}_{\mathcal{F}}\hat{Q}_{l+1}\|_{\mathcal{F}}.$$

6:    **end for**
7: **end for**

A.4   Implementation Details of the FQI Algorithm Over the Deterministic RRE Model

For the deterministic control approach presented in Section 3, 400 tuples have been generated corresponding to a single trajectory. The trajectory has been

Table 5. Tuned LS-SVM parameters to track a set-point of 1500 (a.u.) using the controller trained on the deterministic RRE model

Time Horizon (T)	Squared Bandwith (σ^2)	Regularization Parameter (γ)
700 - 651	40000	10^{-1}
650 - 601	40000	10^{-2}
600 - 551	40000	10
550 - 501	40000	200
500 - 451	40000	1
450 - 401	40000	200
400 - 351	40000	100
350 - 301	40000	200
300 - 251	40000	100
250 - 201	40000	300
200 - 151	40000	100
150 - 1	40000	100

obtained by simulating reactions of the gene expression network using the RRE. A time horizon of 700 minutes has been considered and the regression algorithm has been implemented using the LS-SVM MATLAB toolbox in [15]. The Fixed-Size LS-SVM model provides two parameters for tuning: the squared bandwidth σ^2 and the regularization parameter γ. The parameters have been tuned manually using a trial-and-error method, and the selected ones are reported in Tables 5 and 6 below.

For the deterministic control approach, the deterministic version of the FQI algorithm has been trained and implemented over the RRE model. The simulation results in Figure 5 show that the system is able to track the reference profiles within a maximum deviation of 5%.

Table 6. Tuned LS-SVM parameters to track the sinusoidal reference signal using the controller trained on the deterministic RRE model

Time Horizon (T)	Squared Bandwith (σ^2)	Regularization Parameter (γ)
700 - 651	400000	50
650 - 601	40000	9
600 - 551	20000	1
550 - 501	400000	30
500 - 451	40000	596
450 - 401	40000	800
400 - 351	40000	300
350 - 301	100000	1
300 - 251	100000	11
250 - 201	100000	5
200 - 1	100000	4000

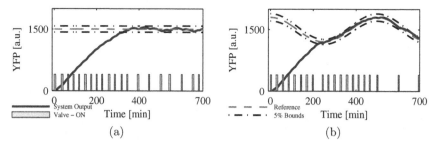

(a)   (b)

Fig. 5. Results for the deterministic control scheme in the full information case. The deterministic controller tracks time-varying and constant profiles within a deviation of 5% from the reference trajectory.

A.5   Hybrid Estimation Algorithm

Algorithm 3. Hybrid Filter for Estimation of the Model States

---

1: Initialize $\hat{x}_{0|-1}^{d,i}(0)$, $\hat{x}_{0|-1}^{i,c}(0)$, and w_i, with $i = 1, \ldots, N$, s.t. $\sum_i w_i = 1$
2: **for** $k = 0, 1, 2, \ldots$ **do**
3:     Acquire new measurement y_k
4:     Compute and normalize weights $w_i \propto \log p\big(y_k | \hat{x}_{k-1}^{c,i}(t_k)\big)$, with $i = 1, \ldots, N$
5:     Compute UKF estimate $\hat{x}_{k|k}^{i,c}$ from $\hat{x}_{k-1}^{i,c}(t_k)$ and y_k, with $i = 1, \ldots, N$
6:     Compute and provide ensemble estimates (12)
7:     Define N new particles $\hat{x}_{k|k}^{i,d}$ by resampling particles $\{\hat{x}_{k-1}^{i,d}(t_k)\}$ with prob. $\{w_i\}$
8:     Simulate $\hat{x}_k^{i,d}(t)$, $t \in [t_k, t_{k+1})$, from $\hat{x}_k^{i,d}(t_k) = \hat{x}_{k|k}^{i,d}$, with $i = 1, \ldots, N$
9:     Compute UKF prediction $\hat{x}_k^{i,c}(t)$ along $\hat{x}_k^{i,d}(t)$, $t \in [t_k, t_{k+1})$, with $i = 1, \ldots, N$
10: **end for**

A Rule-Based Model of Base Excision Repair[*]

Agnes Köhler[1], Jean Krivine[2], and Jakob Vidmar[2]

[1] INRIA-Rocquencourt
Domaine de Voluceau - Rocquencourt B.P. 105 - 78153 Le Chesnay, France
[2] Univ. Paris Diderot, Sorbonne Paris Cité,
Laboratoire PPS, UMR 7126, 75205 Paris, France

Abstract. There are ongoing debates in the DNA repair community on whether the coordination of DNA repair is achieved by means of direct protein-protein interactions or whether substrate specificity is sufficient to explain how DNA intermediates are channeled from one repair enzyme to the other. In order to address these questions we designed a model of the Base Excision Repair pathway in Kappa, a rule based formalism for modeling protein-protein and protein-DNA interactions. We use this model to shed light on the key role of the scaffolding protein XRCC1 in coordinating the repair process.

1 Introduction

A modern trend of Systems Biology sees high-throughput experiments being set up, resulting in an inflation of the publication volume in Biology and medicine[1]. As a consequence it has become impossible for a biologist, specialist of a certain system, to remain up-to-date with all relevant information pertaining to her topic of interest. To counter for this problem, biologists make an intensive use of review papers which are regularly published on a given system[2].

As an alternative to classical reviews, which are static objects with a natural obsolescence, we propose to use rule-based modeling [1, 2] to designing formal updatable reviews that are at the same time executable [3, 4].

More specifically, this paper presents the first executable model of the *Base Excision Repair* (BER) pathway that includes protein-DNA interactions. The outline of the paper is as follows: in Section 2 we briefly present the BER system (reviewed for instance in Ref. [5, 6]), we discuss our tools and methods in Section 3 and we present some results in Section 4.

2 Base Excision Repair

Figure 1 gives a possible *unfolding* of the *abstract* Base Excision Repair (BER) pathway: Various types of damage (A) may modify a nucleotide (oxydation,

---

[*] This work has been partially supported by the French National Research Agency (ANR), project ICEBERG.
[1] In 2000 about 500,000 papers were published in Biology and Medicine. In 2012 this number had escalated to 1,000,000 (source Pubmed.org).
[2] There are 520 review papers mentioning Base Excision Repair in the title or abstract (source Pubmed.org).

P. Mendes et al. (Eds.): CMSB 2014, LNBI 8859, pp. 173–195, 2014.

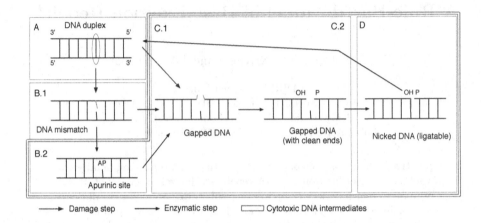

Fig. 1. Abstract Base Excision Repair pathway

deamination) and induce a mismatch in the DNA (B.1) or a single strand break (C.1). Enzymes from the family of DNA Glycosylase may recognize mismatches and excise the modified base, creating an Apurinic (AP) site (B.2). Enzymes with AP endonuclease capacity may open DNA at the lesion locus, generating a single strand break. End cleaning enzymes may prepare the 3' and 5' moieties for the polymerase step (C.2). Eventually DNA ligases can seal the DNA backbone (D) to retrieve a well-formed DNA duplex.

This scenario corresponds to only one possible *unfolding* of BER and the story could diverge at various points: for instance a direct single strand break may induce the loss of more than one nucleotide. Also when the end cleaning enzymes fail to prepare a proper substrate, some polymerases may synthesize more than one new nucleotide and trigger an alternative *long patch repair* pathway.

Furthermore this map is *abstract* as several enzymes may engage in the various catalytic steps that are described. For instance the transition from (B.1) to (B.2) or directly (C.1) is realized by different glycosylases, the identity of which depends on the type of nucleotide modification that has occurred. Ten glycosylases have been found so far in higher eukaryotic cells, we modeled the activity of four of them and used UDG (for uracil excision in U/G mismatches) as default enzyme in our simulations. We give a more concrete description of the BER enzymes in Figure 2.

Together, Figure 1 and Figure 2 give an almost complete view of BER and ODE based models have been proposed to formalize this part [7, 8]. However they only reveal the catalytic steps that transform DNA, and do not take into account important proteins that have no direct enzymatic activity but are important to coordinate the repair process. More importantly, BER enzymes do not behave as typical enzymes that often have little affinity for their products. We will see that most enzymes of Fig. 2 have a non negligible affinity for various DNA intermediates, and this feature is probably critical for channeling DNA products to the next enzyme in the pathway, through protein-protein or protein-DNA interactions [9].

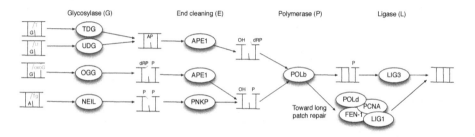

Fig. 2. A more concrete view of BER

This last point is the key to our modeling project, since (1) it is difficult to model as it entails a combinatorial explosion in the number of variables of the model, and (2) these complex interactions can give us insight into the coordination of the repair pathway i.e passing the baton between the different enzymes of Fig. 2.

Coordination is not so much an issue of efficiency, since high enzymatic turnover would probably yield a faster global repair rate. Figure 1 shows that the intermediate substrates (B.2), (C.1), (C.2) and (D) of the repair pathway are cytotoxic. In a nutshell, AP sites (a missing base), gapped and nicked DNA induce genomic instability and BER has probably evolved so as to prevent these substrates from being accessible to enzymes that may trigger apoptosis if such damage is detected (such as Topoisomerases).

The main protein that is believed to act as a coordinator of BER is the X-Ray Cross Complementing protein 1 (XRCC1). Although it has no known catalytic activity, this protein can bind to all BER enzymes that are downstream of the glycosylase. It is noteworthy that proteins interacting with XRCC1 are also those operating on the cytotoxic substrates. It is therefore assumed that XRCC1 acts as a scaffolding protein that coordinates BER, as well as a patch over the lesion to protect it from the environment.

We give Figure 3, the protein-protein and protein-DNA contact map that we inferred from the literature. The strength of the interaction is depicted here through various line widths and the dissociation constant (K_d) is shown. Dotted lines represent known interactions the K_d of which could not be found. This map makes apparent that several proteins compete for the same family of substrates. For instance APE1 and POLβ tend to bind to gapped DNA. Since DNA substrates are complex polymers one cannot assume that binding to a particular DNA substrate is exclusive of any other binding. Notably, it is assumed that XRCC1 can stay connected to gapped DNA throughout the whole repair process.

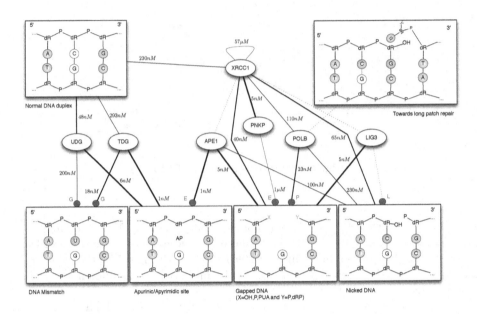

Fig. 3. The contact map of the BER model. Lines ending with a red circle indicate a catalytic activity on the corresponding substrate. The red letters map to the steps of Fig. 2.

3   Methods

Data. We have assembled qualitative (mechanisms of action) and quantitative (concentration, dissociation constant, catalytic rates) data from 59 papers pertaining to BER or to its participants[3]. For lack of space we do not include the complete references in the present paper but they are included in the model repository as an annotated bibtex file.

Quantitative data are particularly difficult to find. For testing the model under plausible conditions, we extracted BER protein copy numbers from Ref.[10] which evaluates protein concentration for HeLa cell extracts. Note that we expect repair accuracy to be robust with respect to variation of protein numbers, as the chromatin state might create local concentration effects on DNA [11].

Catalytic rates for enzymatic activities are easier to find although product inhibition (enzyme with a non negligible binding affinity for its product), which is typical of BER, complicates the interpretation of the rates which are often given in terms of steady state kinetics. More precisely, as can be seen in Fig. 3, most BER enzymes exhibit a scheme of the form:

$$E + S \leftrightarrow_{K_d} ES \rightarrow_{k_{chem}} EP \leftrightarrow_{K'_d} E + P$$

---

[3] We thank Dr. S. Mitra (Houston Methodist Hospital), Dr. D. M. Wilson III (National Institute on Ageing), Dr. S. H. Wilson (NIEHS, NIH) and Dr. K. Caldecott (Univ. Sussex), for direct discussions which directed us to relevant publications.

with a reasonably low K'_d. Experimental catalytic rates k_{cat} are measures of the production speed of P, which, in the above case, is slowed down by product inhibition. Some experiments [12–14] give a measure of k_{chem} for the above scheme, but most paper will only give k_{cat} (which underestimates the hidden k_{chem}). Whenever k_{chem} is not available we assumed the scheme:

$$E + S \leftrightarrow_{K_d} ES \to_{k_{cat}} E + P \leftrightarrow_{K'_d} EP$$

which simply lets the enzyme rebind to its product according to the given K'_d when available.

Yet, dissociation constants for protein-protein and protein-DNA interactions are also complicated to find. Whenever facing unknown data, we used the rate of a similar interaction. For instance we assumed that all glycosylases have the same facilitated diffusion on DNA, using data published for hOGG1 [15].

Importantly, KASIM requires concrete on and off rates for complex formation. When only steady state dissociation constants are known, we used a default k_{on} (randomized in simulations) to deduce k_{off} ($K_d = k_{off}/k_{on}$). Importantly, complex formation occurring in a uni-molecular fashion are assumed to be fast ($k_{uni} = 10^4 s^{-1}$). Whenever the kinetic data was unknown for a given reaction, we used a default kinetic rate k (taken from realistic values for the type of reaction), and randomized it uniformly in the interval $[\frac{k}{10}, 10 * k]$. The list of complete kinetic rates is provided in the Supplementary Data A.2.

Rule-Based Modeling. The input language of KASIM simulator is Kappa [2], a (rule-based) graph rewriting language, the syntax of which is recalled in Supp. Data A.1.

Figure 4 illustrates how DNA polymers are encoded in our model: (A) an apurinic -AP- site and (B) a one nucleotide gapped DNA. The ports on top of DNA nodes allows one to connect various BER enzymes. Internal states are mapped to the corresponding port via a green edge. There are a few key modeling features to notice. Our DNA nodes denote either physical DNA bases, or an empty slot on DNA. Hence a DNA node can either be part of the (physical) DNA backbone, as in the encoding of (A), or be a place holder for enzymes that recognize holes on DNA, as in the red part of the encoding of substrate (B). In the latter substrate, one may read from the Kappa encoding that the middle node is in fact a gap on DNA because it is no longer ligated to the 3' and 5' neighbors (internal state of the e3 and e5 ports set to NA). Notice that the e3 port of upper left DNA node of part (B) is set to P, indicating that the 3' end of the gap bears a phosphate group (that can be for instance recognized by APE1).

Simulation. Simulations of the model were conducted on a dedicated HP server (1.60GHz/4-cores) with 128 GB of RAM. Iterations of simulations under various parameters and randomization of kinetic rates are piloted by a python script (included in the model repository) which requires python 2.7 and simplejson. KASIM 3.5 is necessary to run simulations and is available on github[4].

---

[4] https://github.com/jkrivine/KaSim

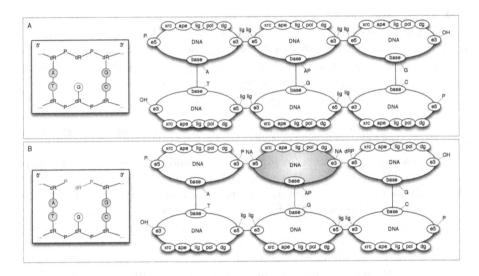

Fig. 4. Two oligonucleotides in Kappa

Unless specified otherwise, simulation results are obtained as the average of 16 simulations ran on a DNA substrate of 100,000 bp randomly generated according to the distribution: $(0.6 : GC, 0.4 : AT)$. In addition to the DNA substrate, initial number of BER enzymes are 2200 UDG; 30,000 APE1; 3,000 POLβ; 400 LIG3 and 1200 XRCC1 (ratios are taken from Ref. [10] and normalized so that the lowest number of potentially modeled enzyme (TDG) is greater than 100). Simulations are run without any damage on DNA for 2 seconds (biological time unit) after which we induced 2% of cytosine deamination, generating U/G mismatches on DNA. Simulations are conducted until complete repair is observed, on average after $t \sim 6$ minutes (biological time). Efficiency of simulations is discussed in the Supp. Data A.3.

Causality Analysis. Causality analysis of Section 4 were performed by enabling KASIM's causal tracking of the ligation rule, on simulations on a 25K bp DNA substrate. From a single simulation, KASIM computed 189 causally ordered traces (causal flows) leading to the ligation steps occurring between t=20s and t=80s (ligation activity is constant after 10s, data not shown). We then performed weak compression [16] that quotiented the number of causal flows to 122 (36% compression) incompressible scenarios[5].

---

[5] Incompressible flows are partial ordering of simulation events with the property that all the events are (transitively) a cause of the final event (the observable) and no trace containing a strict subset of these events may still contain the observable.

4   Results

A Kappa Model of Base Excision Repair Including Protein-DNA Interactions. We have assembled the first executable model of BER that incorporates protein-protein and protein-DNA interactions as well as enzymatic activity on DNA substrates. The actual model contains the interaction rules for 4 glycosylases (UDG, TDG, NEIL and OGG1); the AP-endonuclease (APE1) and the end cleaning enzyme PNKP; the polymerase (POLβ) and the ligase (LIG3). The interactions with scaffold protein XRCC1 is also included. Various DNA substrates for initial conditions can be generated using a python script.

The complete model as well as python and json configuration files are accessible as a github repository[6] and can be tested under the requirements specified in the Methods section. The repository also contains the complete listing of rules with a short description, separated by file[7]. The Kappa syntax, used to define the rules is described in Supp. data.

The rules of the model essentially implement the catalytic activities reported in Fig. 2 as well as the interactions depicted in Fig. 3 using the encoding of DNA shown in Fig. 4.

We used the module sanity.ka to detect invariant violation during the elaboration of the model. We included this file in the model repository because it can be used to test further invariants. The idea is to write rules of the form $I \rightarrow I + Err()$ where I is an invariant violation (for instance an invalid DNA polymer) and Err() is an "error" protein. We can then use the causality analysis features of KASIM to have an explanation on how Err() (and hence the invariant) was created.

DNA Glycosylase. For the simulations presented in the present paper, we focused on the UDG glycosylase the behavior of which is similar to TDG, the other monofunctional glycosylase of the pathway (although rates differ greatly, TDG having a very slow chemical step, followed by a strong product binding).

The rules pertaining to glycosylases interactions are given in the DG.ka and sliding.ka files of the model repository. We describe below the main ones, and we give their graphical description in Fig. 5.

Glycosylases are assumed to use *facilitated diffusion* on DNA (i.e. a random walk on DNA) to find mismatches at a rate that exceeds what can be achieved by mere random binding after diffusion in the nucleus (see for instance [15]).

Facilitated diffusion can be simply modeled by a rule that enables the glycosylase to "jump" to the next base 3' or 5' to its current position:

```
'slide 3' DG(dbd!1, cat), DNA(dg!1,e3!2), DNA(dg , e5!2) -> \
            DG(dbd!1, cat), DNA(dg , e3!2), DNA(dg!1,e5!2) @ 'DG_DNA_slide'
'slide 5' DG(dbd!1, cat), DNA(dg , e3!2), DNA(dg!1,e5!2) -> \
            DG(dbd!1, cat), DNA(dg!1,e3!2), DNA(dg , e5!2) @ 'DG_DNA_slide'
```

---

[6] https://github.com/ramdiv/ber-model
[7] https://github.com/ramdiv/ber-model/blob/master/list.md

Table 1. Kappa files of the model. The notation $(+x)$ indicates additional rules not used in simulations.

File name	rules #	binding and **catalytic activity**
DG.ka	13 (+7)	**Glycosylase** activities for UDG, TDG, NEIL and OGG1
APE1.ka	11	AP, gaped and nicked DNA; **pho'diesterase**; **endonuclease**
POLb.ka	17 (+4)	gaped and nicked DNA; **dRP lyase**; **polymerase**
PNKP.ka	3	gaped DNA; **phosphatase**
XRCC1.ka	29 (+2)	XRCC1 (in xrcc_dimer.ka), APE1, POLb, PNKP, LIG3 and DNA
LIG3.ka	8	gaped and nicked DNA ; **ligase**
sliding.ka	2 (+1)	facilitated diffusion on DNA (approx. in alter_sliding.ka)
damage.ka	3	deamination and direct single strand break
sanity.ka	(+9)	sanity check

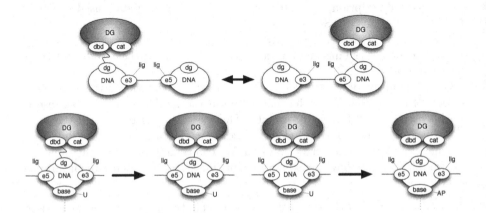

Fig. 5. Main DNA Glycosylase rules

Note that sliding is no longer possible when the cat port (representing the catalytic pocket of the DG) is bound (to a mismatch), as a consequence of the following rule:

```
'UDG anchors DNA mismatch' \
    DNA(e3~lig?, base~U?, dg!1, e5~lig?), DG(dbd!1, cat, type~U) -> \
    DNA(e3~lig?, base~U?, dg!1, e5~lig?), DG(dbd, cat!1, type~U) \
    @ 'DG_DNA_anchors'
```

Once anchored on the mismatch, the DG flip the faulty base into its catalytic pocket for excision:

```
'UDG mismatch excision' \
    DG(cat!1, type~U), DNA(e3~lig!_, dg!1, base~U?, e5~lig!_) -> \
    DG(cat!1, type~U), DNA(e3~lig!_, dg!1, base~AP?, e5~lig!_ ) \
    @ 'UDG_excision'
```

An important point of the above rule is that the DNA node is still ligated to the 3' and 5' neighboring base pairs after the nucleotide excision. This is a key difference with bi-functional glycosylases such as OGG that perform both excision and endonuclease in the same step (not included in simulations):

```
'OGG mismatch excision' \
    DG(dbd!1, cat,   type~OGG), DNA(e5~lig!0), \
    DNA(e3~lig!0,dg!1,base~oxoG?,e5~lig!2),DNA(e3~lig!2) -> \
    DG(dbd,   cat!1, type~OGG), DNA(e5~PUA!0), \
    DNA(e3~NA!0,  dg!1,base~AP?, e5~NA!2 ),DNA(e3~P!2) \
    @ 'OGG_excision'
```

Notice also that in both cases the DG remains bound to its product.

APE1 endonuclease. We present here the main rule pertaining to APE1 activity and give its graphical representation in Fig. 6.

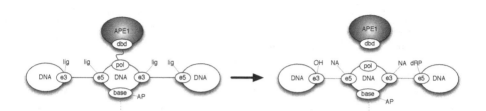

Fig. 6. APE1's endonuclease

Whenever APE1 is bound to DNA and encounters an AP site, it may incise DNA 5' of the damage. The resulting gaped site has a 3' OH and a 5'dRP residues:

```
'APE1 5-endonuclease' \
    APE1(dbd!1), DNA(e3~lig!0), \
    DNA(e5~lig!0, base~AP?, ape!1, e3~lig!2), DNA(e5~lig!2) -> \
    APE1(dbd), DNA(e3~OH !0), \
    DNA(e5~NA!0, base~AP?, ape, e3~NA!2), DNA(e5~dRP!2) \
    @ 'APE1_incision'
```

POLβ gap filling. The main rules of POLβ concern its gap filling activity. There are actually four variants of the rule presented below (see Fig. 7 for the graphical representation), one for each different nucleotide insertion.

```
'POLb polymerase A on gap' \
    POLb(dbd!1), DNA(e3~OH !0), \
    DNA(e5~NA!0,  pol!1, base~AP!2, e3~NA!_), DNA(base~T!2) -> \
    POLb(dbd ), DNA(e3~lig!0), \
    DNA(e5~lig!0, pol , base~A !2, e3~OH!_), DNA(base~T!2) \
    @ 'POLb_polymerase'
```

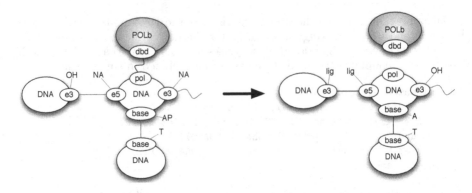

Fig. 7. Polymerase step: insertion of nucleotide with an adenine base

On uni-molecular binding rules. A fundamental assumption of the model is that uni-molecular binding events are fast. The underlying hypothesis is that steric constraints favor complex formation. This enables XRCC1 to reinforce the product inhibition following the scheme presented in Fig. 8 where k_{unary} is the uni-molecular binding rate. With a relatively weak affinity with enzyme A, and an equally moderate affinity with DNA (both have a strong $k_{off}[ns]$ - for non specific), XRCC1 is able to stabilize efficiently A on both its substrate and product, although A has little affinity for its product ($k_{off}[ns] > k_{off}[s]$).

XRCC1 Primarily Impacts Ligation Efficiency. There are a large number of possible *in silico* experiments one can try with our BER model. Since it is the first to incorporate protein-protein interactions, we naturally sought to study the role of the scaffolding protein XRCC1. As a first approach we ran a batch of simulations, under the conditions specified in the Methods section, with and without XRCC1. The simulations are denoted hereafter x^+ (with XRCC1) and x^- (without). Figure 9 shows the average plots for both x^+ and x^-. We first observed that both series of simulations were able to process the totality of initial damage (Fig. 9, left plot). However x^- exhibited a significant decrease in repair speed with respect to x^+, with 10 healed base pairs per second (maximal speed) vs. 17 healed base pairs per second (Fig. 9, left plot, small insert).

We then tried to narrow down the origin of that speed difference by decomposing the global repair into the 4 catalytic steps that follow the gylcosylase reaction (Fig. 9, right). No significant difference between x^- and x^+ could be observed for endonuclease (APE1), lyase and polymerase (POLβ) reactions. Therefore the only possible difference of total repair activity lies in the ligation step. These observations are consistent with the fact that XRCC1 is dispensable for complete repair in vitro [17] and that XRCC1 deficient cells are defective in processing nicked DNA intermediates [18], although the authors propose that this is due to a hypothetical stabilising effect of XRCC1 on LIG3.

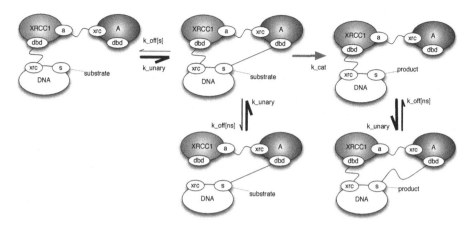

Fig. 8. XRCC1 stabilization ($k_{unary} \gg k_{off}[ns] > k_{off}[s]$)

Mechanistic Insights into the Transition from POLβ to LIG3. XRCC1 is commonly referred to as a scaffolding protein for BER enzymes. The intuitive idea is that XRCC1 maintains APE1, POLβ, and LIG3 at the lesion site, throughout the whole repair process. As suggested by our simulations, the scaffolding role of XRCC1 is unevenly distributed among its potential partners. In order to investigate whether BER enzymes are actually brought to DNA by XRCC1, we used KASIM's causal tracking mode for the ligation step. To do so, we analyzed 122 (compressed) causal flows generated by a ligase event, produced under the conditions described in the Methods section. According to the sample, approximately 80% of ligation events contained an action of XRCC1 in their causal history (data not shown). We therefore sought to analyze more in detail what the exact role of XRCC1 was in the ligation pathways.

The histogram of Fig. 10 indicates that nearly 45% of uni-molecular binding events occurring on DNA, that are in the causal past of a ligation event, correspond to the recruitment of LIG3 to the nicked DNA intermediate by XRCC1 (C). XRCC1 is also found recruiting POLβ to AP sites (B) and nicked DNA (A) in about 15% of the scenarios leading to ligation. Interestingly XRCC1 is recruited to DNA by APE1 (15%) and POLβ (7%) in a significant number of scenarios. The little impact of XRCC1 on the recruitment of APE1 (less than 1% of scenarios) to the lesion sites is likely due to the relatively low turnover rate of UDG coupled with the large amount of APE1 in the system which enable a smooth transition between UDG and APE1 over the AP substrate. A characteristic causal flow obtained from a simulation is given in Fig. 10 (right): nodes correspond to rule applications and arrows represent causality between them. As a labeling convention, X.Y indicates a complex formation between proteins X and Y and rectangular nodes indicate uni-molecular reactions that occurred under the scaffolding of XRCC1. Red nodes correspond to the chemical steps of the pathway.

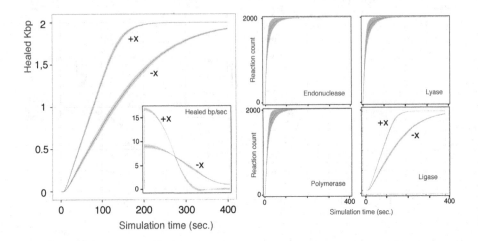

Fig. 9. (Left plot) Simulation efficiency with and without XRCC1, shadows indicate standard deviation. (Right plot) Efficiency of BER catalytic steps, with and without XRCC1. +x denotes the simulation with XRCC1 present and -x denotes the absence of it.

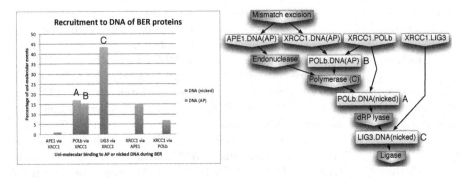

Fig. 10. (Left) Analysis of causal 122 flows exhibiting the intervention of XRCC1 and leading to a ligation event. (Right) A example of causal flow containing POLβ and LIG3 binding events *via* XRCC1.

Overall, causality analysis confirms the key role of XRCC1 in securing the transition between the polymerase step and ligation.

XRCC1 Limits Cytotoxicity BER. Since XRCC1 plays an active role in passing the baton between POLβ's products and LIG3's substrate, we wondered whether this coordination impacts on the cytotoxicity of BER. To do so we analyzed the amount of visible toxic DNA intermediates over time in x^+ and x^-. A toxic substrate is understood here as either an AP site, a gapped DNA or a nicked DNA node that is not bound by any BER protein. Figure 11 shows the amount of total nicked DNA that is present *in silico* over the duration of BER (Left plot). As expected, x^+ and x^- produce approximatively the same amount of nicked intermediates in the pre-steady state phase, since the ligation step is rate limiting ($k_{cat} = 0.04s^{-1}$ for LIG3 which is half the speed of the second slowest reaction, see Supp. Data A.2). However x^+ has an apparent faster rate for processing nicked DNA. Interestingly this results in a much higher cytotoxicity of x^- simulations (Fig. 11, right plot) which is almost entirely caused by unprotected nicked DNA (data not shown). This is consistent with the experimentally observed sensitivity of XRCC1 mutant cells to induced DNA damage [11].

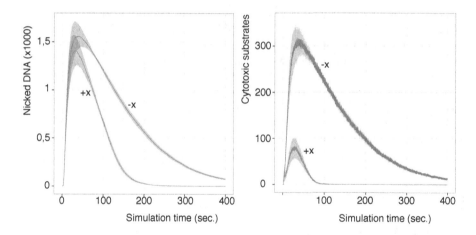

Fig. 11. Nicked DNA intermediates (left) and cytotoxic substrates in the presence and absence of XRCC1. +x denotes the simulation with XRCC1 present and -x denotes the absence of it.

A Tradeoff between Accuracy and Efficiency Under Varying Amount of POLβ. The nucleus is a very crowded medium and the local chromatin state can induce local concentration effects [11]. We thus investigated further the role of XRCC1 under decreasing amount of available POLβ (Figure 12). These experiments showed that when the system is moderately deprived of POLβ (up to 1/4 dilution of the default amount), XRCC1 contributes to maintaining a fast

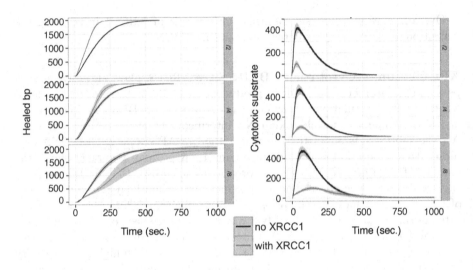

Fig. 12. Repair efficiency and cytotoxicity of simulations with increasing dilution of POLβ (/2, /4 and /8 with respect to default amount)

repair rate by holding LIG3 at lesion sites, waiting for the product of POLβ's reactions. Surprisingly, at higher dilution of POLβ, the price of this coordination becomes rate limiting for the overall repair speed. This is likely due to LIG3 being sequestered too long on AP sites in the absence of POLβ when ligatable substrates are elsewhere available. Importantly, under such extremal conditions XRCC1 still actively limits the amount of cytotoxic substrates (mainly in the form of nicked DNA) available to the environment.

This observation could imply that partial mutant cells (XRCC1$^-$,POLβ^-) would have a global faster repair activity than simple (XRCC1$^+$,POLβ^-) mutants, although with a very likely higher sensitivity to damage.

5   Discussion

A Kappa Model of BER. We have collated a set of mechanisms of action pertaining to BER, as a set of Kappa rules. It results in an executable model of DNA repair that can be used to test various hypothesis on DNA repair mechanisms. As an instance of such applications, we have investigated the impact of the scaffolding protein XRCC1 on repair activity. Consistent with experimental observations, our model shows that complete repair can be achieved in the absence of XRCC1. Furthermore the model successfully predicts the impact of XRCC1 on BER, the absence of which resulted in the accumulation of unprotected nicked DNA intermediates and an impaired repair speed in simulations. Beyond available experiments, our model indicates that XRCC1 might be recruited by APE1 to the damage site. It would afterward proceed with the recruitment of POLβ and, to a higher degree, LIG3. Eventually we showed that

XRCC1 contributes to the robustness of BER with respect to large variation of POLβ's concentration, preserving the repair efficiency up to a certain dilution level (/4), and maintaining a low amount of cytotoxic substrate over time.

Towards a Comprehensive Model of BER. More studies on the dynamics of the present model can be performed, beyond the scope of this paper. Also more biological facts need to be incorporated. Among them, the addition of PARP1, that plays an important role in an alternative way to recruit BER enzymes to the damaged site, seems to be a priority. We would also like to model the alternative *long patch repair* which occurs when BER is unable to produce a ligatable substrate. It will be particularly interesting to see how XRCC1 can regulate the switch between long patch and short patch BER, as experimental studies indicate.

Perspectives. Our stance is to take both qualitative and quantitative data seriously and collate them into a comprehensive model. We believe this model can be used to raise challenges to the biologist community about missing information and also highlight key points where the DNA repair community agrees or disagrees. We also believe that a comprehensive model can be used to make predictions on possible experiments and help the biologists to explore the wet lab perturbation space in a rational manner.

References

1. Faeder, J.R., Blinov, M.L., Hlavacek, W.S.: Rule based modeling of biochemical networks. In: Complexity, pp. 22–41 (2005)
2. Danos, V., Feret, J., Fontana, W., Harmer, R., Krivine, J.: Rule-based modelling of cellular signalling. In: Caires, L., Vasconcelos, V.T. (eds.) CONCUR 2007. LNCS, vol. 4703, pp. 17–41. Springer, Heidelberg (2007)
3. Danos, V., Feret, J., Fontana, W., Krivine, J.: Scalable simulation of cellular signaling networks. In: Shao, Z. (ed.) APLAS 2007. LNCS, vol. 4807, pp. 139–157. Springer, Heidelberg (2007)
4. Sneddon, M.W., Faeder, J.R., Emonet, T.: Efficient modeling, simulation and coarse-graining of biological complexity with NFsim. Nature Methods 8, 177–183 (2011)
5. Hedge, M.L., Hazra, T.K., Mitra, S.: Early steps in the DNA base excision/single-strand interruption repair parthway in mammalian cells. Cell Research 18, 27–47 (2008)
6. Kim, Y.J., Wilson III., D.M.: Overview of base excision repair biochemistry. Curr. Mol. Pharamacol. 5, 3–13 (2012)
7. Sokhansanj, B.A., Rodrigue, G.R., Fitch, J.P., Wilson III., D.M.: A quantitative model of human DNA base excision repair. i. mechanistic insights. Nucleic Acids Research 30, 1817–1825 (2002)
8. Sokhansanj, B.A., Wilson III, D.M.: Estimating the effect of human base excision repair protein variants on the repair of oxidative DNA base damage. Cancer Epidemiol. Biomarkers Prev. 15, 1000–1008 (2006)

9. Prasad, R., Shock, D.D., Beard, W.A., Wilson, S.H.: Substrate channeling in mammalian base excsion repair pathways: Passing the baton. Journal of Biological Chemistry 285, 40479–40488 (2010)

10. Nagaraj, N., Wisniewski, J.R., Geiger, T., Cox, J., Kircher, M., Kelso, J., Pääbo, S., Mann, M.: Deep proteome and transcription mapping of a human cancer cell line. Mol. Sys. Biol. 7 (2011)

11. Lan, L., Nakajima, S., Oohata, Y., Takao, M., Okano, S., Masutani, M., Wilson, S.H., Yasui, A.: In: situ analysis of repair processes for oxidative DNA damage in mammalian cells. PNAS 101, 13738–13743 (2004)

12. Strauss, P.R., Beard, W.A., Patterson, T.A., Wilson, S.H.: Substrate Binding by Human Apurinic/Apyrimidinic Endonuclease Indicates a Briggs-Haldane Mechanism. Journal of Biological Chemistry 272, 1302–1307 (1997)

13. Maher, R.L., Bloom, L.B.: Pre-steady-state Kinetic Characterization of the AP Endonuclease Activity of Human AP Endonuclease. Journal of Biological Chemistry 282, 30577 (2007)

14. Fitzgerald, M.E., Drohat, A.C.: Coordinating the initial steps of base excision repair. Journal of Biological Chemistry 47, 32680 (2008)

15. Blainey, P.C., Oijen, A.M.c., Banerjee, A., Verdine, G.L., Xie, X.S.: A base-excision DNA-repair protein finds intrahelical lesion bases by fast sliding in contact with DNA. PNAS 103, 5752–5757 (2006)

16. Danos, V., Feret, J., Fontana, W., Harmer, R., Hayman, J., Krivine, J., Thompson-Walsh, C., Winskel, G.: Graphs, Rewriting and Pathway Reconstruction for Rule-Based Models. In: FSTTCS 2012. LIPIcs (2012)

17. Kubota, Y., Nash, R.A., Klungland, A., Schär, P., Barnes, D.E., Lindahl, T.: Reconstitution of DNA base excision-repair with purified human proteins: interaction between DNA polymerase β and the XRCC1 protein. The EMBO Journal 15, 6662–6670 (1996)

18. Cappelli, E., Taylor, R., Cevasco, M., Abbondandolo, A., Caldecott, K.W., Frosina, G.: Involvement of XRCC1 and DNA ligase III gene products in DNA base excision repair. Journal of Biological Chemistry 272, 23970–23975 (1997)

A   Supplementary Data

A.1   The Kappa Language

We adapt here the presentation of Kappa, given in KaSim's manual[8].

General Remarks. The *Kappa File* (KF) is the formal representation of a model. We use KF to denote the union of the files that are given as input to KaSim (argument -i). Each line of the KF is interpreted by KaSim as a *declaration*. If the line is ended by the escape character '\' the continuation of the declaration is parsed onto the next line. Declarations can be: agent *signatures, rules, variables, initial conditions, perturbations* and *parameter configurations*. The KF's structure is quite flexible and can be divided in any number of subfiles in which the order of declarations does not matter (to the exception of variable declarations). Comments can be used by inserting the marker # that tells KaSim to ignore the rest of the line.

Agent Signature. In Kappa there are two entities that can be used for representing biological elements: *agents* and *tokens* (we don't consider token here). Agents are used to represent complex molecules that may bind to other molecules on specific sites.

In order to use agents in a model, one needs to declare them first. *Agent signatures* constitute a form of typing information about the agents that are used in the model. It contains information about the name and number of interaction sites the agent has, and about their possible internal states. A signature is declared in the KF by the following line:

%agent: *signature_expression*

according to the grammar given Table 2 where terminal symbol are denoted in (blue) typed font. Symbol Id can be any string generated by regular expression $[a-z\,A-Z][a-z\,A-Z\,0-9\,\_\,-\,+]^*$. Terminal symbol ε stands for the empty symbol.

Table 2. Agent signature expression

signature_expression	::= Id(*sig*)
sig	::= Id *internal_state_list, sig* | ε
internal_state_list	::= ~Id *internal_state_list* | ε

For instance the line:

%agent: A(x,y~u~p,z~0~1~2) # Signature of agent A

---

[8] http://www.pps.univ-paris-diderot.fr/~jkrivine

will declare an agent A with 3 *(interaction) sites* x,y and z with the site y possessing two *internal states* u and p (for instance for the unphosphorylated and phosphorylated forms of y) and the site z having possibly 3 states respectively 0, 1 and 2. Note that internal states values are treated as untyped symbols by KASIM, so choosing a character or an integer as internal state is purely matter of convention.

Rules. Once agents are declared, one may add to the KF the rules that describe their dynamics through time. A *pure rule* looks like:

'my rule' *kappa_expression* → *kappa_expression* @ *rate*

where 'my rule' can be any name that will refer to the subsequent rule that can be decomposed into a *left hand side* (LHS) and a *right hand side* (RHS) kappa expressions together with a *kinetic rate expression* . One may also declare a *bi-directional rule* using the convention:

'bi-rule' *kappa_expression* ↔ *kappa_expression* @ *rate⁺,rate⁻*

Note that the above declaration corresponds to writing, in addition of 'my-rule', a backward rule named 'my rule_op' which swaps left hand side and right hand side, and with rate *rate⁻*.

Kappa and rate expressions are generated by the grammar given in Table 3.

Table 3. Kappa expressions

kappa_expression	::= *agent_expression* , *kappa_expression* | ε
agent_expression	::= Id(*interface*)
interface	::= ε | Id *internal_state link_state*
internal_state	::= ε | ˜Id
link_state	::= ε | !n | !_ | ?
token_name	::= Id
rate_expression	::= *algebraic_expression*
	| *algebraic_expression* (*algebraic_expression*)

Table 4. Algebraic expressions

algebraic_expression	::= $x \in \mathbb{R}$ | variable
	| *algebraic_expression* binary_op *algebraic_expression*
	| unary_op (*algebraic_expression*)

Simple Rules. With the signature of A defined in the previous section, the line

'A dimerization' A(x),A(y~p) → A(x!1),A(y~p!1) @ γ

denotes a dimerization rule between two instances of agent A provided the second is phosphorylated (say that is here the meaning of p) on site y. Note that the bond between both As is denoted by the identifier !1 which uses an arbitrary integer (!0 would denote the same bond). In Kappa, a bond may connect exactly 2 sites so any occurrence of a bond identifier !n has to be paired with exactly one other sibling in the expression. Note also the fact that site z of A is not mentioned in the expression which means that it has no influence on the triggering of this rule. This is the *don't care don't write convention* (DCDW) that plays a key role in resisting combinatorial explosion when writing models.

Adding and Deleting Agents. Sticking with A's signature, the rule

'budding A' A(z) → A(z!1),A(x!1) @ γ

indicates that an agent A free on site z, no matter what its internal state is, may beget a new copy of A bound to it *via* site x. Note that in the RHS, agent A's interface is not completely described. Following the DCDW convention, KaSim will then assume that the sites that are not mentioned are created in the *default state*, *i.e* they appear free of any bond and their internal state (if any) is the first of the list shown in the signature (here state u for y and 0 for z).

Importantly, KaSim respects the *longest prefix convention* to determine which agent in the RHS stems from an agent in the LHS. In a word, from a rule of the form $a_1, \ldots, a_n → b_1, \ldots, b_k$, with a_is and b_js being agents, one computes the biggest indices $i \leq n$ such that the agents $a_1, \ldots, a_i$ are pairwise consistent with $b_1, \ldots, b_i$, *i.e* the a_js and b_js have the same name and the same number of sites. In which case we say that the for all $j \leq i$, a_j is *preserved* by the transition and for all $j > i$, a_j is *deleted* by the transition and b_j is *created* by the transition. This convention allows us to write a deletion rule as:

'deleting A' A(x!1),A(z!1) → A(x) @ γ

which will remove the A agent in the mixture that will match the second occurrence of A in this rule.

Side Effects. It may happen that the application of a rule has some *side effects* on agents that are not mentioned explicitly in the rule. Consider for instance the previous rule:

'deleting A'A(x!1),A(z!1) → A(x) @ γ

The A in the graph that is matched to the second occurrence of A in the LHS will be deleted by the rule. As a consequence all its sites will disappear together with the bonds that were pointing to them. For instance, when applied to the graph

G =A(x!1,y~p,z~2),A(x!2,y~u,z~0!1),C(t!2)

the above rule will result in a new graph G' = A(x!1,y~p,z~2),C(t) where the site t of C is now free as side effect.

Wildcard symbols for link state ? (for bound or not), !_ (for bound to someone), may also induce side effects when they are not preserved in the RHS of a rule, as in

'Disconnect A'   A(x!_) → A(x) @ γ

or

'Force bind A'   A(x?) → A(x!1),C(t!1) @ γ

Rates As said earlier, Kappa rules are equipped with *kinetic rate(s)*. A rate is a real number, or an algebraic expression evaluated as such, called the *individual-based or stochastic rate constant*, it is the rate at which the corresponding rule is applied per instance of the rule. Its dimension is the inverse of a time $[T^{-1}]$.

The stochastic rate is related to the *concentration-based rate constant* k of the rule of interest by the following relation:

$$k = \gamma(\mathcal{A}\,V)^{(a-1)} \tag{1}$$

where V is the volume where the model is considered, $\mathcal{A} = 6.022 \cdot 10^{23}$ is Avogadro's number, $a \geq 0$ is the arity of the rule (*i.e* 2 for a bimolecular rule).

In a modeling context, the constant k is typically expressed using *molars* $M := moles\,l^{-1}$ (or variants thereof such as μM, nM), and seconds or minutes. If we choose molars and seconds, k's unit is $M^{1-a}s^{-1}$, as follows from the relation above.

Concentration-based rates are usually favored for measurements and/or deterministic models, so it is useful to know how to convert them into individual-based ones used by KASIM.

A.2   The Kinetic Rates of the BER model

Complex Formation Rates. The contact map illustrated Fig. 3 is derived from the papers the references of which are listed in Table 5. Italic fonts denote qualitative studies. Question marks denote postulated interactions without explicit references.

Catalytic Rates. As pointed out in Section 3, the catalytic rate of an enzyme, k_{cat}, is usually given in terms of steady state kinetics, i.e following the scheme:

$$E + S \leftrightarrow ES \rightarrow_{k_{cat}}^* E + P$$

But as we pointed out, in the presence of non negligible product inhibition, enzymatic activity is better accounted for using the scheme:

$$E + S \leftrightarrow ES \rightarrow_{k_{chem}} EP \leftrightarrow E + P$$

Table 6 gives the catalytic rates and the reference paper(s) that were used in our model. Both k_{cat} and k_{chem} are given when available.

Table 5. Protein-DNA and protein-XRCC1 interactions. (†) Private conversation with Dr. S. H. Wilson (NIH).

Protein	DNA duplex	Mismatch	AP site	Gaped DNA	Nicked DNA	XRCC1
XRCC1	[1]	?	*[2]*	[1]	[1]	[1]
APE1			[3], [4], [5]	[5]	(†)	*[6]*
UDG	[7]	[7]	[7]			
TDG	[8]	[8]	[8]			
PNKP				[9]		[10]
POLβ				[11]	(†)	[12]
LIG3				[13]	?	*[14]*

Table 6. Catalytic rates used in the model. The notation [15] (from Ref X) indicates that the number comes from reference X of paper [15].

Protein	k_{chem} (s^{-1})	k_{cat} (s^{-1})	Reference
APE1 (3' PUA cleaning)		0.05	[15] (from Ref. 60)
APE1 (Endonuclease)	1000	3	[15], [16]
PNKP		0.14	[17]
LIG3 (Ligase)		0.04	[15] (from Refs. 56 and 63)
POLβ (3' dRP cleaning)		0.075	[15] (from Ref. 62)
POLβ (gap filling)	10	0.45	[11], [15] (from Refs. 29 and 61)
TDG	0.03		[8]
UDG		15	[18]

Default Rates. When no quantitative data is known, we used "realistic" default rates that are randomized at each simulation from the intervals presented in the table below:

process	interval rate	
general bi-molecular binding	$[10^7 - 10^9]$	$M^{-1}s^{-1}$
general uni-molecular binding	$[10 - 10^4]$	s^{-1}
general unbinding	$[10^{-3} - 10^{-1}]$	s^{-1}

A.3   Simulation Efficiency

The efficiency of an *in silico* experiment with respect to a wet lab experiment is usually measured in terms of time and money consumption. It is interesting to check, for a given model, how long it takes (in CPU seconds) to simulate one (biological world) second of the real system. Fig. 13 shows the evolution of the efficiency of one simulation running with the parameters specified in the Methods section.

A good measure of the efficiency of a simulation at time t can be given by $eff(t) \stackrel{def}{=} \frac{dCPU_{time}(t)}{dt}$. As can be seen in Fig. 13, $eff(t)$ has three distinguished phases during which it becomes quasi-linear. The three phases correspond to the pre-steady state ($eff(t) \sim 15$ CPU seconds for 1 bio second) and steady

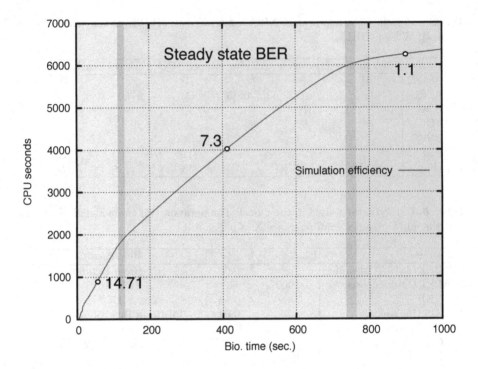

Fig. 13. A global picture of the CPU cost of a BER simulation. The derivative at distinguished points are given (pre-steady state, steady state and post BER).

state of BER ($\textit{eff}(t) \sim 7$ CPU seconds for 1 bio second) and the phase at which no more repair is conducted ($\textit{eff}(t) \sim 1$ CPU seconds for 1 bio second).

Importantly *in silico* BER efficiency varies quite a lot as different simulations are run with randomized dissociation constants, though they still exhibit the same distinct phases (data not shown).

References Listed in Supp. Data A.2

1. Mani, R.S., et al.: Biophysical characterization of Human XRCC1 and its binding to damaged and undamaged DNA. Biochemistry 43, 16505–16514 (2004)
2. Nazarkina, Z.K., et al.: XRCC1 interactions with base excision repair DNA intermediates. DNA Repair 5, 254–264 (2007)
3. Wilson III, D.M., et al.: Abasic site binding by the human apurinic endonuclease, Ape, and determination of the DNA contact sites. Nucleic Acids Research 25(5), 933–939 (1997)
4. Maher, R.L., Bloom, L.B.: Pre-steady-state Kinetic Characterization of the AP Endonuclease Activity of Human AP Endonuclease. J. Biol. Chem. 282(42), 30577 (2007)
5. Strauss, P.R., et al.: Substrate Binding by Human Apurinic/Apyrimidinic Endonuclease Indicates a Briggs-Haldane Mechanism. J. Biol. Chem. 272(2), 1302–1307 (1997)
6. Vidal, A.E., et al.: XRCC1 coordinates the initial and late stages of DNA abasic site repair through protein-protein interactions. EMBO 20(22), 6530–6539 (2001)
7. Parikh, S.S., et al.: Base excision repair initiation revealed by crystal structures and binding kinetics of human uracil-DNA glycosylase with DNA. EMBO 17(17), 5214–5226 (1998)
8. Morgan, M.T., et al.: Stochiometry and affinity for the thymine DNA glycosylase binding to specific and non specific DNA. Nucleic Acids Research 39(6), 2319–2329 (2011)
9. Weinfeld, M., et al.: Tidying up loose ends: the role of polynucleotide kinase/phosphatase in DNA strand break repair. Trends Biochem. Sci. 36(5), 262–271 (2011)
10. Ammar, A.E.A., et al.: Specific recognition of a multiply phosphorylated motif in the DNA repair scaffold XRCC1 by the FHA domain of human PNK. Nucleic Acids Research 37(5), 1701–1712 (2009)
11. Vande Berg, B.J., et al.: DNA Structure and Aspartate 276 Influence Nucleotide Binding to Human DNA Polymerase β. J. Biol. Chem. 276(5), 3408–3416 (2001)
12. Cuneo, M.J., London, R.E.: Oxidation state of the XRCC1 N-terminal domain regulates DNA polymerase β binding affinity. PNAS 107(15), 6805–6810 (2010)
13. Leppard, J.B., et al.: Physical and Functional Interaction between DNA Ligase III and Poly(ADP-Ribose) Polymerase 1 in DNA Single-Strand Break Repair. Mol. Cell. Biol. 23(16), 5919–5927 (2003)
14. Caldecott, K.W., et al.: An Interaction between the mammalian DNA repair protein XRCC1 and DNA ligase III. Mol. Cell. Biol. 14(1), 68–76 (1994)
15. Sokhansanj, B.A., et al.: A quantitative model of human DNA base excision repair. I. Mechanistic insights. Nucleic Acids Research 20(8), 1817–1825 (2002)
16. Carey, D.C., Strauss, P.R.: Human Apurinic/Apyrimidic Endonuclease is processive. Biochem. 38, 16553–16560 (1999)
17. Das, U., Shuman, S.: Mechanism of RNA 2',3'-cyclic phosphate end healing by T4 polynucleotide kinase–phosphatase. Nucleic Acids Research 41(1), 355–365 (2012)
18. Jiang, Y.L., Stivers, J.T.: Reconstructing the Substrate for Uracil DNA Glycosylase: Tracking the Transmission of Binding Energy in Catalysis. Biochem. 40, 7710–7719 (2001)

Using Process Algebra to Model Radiation Induced Bystander Effects

Rachel Lintott[1], Stephen McMahon[2], Kevin Prise[2] Celine Addie-Lagorio[1], and Carron Shankland[1]

[1] Computing Science and Mathematics, University of Stirling, Stirling FK9 4LA, UK
[2] Centre for Cancer Research and Cell Biology, Queen's University Belfast, Belfast B79 7BL, UK

Abstract. Radiation induced bystander effects are secondary effects caused by the production of chemical signals by cells in response to radiation. We present a Bio-PEPA model which builds on previous modelling work in this field to predict: the surviving fraction of cells in response to radiation, the relative proportion of cell death caused by bystander signalling, the risk of non-lethal damage and the probability of observing bystander signalling for a given dose. This work provides the foundation for modelling bystander effects caused by biologically realistic dose distributions, with implications for cancer therapies.

1 Introduction

Radiation is often referred to as a double edged sword [14]. Whilst it is one of the most effective treatments for several forms of cancer, exposure to radiation can also cause damage to healthy cells leading to long term side effects for the patient. Radiotherapy has been used to treat cancer for over a century, with over half of all modern day patients receiving this treatment at some point [15]. Due to the inherent risks associated with radiotherapy treatments, there is a constant drive to understand the resulting physical and biological processes in order to reduce exposures, both in terms of area exposed, and dose delivered. For many years it has been thought that radiation causes damage to biological cells through direct damage to the targeted DNA. However, over the past 20 years, experimental evidence has been shown to suggest subsequent, non-targeted effects of radiation. It has been suggested recently [1] that these non-targeted effects are mediated through cellular signalling. These effects fall into three distinct groups: cohort effects, mediated via gap junctions between neighbouring cells, bystander effects between nearby cells mediated by release of signals into a shared medium, and abscopal, or long range effects in distant tissues [2]. In this study we consider only the second of these effects, radiation induced bystander effects (RIBEs). Cells which have suffered DNA damage due to radiation release signals, in the form of reactive oxygen or nitrogen species [1]. These can be transmitted to surrounding cells via dispersal into the extracellular medium. These signalling molecules are then able induce damage responses in

P. Mendes et al. (Eds.): CMSB 2014, LNBI 8859, pp. 196–210, 2014.

surrounding cells (Figure 1). It has been suggested that RIBEs may be used to amplify the cell killing effect of radiation [23] and hence lead to a reduction in potentially dangerous levels of exposure during treatment. However, RIBEs may also contribute to the increased risk faced by low level exposures to radiation such as those experienced during space travel [1]. These effects therefore have a significant impact on our understanding of the biological response to radiation. Mathematical and computational modelling is a useful tool in understanding the mechanisms at work.

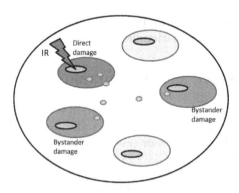

Fig. 1. Basic mechanisms of radiation induced bystander effects. Direct irradation causes damage to the cellular DNA which leads to the production of reactive oxygen or nitrogen species (ROS/RNS). These are released by the cell and can cause damage to neighbouring cells.

In recent years, several mathematical and computational models have been developed to describe aspects of this phenomenon [4,9,10,21]. Faria and Dickman [9] present an epidemic-type model of the damage caused to a population of cells by the spread of a decaying diffusing signal. Whilst this model has some interesting analytic results, such as the discovery of a critical threshold between the spreading and non-spreading phases of the signal, the assumptions used to develop it are necessarily simplifying, and their paper provides little biological verification. McMahon *et. al.* [21], on the other hand, presented a model based largely on the experimental approaches used to assess the impact of bystander effects.

In this paper we present a process algebra model, developed using Bio-PEPA [7], which combines the epidemic-type structure of Faria and Dickman [9] with the biologically verified mechanisms of McMahon *et. al.* [21]. Over recent years, process algebra has been increasingly used in a wide range of biological applications [25,13,18]. Bio-PEPA has been specifically developed for use in biological applications, and allows the user to define rates of reaction between distinct species in terms of their concentrations. The underlying continuous time Markov-Chain (CTMC) semantics allows this type of model to be solved either through stochastic simulation, for example using Gillespie's algorithm, or via conversion of the model to a deterministic system of ordinary differential equations. The level of abstraction afforded by Bio-PEPA is a key attraction to this computational method, allowing biological systems to be defined by a number of species, and the reactions between these species. The versatility in analysis techniques allows Bio-PEPA models to capture properties determined at an individual level, in a computationally less intensive way than traditionally defined individual based models. The range of applications for which this framework has been used has been expanded from its roots in biochemical networks [6] to areas such as epidemic modelling [18], crowd dynamics [19], and population modelling of aquatic invertebrates [25]. In this paper, we further extend this range to model cellular damage caused by radiation induced bystander effects.

2   Model Structure and Assumptions

The effectiveness of a treatment such as radiation is experimentally tested using a clonogenic survival assay [12], whereby a collection of cells is irradiated and then split up into individual cells or small clusters of cells and left to divide to form colonies. After a number of days (a week or more) the number of successful colonies (those which have grown to a stated size, usually around 50 cells) is counted and the survival fraction is calculated as

$$SF = \frac{\text{fraction of successful, treated colonies}}{\text{fraction of successful (untreated) colonies in control}}.$$

This ratio of treated to untreated colonies accounts for the fact that some colony death may be due to the experimental techniques. The structure our model is based on this experimental set-up and aims to predict the surviving fraction of cell colonies for a given radiation dose. In addition to this, we predict the proportion of cell colonies which have suffered some level of damage, but have maintained their capacity for division. It is these cells which have the potential to lead to long term problems for the patient, and hence this prediction gives a measure of risk for a given dose.

Modelling in Bio-PEPA requires the definition of three components; the model compartments, rates of reaction between these compartments, and the initial concentrations within each compartment. In the following sections we describe each of these components, with the Bio-PEPA model in full given in the appendix.

2.1   Model Compartments

The population considered is the fixed number, N, of possible cell colonies in an experiment. The colonies are subdivided depending on their damaged status. Prior to treatment, all cells will be classed as healthy (H), having suffered no damage from irradiation. Upon treatment, cells sustain a level of damage, as described in section 2.3. Depending on the initial dose, the direct damage suffered by a cell will be either a) sufficient to induce cell death, or b) sufficient to cause a 'recoverable' level of damage. Those cells which fall into category a) will enter the apoptotic class (A). Apoptosis is the process of controlled cell death or cell suicide. This leads the cell to break down and disperse in a controlled, and non-toxic way. Cells which suffer a recoverable level of damage enter into the infectious class (I). Both infectious and apoptotic cells are able to produce and emit bystander signal. The period of emission is limited by the processes of death and recovery. Apoptotic cells will cease signalling once they have broken down and enter the dead class (D), and infectious cells will cease emitting once they have entered the recovered class (R).

The compartments of the Bio-PEPA model are therefore given by the five damaged classes H,I,A,R,D, along with a class tracking the concentration of bystander signal, C. These compartments and the network of transitions between classes are shown in figure 2

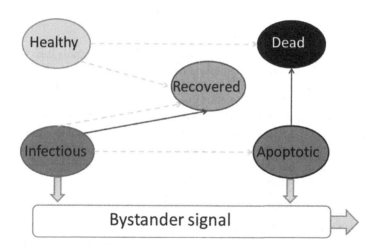

Fig. 2. Schematic showing compartments of the model and transitions between them. Thick arrows show signalling kinetics, dashed arrows show signal mediated transitions, and solid thin arrows show signal independent transitions.

2.2   Model Reactions

Cell colonies change state from one compartment to another due to a number of reactions. The functions defining these reactions are summarised in Table 1 and can be divided into three categories; bystander signal dynamics, damage suffered as a result of bystander signalling and natural cellular processes.

Table 1. Table of model reactions with their corresponding rates

Process Label	Description	Definition	Rate
emitI	emission of bystander signal by infectious cells	$C \to C + 1$	$\frac{\epsilon I}{V}(100 - C)$
emitA	emission of bystander signal by apoptotic cells	$C \to C + 1$	$\frac{\epsilon A}{V}(100 - C)$
decay	decay of bystander signal	$C \to C - 1$	δC
damageH	damage of healthy cells due to bystander signal	$H \to H - 1, R \to R + 1$	$\rho\beta f(C)H$
deathH	death of healthy cells due to bystander signal	$H \to H - 1, D \to D + 1$	$(1 - \rho)\beta f(C)H$
damageI	damage of infectious cells due to bystander signal	$I \to I - 1, R \to R + 1$	$\phi\beta f(C)I$
deathI	death of infectious cells due to bystander signal	$I \to I - 1, A \to A + 1$	$(1 - \phi)\beta f(C)I$
recovery	recovery of infectious cells	$I \to I - 1, R \to R + 1$	γI
apoptosis	death of apoptotic cells	$A \to A - 1, D \to D + 1$	αA

Bystander Signalling Dynamics

Experimental evidence suggests that irradiated cells emit signals into their environment which induce damage responses in nearby cells (refs). In the context of this model, irradiated cells are those which are either infectious (I) or apoptotic (A). The process of emission is defined in Table 1 as emitX (where X={I,A}), and is a decreasing function of the signal concentration, C. The concentration of bystander signal modelled here is not in fact the measurable signal quantity, rather the percentage of the saturated quantity. This normalised concentration brings our model in line with that of McMahon *et. al.*, and enables us to use published parameter values. Bystander signal decays at a rate proportional to its concentration, and defined by *decay* in Table 1.

Damage Induced by Contact with Signal

It has been suggested in the literature that bystander effects are an 'all or nothing' response [24], with damaging effects only observed at concentrations above a threshold $\hat{C}$. McMahon *et. al.* estimates this threshold to be around 21% of the saturated concentration. In order to capture this 'all or nothing' property, we use the Heaviside step function

$$\mathrm{h}(C) = \begin{cases} 0 & \text{if} \quad C < \hat{C} \\ 1 & \text{if} \quad C \geq \hat{C}. \end{cases} \tag{1}$$

This on/off switch activates the bystander signal, with no cellular response possible if the concentration is below this threshold. Active concentrations of bystander signal ($C \geq \hat{C}$) may cause novel damage to healthy cells (H) or subsequent damage to those directly damaged by irradiation (I). Damaging interactions occur at a rate $\beta X \mathrm{h}(C)$ (X={H,I}). If the signal concentration falls below this threshold quantity, the reaction rate falls to zero, and no further damage is done. Both healthy and infectious cells, when challenged with bystander signal, may either suffer sufficient damage to cause cell death, or may survive this challenge. When a healthy cell colony comes into contact with active bystander signal, a proportion, ρ, of these reactions will result in the survival of the colony, and these colonies will move to the recovered class. The rest of the reactions $(1 - \rho)$ will result in colony death, and the healthy colony will move to the dead class. A similar proportion, ϕ, of reactions between the infectious colonies and bystander signal, will lead to infectious colony survival, with $(1 - \phi)$ of these reactions resulting in colony death.

The proportion of bystander reactions resulting in colony death/survival will depend on the amount of direct damage the colony has suffered. Many recent modelling studies have formulated both direct, and bystander damage as a number of 'hits' [20] [21] [22] with cell death occurring at 5 'hits' or more. This formulation, whilst not strictly measurable is correlated with the number unrepaired or misrepaired DNA double-strand breaks suffered by cells, with approximately 5 or more of these lesions being fatal to the cell. The number of 'hits' suffered by a cell upon contact with bystander signal is assumed to be taken from a Poisson distribution, with mean given by the parameter H_b. This probabilistic approach to determining damage enables the model to capture the distribution of damaged or dead cells, and to account for the stochasticity and uncertainty in the underlying chemical kinetics without modelling these processes explicitly. This approach also allows us to reduce the parameters ρ and ϕ to functions of a single parameter H_b. These functions are given explicitly by

$$\rho = \frac{P(1) + P(2) + P(3) + P(4)}{1 - P(0)} \tag{2}$$

$$\phi = \frac{P(1)(I(1) + I(2) + I(3)) + P(2)(I(1) + I(2)) + P(3)(I(1))}{(1 - P(0))I_{init}/N} \tag{3}$$

where $P(i)$ is the probability of receiving i hits from a reaction with bystander signal, and $I(i)$ is the probability of receiving i hits from direct damage.

Natural Cellular Processes. The remaining reactions from Table 1 are 'apoptosis' and 'recovery'. These processes are not dependent on the concentration of bystander signal. The reaction 'apoptosis' describes the change of state of a cell colony from apoptotic to dead. The process of apoptosis or cell suicide can take several days to complete, and whilst the colony is clonogenically dead (i.e. unable to continue to grow and produce clones) these cells are still able to emit bystander signal. Without the inclusion of this class, the amount of signal produced, and the duration when the signal is active are underestimated. The reaction 'recovery' enables infectious cells to cease emitting bystander signal. The emission period has been shown to be limited, as damaged cells will begin repair processes. The emission period here is $1/\gamma$, where γ is dependent on the initial dose of radiation delivered. A high initial dose leads to a high γ, and therefore a slow rate of recovery. Higher doses allow bystander signal to be produced for longer periods by damaged cells.

2.3 Initial Concentrations

The final specifications necessary to complete our model are the initial species concentrations. In order to reduce complexity, and to focus our model on the dynamics of bystander effects, we assume that the direct damage due to the initial dose of radiation is fixed. Ionising radiation exposures such as those used in radiotherapy treatments fall into two main categories, electromagnetic (such as γ- or X-rays) or particulate (such as α particles). In both cases, damage to cells occurs due to the energy deposited within the cell, causing ionisations which can lead to either single or double strand breaks in the DNA. This energy deposition is dependent on the type of radiation to which cells are exposed. In order to provide a generic model of a non-specific radiation treatment, we again describe the damage done by a number of 'hits', λ, proportional to the radiation dose, D, delivered to the cells, hence $\lambda = sD$. If a population of cells is exposed to a uniform dose, then the number of 'hits' suffered by an individual cell is Poisson distributed with mean λ.

The initial states of the colonies, immediately following irradiation (time=0) is therefore determined as follows

$$
\begin{aligned}
H_{init} &= Ne^{-\lambda}, \\
I_{init} &= N\lambda e^{-\lambda}\left(1 + \tfrac{\lambda}{2!} + \tfrac{\lambda^2}{3!} + \tfrac{\lambda^3}{4!}\right), \\
A_{init} &= N - H_{init} - I_{init}, \\
R_{init} &= 0, \\
D_{init} &= 0.
\end{aligned}
\tag{4}
$$

These states are therefore inherently dependent on the dose delivered, with high initial doses leading to a high number of cells which are initially apoptotic and destined to die. Initially low doses lead to a higher number of initially healthy cells with some infected cells and fewer apoptotic. We would expect therefore that at high doses, the majority of the cell death observed will be due to the initial dose, with bystander effects being a predominantly low dose phenomenon.

2.4 Parameter Estimates

Our model requires the estimation of 7 parameters, given in Table 2. Since the dynamic behaviour of this model is based on the work of McMahon *et. al* [21], there is a correspondence between the parameters described, and the relevant estimates. In that paper, the model was parameterised against a number of cell lines, and experiments. Here we focus on the data obtained in [5] for the human prostate cancer cell line DU145. Where there is a correspondence in kinetic rates between our model and McMahon's, we use the estimate found in that paper. The parameter s gives the average number of hits per Gy obtained from the direct effect of radiation. Here we set this value to 1 which offers a better fit to the obtainable data than the McMahon *et. al.* estimate $s = 0.78$. This parameter is not easily measurable directly, and its closest biological interpretation is the

Table 2. Table of parameter values used in simulation results

Parameter Name	Symbol	Estimate
Hits per Gray	s	1 hit
Rate of bystander damage	β	0.0028 min^{-1}
Signalling duration	$1/\gamma$	61/Gy min Gy^{-1}
Rate of signal emission (at low concentration)	ϵ	0.00011 min^{-1}
Rate of signal decay	δ	0.019 min^{-1}
Hits per interaction with bystander	H_b	3.9 hits
Rate of death due to apoptosis	α	0.000417 min^{-1}

number of unrepaired DNA double strand breaks (DSB). For each Gray, this figure is of the order 1 DSB/Gy, and hence $s \approx 1$ is reasonable. The difference in s between our model and that of McMahon *et. al.* is due to the structure of the model. McMahon *et. al.* accounts for a heterogeneity in cell response to direct irradiation based on the cell cycle. We avoid this level of complexity by considering a coarser-grained, colony-level model. This difference in complexity accounts for the slight difference in this parameter estimate.

The rate of bystander damage, β is determined from the probability that a cell will be damaged through an encounter with bystander signal in a time interval δt. This probability is given in McMahon *et. al.* as

$$P_B = 1 - e^{-\beta \delta t}. \tag{5}$$

In order to translate this probability into a rate of damage over time, the exponential term is expanded following [16], both sides divided by the time interval δt and the limit taken as $\delta t \to 0$. This gives the rate of damage over time as

$$\frac{dP_B}{dt} = \beta \tag{6}$$

and the numerical estimate of this parameter can be taken directly from McMahon *et. al.*

The signalling duration, $1/\gamma$ translates to a rate of recovery for infectious cells of γ. This parameter is dependent on the initial dose of radiation: large initial doses lead to longer signalling duration and hence slower rate of recover. Signal emission is assumed to be a feature of cells which have been directly irradiated, hence both the signalling duration and the rate of emission, ϵ, should be the same for this model as for the McMahon model, regardless of scale, since the initial number of irradiated cells is the same. Hence we are justified in using the parameters found in McMahon *et. al.* The rate ϵ is the rate of emission at low signal concentrations, with this declining linearly to zero as the concentration reaches saturation. The mean number of 'hits' received by an encounter with bystander signal is given by $H_b = 3.9$. This parameter is the equivalent to the damage induced by direct effects, given by s, and is similarly slightly higher than that used by McMahon *et.al.* This discrepency is once again due to the difference in scales of the models. The final parameter to discuss is the rate of death due to apoptosis. We have estimated the duration of apoptosis to last an average of approximately 40 hours, leading to a rate of death of $\alpha = 0.000417$ min^{-1}.

3   Results

The model was simulated using Gillespies stochastic algorithm, for initial doses of between 0 and 8Gy, at intervals of 0.5 Gy. This range of initial doses relates to a range of initial conditions for the model to be tested under. For each initial dose, fifty simulations were run, and the 95% confidence intervals on the survival fraction were calculated. From these experiments, we found that the largest confidence interval of 0.95 ± 0.0058 was found at a dose of 1Gy. This is equivalent to a confidence interval of around 0.1% of the mean. Since the variation in simulation outputs is so low, the mean output of the simulations is used.

In order to validate the model, the output was compared to survival data from experiments published in [5]. This data is in the form of the surviving fraction of cell colonies, taken several days after treatment with ionising radiation. In order to simulate this, we ran the model for 10000 time steps, simulating approximately 7 days. By this point, the system had reached equilibrium, and simulation over longer periods provided negligibly different results. The survival fraction predicted by the model shows a good fit to the available data, with the predicted output falling within experimental confidence limits (Figure 3). Comparing the bio-PEPA model to that of McMahon *et. al.* we see that the McMahon *et. al.* predicts a lower survival fraction at low doses, and a higher survival fraction at higher doses.

Having validated the model, against experimental data, we are able to produce a number of predictive results. Figure 4 (a) shows the proportion of cell colonies killed by the relative effects of either direct irradiation or bystander effect. It is clear from this plot that for high doses, the proportion of cells killed by bystander signalling is small when compared to direct effects. Doses up to around 3Gy see a higher proportion of colony death being attributed to bystander effects than the direct effect of irradiation, hence it is in this range of exposures that bystander effects are most important. This result is significant since it is this range of dosages that is used in the clinical setting. High exposures of radiotherapy treatments are often delivered as fractionated doses with a typical fraction being around 2Gy. This same range of doses predicts the peak levels in the number of cell colonies which are assumed to be 'recovered' (Figure 4 (b)). These colonies have been damaged by the effects of radiation but maintain the ability to divide. This class of cells is particularly important when considering the negative, knock-on effects of radiotherapy, since these cells are more prone to detrimental mutations, possibly leading to long term problems for the health and well-being of the patient.

The Bio-PEPA Eclipse plug-in [8] allows direct calculation of probability and cumulative distribution functions for a specified target output. Due to the on/off switch in signal activity present in this model, these functions allow us to quantify the risk of seeing an actively damaging bystander effect for a given initial dose.

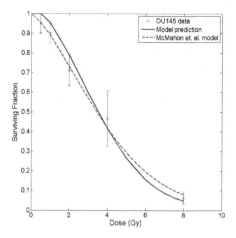

Fig. 3. Survival curve predicted by model and compared with data from Butterworth et al [5] with error bars, and model output from McMahon *et.al.* [21]. Model has been parameterised using McMahon *et. al.* for the human prostate cancer cell line DU145.

By specifying the target output as $h(C) = 1$, the PDF output gives the time at which each simulation reached this condition. Once more, we consider the range 0-8Gy, at intervals of 0.5Gy. For each dose considered, 1000 simulations were run for a length of 10000 time steps (minutes). The number of simulations performed in this case has been increased to allow greater accuracy, for each dose the time taken to simulate the model 1000 times was of the order of around 1 second. The time to bystander onset was calculated for each simulation, as well as the proportion of all simulations where bystander effect was observed. The median

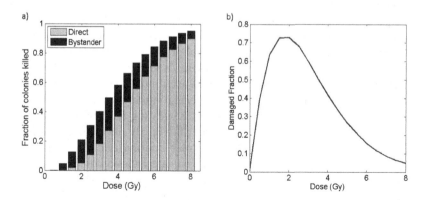

Fig. 4. Left hand plot: Proportion of colonies killed through direct effects and bystander signalling. Right hand plot: Recovered fraction of cell colonies. A peak in recovered colonies is observed at low initial doses.

time to bystander onset, along with 25% and 75% quartiles are shown in Figure 5. For the lowest dose considered (0.5Gy), only 0.6% of all simulations predicted a bystander response. At this dosage, the risk of producing enough bystander signal to cause damage or death to surrounding cells is therefore very low. At doses above 1.5Gy however, all simulations predicted a bystander response.. The median time between direct irradiation and bystander onset reduces with increasing dose, before saturating at a lower limit of around 12-13 minutes.

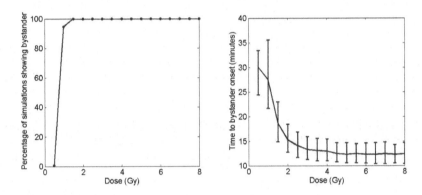

Fig. 5. Left plot shows the percentage of simulations resulting in the onset of bystander effect. Right plot shows median time to bystander onset, with error bars showing upper and lower quartiles

4   Discussion

In this paper we have presented a model of the *in vitro* consequences of radiation induced bystander effects. By basing our model structure on that of an epidemic type model, considering the transitions between 'infectious' classes, in combination with the biologically relevant assumptions of previous modelling work [21], we have been able to identify a number of key results. We have shown that the dose region up to around 3Gy presents a significant risk of bystander effect, both in terms of the proportion of death attributed to these effects (Figure 4) and the number of stable cell colonies suffering minor damage (Figure 4(b)). This result is in line with current literature, suggesting that bystander effects are potentially a major problem at low doses [3] [23]. By developing this model in Bio-PEPA we have easily been able to extract information about the risk of observing bystander effect for a given dose, given that emission of these signals is inherently a stochastic process. Along with an evaluation of the risk of bystander signalling, we have predicted the time between direct irradiation and bystander onset for a range of doses, showing that this time lag reaches at a minimum of around 12 minutes for doses above around 2Gy. Whilst this predictive result has not been verified experimentally, experimental data using different cell lines and radiotherapeutic delivery have shown that bystander signal can be detected on the order of 30 minutes or less [26].

The discovery of radiation induced bystander effects, in both the *in vitro* and *in vivo* settings, has important consequences for the future of radiation research. These effects must be taken into account when evaluating the risk of low level background exposures, and also when developing radiotherapeutic strategies for the treatment of cancer. Delivery of radiation doses and their effects can no longer be thought of as targeted point in time damage to specific cells. The number of cells killed by a dose of radiation has been modelled effectively for many years, primarily through the linear-quadratic model (LQ), and the biologically effective dose (BED) [11]. The LQ model offers a prediction of the fraction of cell kill for a single given dose of irradiation, whilst the BED converts the total dose delivered over a number of fractions (typically of around 2Gy) given at discrete times into the single dose required to achieve the equivalent cell kill. Whilst these methods of modelling cell death has been clinically effective, this approach neglects the non-lethal effects of bystander signalling. The model presented here shows a peak in the risk of bystander effects, leading to both cell death and damage, at around 2-3Gy. This result has been observed in previous experimental work [27].

Our prediction that the number of 'recovered' cell colonies peaks at clinically relevant doses suggests that dose fractionation may increase the risk of side effects due to bystander damage. These side effects may include long term mutations, leading, for example, to the development of tumours. The model presented here takes into account both the lethal and non-lethal effects of bystander signalling, this key property is novel to this approach, and has been facilitated by the use of Bio-PEPA in this modelling work.

The model presented here is a step in investigating of the impacts of bystander effects in clinical situations. To develop this work further, an in-depth analysis of the parameter estimates should be done with further experimentation done to validate our predictions, in particular those shown in Figure 5. In order to estimate experimentally unobtainable parameters, modelling in Bio-PEPA facilitates the use of evolutionary computation tools for model fitting [18].

The results shown here are specific to the treatment of the human prostate cancer cell line, with photon beam irradiation. Whilst this model and the mechanisms hypothesised fit this system relatively well, we can say nothing more general about the mechanisms acting in different cell lines, or in response to different radiotherapies such as targeted radionuclide therapy [17]. This model provides a first approximation of the mechanisms of bystander effects. This style of modelling enables us to test these assumed mechanisms against these other experimental settings ultimately enabling us to deduce general trends in the mechanisms of radiation induced bystander effects.

Acknowledgements. R. Lintott and C. Shankland are funded by the EPSRC project EP/K039342/1. C. Addie-Lagorio was funded by the Wellcome Trust Biomedical Vacation Scholarship.

References

1. Azzam, E.I., Toledo, S.M.d., Harris, A.L., Ivanov, V., Zhou, H., Amundson, S.A., Lieberman, H.B., Hei, T.K.: The ionizing radiation-induced bystander effect: evidence, mechanism, and significance. In: Pathobiology of Cancer Regimen-Related Toxicities, pp. 35–61. Springer (2013)
2. Blyth, B.J., Sykes, P.J.: Radiation-induced bystander effects: what are they, and how relevant are they to human radiation exposures? Radiat. Res. 176(2), 139–157 (2011)
3. Bonner, W.M.: Low-dose radiation: thresholds, bystander effects, and adaptive responses. Proceedings of the National Academy of Sciences 100(9), 4973–4975 (2003)
4. Brenner, D.J., Little, J.B., Sachs, R.K.: The bystander effect in radiation oncogenesis: Ii. a quantitative model. Radiat. Res. 155(3), 402–408 (2001)
5. Butterworth, K.T., McGarry, C.K., Trainor, C., O'Sullivan, J.M., Hounsell, A.R., Prise, K.M.: Out-of-field cell survival following exposure to intensity-modulated radiation fields. Int. J. Radiat. Oncol. 79(5), 1516–1522 (2011)
6. Ciocchetta, F., Duguid, A., Guerriero, M.L.: A compartmental model of the camp/pka/mapk pathway in bio-pepa. In: The 3rd Workshop on Membrane Computing and Biologically Inspired Process Calculi, Electronic Proceedings in Theoretical Computer Science. Citeseer (2009)
7. Ciocchetta, F., Hillston, J.: Bio-pepa: A framework for the modelling and analysis of biological systems. Theor. Comput. Sci. 410(33), 3065–3084 (2009)
8. Duguid, A., Gilmore, S., Guerriero, M.L., Hillston, J., Loewe, L.: Design and development of software tools for bio-pepa. In: Winter Simulation Conference, pp. 956–967 (2009)
9. Faria, F.P., Dickman, R.: An epidemic process mediated by a decaying diffusing signal. J. Stat. Mech-Theory E (06), P06006 (2012)
10. Faria, F.P., Dickman, R., Moreira, C.H.: Models of the radiation-induced bystander effect. International Journal of Radiation Biology 88(8), 592–599 (2012)
11. Fowler, J.F.: 21 years of biologically effective dose. Brit. J. Radiol. 83, 554–568 (2010)
12. Franken, N.A., Rodermond, H.M., Stap, J., Haveman, J., Van Bree, C.: Clonogenic assay of cells in vitro. Nature Protocols 1(5), 2315–2319 (2006)
13. Guerriero, M.L., Pokhilko, A., Fernández, A.P., Halliday, K.J., Millar, A.J., Hillston, J.: Stochastic properties of the plant circadian clock. Journal of The Royal Society Interface, rsif20110378 (2011)
14. Hei, T.K., Zhou, H., Ivanov, V.N., Hong, M., Lieberman, H.B., Brenner, D.J., Amundson, S.A., Geard, C.R.: Mechanism of radiation-induced bystander effects: a unifying model. Journal of Pharmacy and Pharmacology 60(8), 943–950 (2008)
15. Joiner, M.C., van der Kogel, A.: Basic Clinical Radiobiology, 4th edn. CRC Press (2009)
16. Keeling, M.J., Rohani, P.: Modeling infectious diseases in humans and animals. Princeton University Press (2008)
17. Mairs, R.J., Fullerton, N.E., Zalutsky, M.R., Boyd, M.: Targeted radiotherapy: microgray doses and the bystander effect. Dose-Response 5(3), 204–213 (2007)
18. Marco, D., Scott, E., Cairns, D., Graham, A., Allen, J., Mahajan, S., Shankland, C.: Investigating co-infection dynamics through evolution of bio-PEPA model parameters: A combined process algebra and evolutionary computing approach. In: Gilbert, D., Heiner, M. (eds.) CMSB 2012. LNCS, vol. 7605, pp. 227–246. Springer, Heidelberg (2012)

19. Massink, M., Latella, D., Bracciali, A., Harrison, M.D., Hillston, J.: Scalable context-dependent analysis of emergency egress models. Formal Aspects of Computing 24(2), 267–302 (2012)

20. McMahon, S.J., Butterworth, K.T., McGarry, C.K., Trainor, C., O'Sullivan, J.M., Hounsell, A.R., Prise, K.M.: A computational model of cellular response to modulated radiation fields. Int. J. Radiat. Oncol. 84(1), 250–256 (2012)

21. McMahon, S.J., Butterworth, K.T., Trainor, C., McGarry, C.K., O'Sullivan, J.M., Schettino, G., Hounsell, A.R., Prise, K.M.: A kinetic-based model of radiation-induced intercellular signalling. PloS One 8(1), e54526 (2013)

22. Partridge, M.: A radiation damage repair model for normal tissues. Physics in Medicine and Biology 53(13), 3595 (2008)

23. Prise, K.M., O'Sullivan, J.M.: Radiation-induced bystander signalling in cancer therapy. Nature Reviews Cancer 9(5), 351–360 (2009)

24. Schettino, G., Folkard, M., Michael, B.D., Prise, K.M.: Low-dose binary behavior of bystander cell killing after microbeam irradiation of a single cell with focused ck x rays. Radiat Res. 163(3), 332–336 (2005)

25. Scott, E., Hoyle, A., Shankland, C.: Pepa'd oysters: Converting dynamic energy budget models to bio-pepa, illustrated by a pacific oyster case study. Electronic Notes in Theoretical Computer Science 296, 211–228 (2013)

26. Seymour, C.M., C: Medium from irradiated human epithelial cells but not human fibroblasts reduces the clonogenic survival of unirradiated cells. International Journal of Radiation Biology 71(4), 421–427 (1997)

27. Seymour, C.B., Mothersill, C.: Relative contribution of bystander and targeted cell killing to the low-dose region of the radiation dose-response curve. Radiat. Res., 508–511 (2000)

A   Full Bio-PEPA Model

$$N = 1000;$$
$$Gy = 1;$$
$$s = 1;$$
$$dose = s * Gy;$$
$$initH = floor(N * exp(0 - dose));$$
$$initI = floor((N * (dose + ((dose^2)/2) + ((dose^3)/6) + ((dose^4)/24))) * exp(0 - dose));$$
$$initA = N - initH - initI;$$
$$beta = 0.0028;$$
$$gamma = 0.0164/Gy;$$
$$eps = 0.00011/5;$$
$$delta = 0.019;$$
$$apop = 1/2400;$$
$$thresh = 21;$$
$$fW = H(W - thresh);$$
$$Hb = 4;$$
$$Bh = 1 - exp(0 - Hb);$$
$$p = (exp(0 - Hb)) * (Hb + ((Hb^2)/2) + ((Hb^3)/6) + ((Hb^4)/24))/Bh;$$
$$rad = initI/N;$$
$$q = exp(0 - dose - Hb) * (dose * (Hb + ((Hb^2)/2) + ((Hb^3)/6))$$
$$+((dose^2)/2) * (Hb + ((Hb^2)/2)) + ((dose^3)/6) * Hb)/(rad * Bh);$$

$$SF = (N - D)/N;$$
$$damageH = [p * beta * H * fW];$$
$$killingH = [(1 - p) * beta * H * fW];$$
$$damageI = [q * beta * I * fW];$$
$$killingI = [(1 - q) * beta * I * fW];$$
$$recover = [gamma * I];$$
$$emitI = [(eps/V) * I * (100 - W)];$$
$$emitA = [(eps/V) * A * (100 - W)];$$
$$deathA = [apop * A];$$
$$decay = [delta * W];$$

$$H = damageH << +killingH <<;$$
$$I = killingI << +damageI << +recover << +emitI(.);$$
$$A = deathA << +killingI >> +emitA(.);$$
$$R = damageH >> +damageI >> +recover >>;$$
$$D = deathA >> +killingH >>;$$
$$W = emitI >> +emitA >> +decay << +damageH(.) + damageI(.) + killingH(.) + killingI(.);$$

$$H[initH] < * > I[initI] < * > A[initA] < * > R[0] < * > D[0] < * > W[0]$$

Fig. 6. Bio-PEPA model in full

Exploring the Cellular Objective in Flux Balance Constraint-Based Models

Rafael S. Costa[1,2], Son Nguyen[3], Andras Hartmann[1,2], and Susana Vinga[2]

[1] Instituto de Engenharia de Sistemas e Computadores, Investigacão e
Desenvolvimento (INESC-ID),
R Alves Redol 9, 1000-29 Lisboa Portugal
{rcosta,ahartmann}@kdbio.inesc-id.pt
[2] Center for Intelligent Systems, LAETA, IDMEC, Instituto Superior Técnico,
Universidade de Lisboa, Av. Rovisco Pais 1, 1049-001 Lisboa, Portugal
svinga@dem.ist.utl.pt
[3] Instituto Superior Técnico, Universidade de Lisboa
Av. Rovisco Pais 1, 1049-001 Lisboa, Portugal
nguyen.son@ist.utl.pt

Abstract. Genome-scale reconstructions are usually stoichiometric and analyzed under steady-state assumptions using constraint-based modelling with flux balance analysis (FBA). FBA requires not only the stoichiometry of the network, but also an appropriate cellular objective function and possible additional physico-chemical constraints to predict the set of resulting flux distributions of an organism.

To compute the metabolic flux distributions in microbes, the most common objective is to consider the maximization of the growth rate or yield. However, other objectives may be more accurate in predicting phenotypes. Since in general objective function selection is highly dependent on the growth conditions, the quality of the constraints and the dataset, further investigation is required for better understanding the universality of the objective function. In this work, we explore the validity of different classes of optimality criteria and the effect of single (or combinations of) standard constraints in order to improve the predictive power of intracellular flux distribution. These were evaluated to compare predicted fluxes to published experimental ^{13}C-labelling fluxomic datasets using two metabolic systems with different conditions and comparison datasets.

It can be observed that by using different conditions and metabolic systems, the fidelity patterns of FBA can differ considerably. However, despite of the observed variations, several conclusions could be drawn. First, the maximization of biomass yield achieves one of the best objective function under all conditions studied. For the batch growth condition the most consistent optimality criteria appears to be described by maximization of the biomass yield per flux or by the objective of maximization ATP yield per flux unit. Moreover, under N-limited continuous cultures the criteria minimization of the flux distribution across the network or by the maximization of the biomass yield was determined as the most significant. Secondly, the predictions obtained by flux balance analysis

P. Mendes et al. (Eds.): CMSB 2014, LNBI 8859, pp. 211–224, 2014.
© Springer International Publishing Switzerland 2014

using additional combined standard constraints are not necessarily better than those obtained using only one single constraint.

Keywords: metabolic networks, constraint-based models, flux balance analysis, objective functions, constraints, flux distributions prediction.

1   Introduction

Together with the development of high-throughput technologies in molecular biology, the applications of system biology are becoming increasingly significant. By integrating abundant experimental data with modelling approaches, one could expand the field of research and also improve the models' accuracies. A wide class of metabolic modelling frameworks applied to study the metabolism of organisms are stoichiometric models which do not require a large set of parameters as kinetic models do [6,18].

With the technological developments in the post genomic era there has been an increasing focus of metabolic reconstructions for a large number of modelled organisms [8,25]. The metabolic network usually contains only stoichiometric information and are analyzed under steady-state assumption using constraint-based modelling with flux balance analysis (FBA) [16,26]. The flux balance and the constraints involving prior knowledge about the biological system could be rewritten as mathematical expressions in FBA and used to predict possible metabolic flux distributions when a certain cellular objective is defined [35]. One approach usually used to address this optimization problem is the linear programming (LP) framework. The consistency of the predictive solutions obtained by FBA for an organism depends on a number of important assumptions. First, it assumes that evolutionary forces have shaped the metabolism towards optimal cellular criteria to represent natural selection. Second, simple constraints refer to the different limited conditions that a given biological system must satisfy, such as physico-chemical (e.g. fixing known as upper and lower bounds of individual input and output fluxes capacities), topological (e.g. connection between metabolites) and environmental factors [16]. Moreover, gene expression data, reactions reversibility and other constraints can be added to restrict the solution space of possible phenotypes fluxes [20,31]. Thus, the task of FBA is to find a solution that optimizes an objective function satisfying the imposed constraints.

Although the most common objective in microbes is to consider the maximization of the growth yield (i.e. assumed "biomass" equation) whose validity has been experimentally tested under some conditions [35] there are cases where this assumption is not valid (e.g. overflow metabolism), and other objective criteria can actually achieve better results [14,28]. Various alternative cellular objective criteria have been suggested, such as the generation of ATP per substrate [32], the minimization of the production rate of redox potential [17], or the maximization of the biomass yield per flux unit. Consequently, the universal principle of the biomass objective remains an open question [12,24]. Over the last years, some works have been carried out to test the use of objective function optimization by FBA. This set of works can be divided into: (i) studies examining hypotheses on presumed cellular objective functions through comparison to measured fluxes generated usually by ^{13}C-labeling methods [28,33,34], and (ii) studies developing algorithms to predict biological objective functions from experimental data [5,11,17].

As part of the first study, an evaluation of objective functions for *Escherichia coli* central metabolism model has been established by Schuetz and co-workers [33]. In total 11 objective functions, together with 8 physico-chemical constraint forming 99 simulations has been evaluated in attempt of finding the combination that could best predict the ^{13}C-determined *in vivo* intracellular flux data from *E. coli* grown under different environmental conditions. The results indicate that, in unlimited resource condition (batch culture), the best objective is the maximization of ATP yield per unit of flux while in chemostat culture with limited nutrient, the linear maximization of ATP or biomass yield proved to be a more suitable objective. However, in the cases tested, there was still significant variation between predicted and experimental fluxes. It is also important to mention that the additional constraints become unimportant if appropriate objective function is chosen in a given condition. Recently, an expanded study tested a large class of possible single objective functions (maximization biomass, ATP yield and the production of each metabolite across the network, and minimizing reaction fluxes). It has been shown that a combination of three efficiency objectives, maximum ATP yield and maximum biomass yield with minimum sum of absolute fluxes, are consistent to the experimental datasets (i.e. operate close to *Pareto* fronts [22]). Alternatively, other studies proposed algorithms to systematically identify or predict a relevant objective function using experimental data. For instance, *ObjFind* [5] is an optimization-based algorithm that was created for that purpose. This *in silico* procedure attempts to solve the coefficients of importance (CoIs) on reaction fluxes while keeping the divergence between resultant and experimental flux distribution to be as small as possible. A bi-level optimization problem is used to describe this task: minimize the error between *in vivo* and *in silico* fluxes by quadratic programming, subject to the fundamental FBA problem. Biological objective solution search (*BOSS*) algorithm [11] is another tool that is claimed to be able to recapitulate the actual objective including even excluded reaction from reconstruction model.

Since objective function selection seems to be, in general, highly depend on the growth conditions, quality of the constraints, size of the metabolic models and comparison datasets specific, more investigation should be established for better understanding the universality of the objective function. This need has been recognized also by the systems biology community [2,9]. In line with that, the main goal of the current work is to explore different optimality principles and the effect of single (or combinations) empirical constraints in order to improve the predictive power of intracellular flux distribution in metabolic networks. We test this with two *E. coli* metabolic subsystems in different cultures and novel comparison datasets, expanding previous evaluations [17,28,33].

2   Methods

The flux balance analysis (FBA) is formulated based on steady-state assumption which constrains the system to follow the mass balance condition. The mass balance equation together with specific environmental (or physico-chemical) restrictions form the solution space of all possible fluxes distribution. Then FBA typically solves an optimization problem on this space, with objective functions linearly expressed as:

$$Z = c.v$$

where v denotes the vector of flux values and c is coefficient vector that defines the weight of each flux v_i in the objective function. c can take the form of simple sparse vector with only one non-zero element in case of objective involving single reaction, such as maximization of biomass or ATP yield. In case of multiple fluxes involved, such as redox potential [24], objective function Z is a linear combination of v_i with varied coefficients c_i. In case of the more complex functions, mixed-integer or non-linear programming have to be used to solve the optimization problem. The general expression for simple FBA problem is:

$$\min/\max(Z) \tag{1}$$
$$\text{s.t.} \quad S.v = 0$$
$$v^{ub} \geq v \geq v^{lb}$$

where S is the stoichiometric matrix of the model and the flux rates are limited by upper (v^{ub}) and lower bounds (v^{lb}) of each of the fluxes in v.

2.1   Metabolic Models and Experimental Flux Datasets

For this work two published metabolic reconstructions of *Escherichia coli* was selected as the metabolic model for flux analysis. The first is a condensed stoichiometric model [27] contains 72 metabolites forming in total a set of 95 central chemical reactions, which consist of 75 internal and 20 drain fluxes (*Core* model). The second model is the genome-scale metabolic reconstruction iAF1260 of *E. coli* [10], which contains in total 1668 metabolites and 2382 reactions (*Genome-scale* model).

The mapping of available measured fluxes to corresponding reactions in FBA model is given as in Supporting information 1. There are six experimental datasets used to compare with model predictions. They consist of metabolic fluxes in different cultures extracted from published phenotype analyses on *E. coli*. The datasets come from various sources of literature (Table 1) to diversify our testing environment (chemostat aerobic N- and C-limited, batch aerobic).

2.2   Objective Functions and Cellular Constraints Examined

For the sets of experimental data selected we evaluated the predictive ability of eight different objective functions: maximization of biomass yield (max BM)

Table 1. Experimental datasets from different sources used in this study

Reference (data source)	Dilution rate (D)	Culture conditions	Number of mapped reactions
Emmerling et al. [7]	$0.09h^{-1}$	chemostat, aerobic, N-limited	33
Ishii et al. [15]	$0.1h^{-1}$	chemostat, aerobic, C-limited	47
	$0.4h^{-1}$	chemostat, aerobic, C-limited	47
	$0.7h^{-1}$	chemostat, aerobic, C-limited	47
Perrenoud et al. [29]	$0.62h^{-1}$	batch, aerobic	36
Holm et al. [13]	$0.67h^{-1}$	batch, aerobic	37

and ATP yield (max ATP); optimization of enzymatic efficiency for cellular growth (i.e. the minimization of the absolute sum of internal fluxes) which was formulated by linear programming with Manhattan-norm , namely min Flux ($\min_{i=1}^{n} \sum | v_i |$); maximization of ATP and biomass yield per flux unit: max ATP/flux ($\max \frac{v_{ATP}}{\sum_{i=1}^{n} v_i^2}$) and max BM/flux ($\max \frac{v_{biomass}}{\sum_{i=1}^{n} v_i^2}$), respectively; minimization of redox potential (min Rd); and minimization and maximization of ATP producing fluxes per unit substrate (min ATPprod and max ATPprod respectively). These objective functions correspond to the most significant performed in [33].

For the two nonlinear objective functions max ATP/flux and max BM/flux, the l_2-norm (Euclidean norm) of the fluxes vector was used in accordance with [33] while min Flux refers to the l_1-norm (Manhattan norm) of the same vector as in [34].

For linear optimization, there are usually non-unique sets of flux values v that give the same optimal value of objective function [23]. To avoid typical degeneracy of FBA solutions, the principle of parsimonious enzyme usage (a derivative of FBA called pFBA) was used [19]. This approach finds a flux distribution with minimum absolute values among the alternative optima, assuming that the cell attempts to achieve the selected objective function while allocating the minimum amount of resources (i.e. minimal enzyme usage). Mathematically, the optimization problem is described as follows:

$$\min \sum_{i=1}^{n} | v_i | \qquad (2)$$
$$\text{s.t.} \quad Z = Z_{optima}$$
$$S.v = 0$$
$$v^{ub} \geq v \geq v^{lb}$$

The cellular constraints on the biological systems were introduced together with objective functions to better locate the ultimate solution. We systematically tested 6 single (or combined) constraints [33] P-to-O (P/O) ratio was set to 1; the upper limit bound for the maximal oxygen consumption rate (qO$_2$max) was

set to 15 mmol/gh as experimentally reported for chemostat cultures [15]; the
bounds on cellular maintenance energy (ATPM reaction) as the requirement
for growth-independent was left at the default model value of 8.39 mmol/gh
[27]; bound all fluxes (bounds) to maximal 200% of the glucose uptake rate as
observed experimentally [37] and 35% of NADPH overproduction compared to
the NADPH requirement for biomass production. Finally the combination of all
above constraints is also included.

Empirical constraints related to the bounds of fluxes given in the original
models were discarded and at least the relative substrate uptake rate (glucose)
was constrained in each case.

2.3   Implementation

All calculations were implemented in Matlab 2012b (Mathworks Inc. Software)
and simulations were performed using the Constraint-Based Reconstruction and
Analysis (COBRA) toolbox (v. 2.0.5) [1]. In terms of optimization solvers, GLPK
[21] was used for linear problems. For the two non-linear non-convex objective
functions, a numeric approximation method was used. For example, with the
objective function max `ATP/flux`, firstly the range of `ATP` flux is calculated then
1000 value points are uniformly selected along the distance. After that, for each
point, the `ATP` reaction rate is constrained to this value and the l_2-norm of all
fluxes in this system $\sum v_i^2$ is minimized simultaneously. The maximum of all
1000 ratios between `ATP` fluxes with their corresponding minimized flux norms
is a good approximation for the objective max `ATP/flux`. Similar approach was
used for max `BM/flux`.

Simulations were executed in parallel on a server machine with 8 AMD processors of 2.3GHz each. The libSBML [4] was the package used for reading SBML
model files. Experimental datasets, metabolic models and ready-to-run matlab
scripts are freely available at `https://github.com/hsnguyen/ObjComparison`.

2.4   Calculating Distance between Predicted and Experimental Data

To evaluate the prediction ability, the definition of predictive fidelity [33] was
used. This error is defined by:

$$\mathsf{d}(v_{comp}, v_{exp}) = \epsilon^T W \epsilon \tag{3}$$

$$\epsilon = \frac{v_{comp} - v_{exp}}{v_{glucose}} \tag{4}$$

$$W_{i,i} = \frac{1}{\sigma_i^{exp}} \left(\sum_i \frac{1}{\sigma_i^{exp}} \right)^{-1} \tag{5}$$

where $\mathsf{d}(v_{comp}, v_{exp})$ is the standardized Euclidean distance between predicted
fluxes v_{comp} and the experimental *in vivo* v_{exp} fluxes weighted by their variances
σ^{exp}. The set of compared reactions are given as Supporting information 1.
Smaller predictive fidelity represent a better agreement between computational
predicted and experimental fluxes.

3 Results and Discussion

The cellular objective functions for an organism are strongly dependent on the comparison-data and size/type of the metabolic models. Thus, across two different metabolic systems and using available intracellular flux measurements for *E. coli*, we examine the optimal criteria derived in a previous study [33]. Furthermore, the effect of single (or pairs) standard constraints are also evaluated. The metabolic models that we used are two stoichiometric models of *E. coli* metabolism [27,33] and the flux distributions for eight different objective functions. The experimental conditions included batch cultures but also chemostat cultures under aerobic glucose and ammonium limitation.

The results obtained in the FBA simulations for each case were ordered according to the fidelity error between predicted and experimental fluxes. Thus the estimation of predictive result is based on two perspectives: (i) the predictive errors represented by stem plots for different simulations and (ii) scatter plots of separated flux-by-flux comparisons to see how each experimental flux distribution matched to corresponding predictions.

3.1 The Impact of Objective Functions by FBA in Different Conditions

The predictive fidelities for various objective and constraint combinations in different experimental conditions are shown in Figure 1 (all the remaining datasets are given in supporting information). It can be observed that for simulations using different growth conditions, the corresponding error patterns can differ considerably. The 8 objective functions showed a great difference in the accuracy. Under nutrient scarcity (chemostat cultures) in glucose-limited, the linear maximization of the biomass yields achieved the best predictive accuracy (Figure 1e and 1f and for other conditions see Supporting information). This result agrees with previous works that supported the use of maximization of biomass as the best objective function for FBA in continuous cultures [7,33,36]. An interesting criteria also is the parsimony criteria ($\min \sum | \mathbf{v} |$). The finding that minimization of the overall intracellular flux plays a key role, as shown for the genome-scale model (e.g. Fig 1f), is in line with a previous work for hybridoma cells in continuous culture [3]. Our analysis for this growth condition indicated that the single standard constraints namely *none, qO2, maintenance energy* or *NADPH* improved their predictive fidelities. Although common properties with the previous work [33], some relative alterations in our results could be observed. Probably because the use of the relative flux values instead of the split ratios used by Schuetz et al [33] between fluxes when comparing computational and experimental results.

The best objective for the ammonium-limited continuous culture was obtained by minimization of flux distribution or by maximization the biomass (Figure 1a and 1b). On the other hand, for batch cultures the results suggest that the cells "prefer" the maximization of biomass yield per flux unit (for example Figure 1c and 1d). In contrast to a previous work [17], this result suggest that the

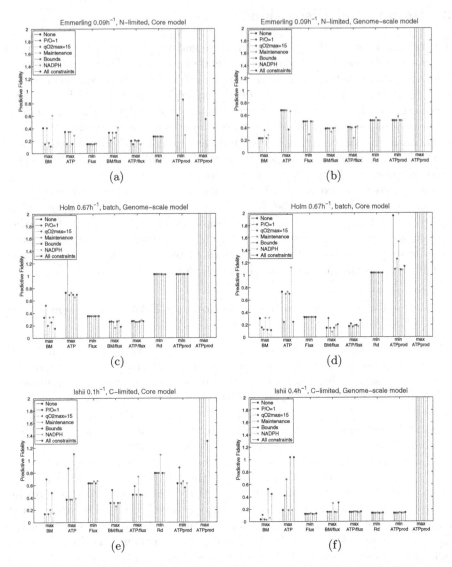

Fig. 1. Predictive fidelities for all objective functions and constraints of different culture conditions

minimization of redox potential is not a good choice as objective function for *E. coli* batch cultures. In addition, the max ATP/flux also gave quite consistent and promising fidelities on the two batch cultures (Holm and Perrenoud dataset) in accordance with Schuetz work [33].

Additionally, in accordance with the finding of Schuetz et al [33], we found that the effect of choice of constraints appear to be mostly insignificant to the choice of the objective function.

3.2 The Impact of Objective Functions by FBA in Different Metabolic Systems

We next examined whether of these findings are consistent across different metabolic systems. Here, the objective functions were tested using the reconstruction models named as *Core* model and `iAF1260` *Genome-scale* model respectively for the same experimental dataset (Figure 2).

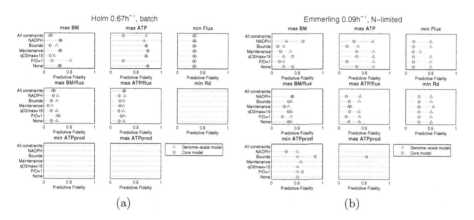

(a)   (b)

Fig. 2. Predictive fidelities of pFBA simulation for each objective function in Holm dataset (2a) and Emmerling dataset (2b) with two different metabolic systems. Values above 1 are out of range.

Figure 2 shows the effect of the metabolic models type and size on the qualitative and quantitative predictions for the same dataset (Holm and Emmerling dataset). While max BM and max BM/flux reported similar fidelities, simulation with max ATP as objective function acted oppositely in the two metabolic systems. This difference reflects the fact that beside agreements on the most effective functions for certain situations, using different metabolic systems could significantly affect the conclusions for certain cases. The factors related to the model reconstructions such as the biomass composition can be one possible reason for this difference. For results of all other datasets see Supporting information.

3.3 Evaluation of Pairwise Constraints

Here FBA was run for all pairwise constraints to predict the fluxes. In order to understand how the phenotype predictions vary across the different constraints, a particular case is selected, namely the chemostat fermentation under the highest dilution rate of $0.7h^{-1}$. This is a typical case where FBA simulations are less accurate, since the cells were sub-optimally grown due to overflow metabolism [14]. For this case the flux predictions behaved qualitatively well for the same optimal criteria and well-described by maximizing biomass yield (Figure 3).

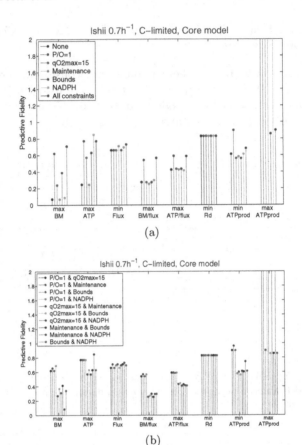

Fig. 3. Comparison between pFBA predictive fidelities using single (3a) and pairwise (3b) constraints for the dataset from Ishii under the dilution rate of $0.7h^{-1}$

Despite some differences, in general there is no improvement in predictive fidelity for any of the single objective functions when the combination of two objective functions are used instead. On the other hand, when the complete set of constraints "*all constraints scenario*" are used simultaneously, its contribution was not significant compared to the "*none*" constrain scenario as reported previously in [33]. Identical conclusions were obtained when repeating all these simulations with the other datasets (data not shown).

4   Conclusions

This paper explored the effect of different optimality principles in FBA for two metabolic systems using different conditions and comparison dataset than previous evaluations. Although the fidelity patterns of FBA can differ considerably under different conditions, the classical biomass optimization was shown as one of the best objective. Moreover, our results show that the metabolic model type/size could have a significant impact on the predictive fidelities.

Despite the observed variations, several generalities emerged. In agreement with previous studies, the single objective of maximization of biomass yield achieves the best predictive accuracy for a wide range of experimental conditions and different models. For the batch growth condition the most consistent optimality criteria appears to be described by the maximization of the biomass yield per flux and also by the objective of maximization of ATP yield per flux. Moreover, under N-limited continuous cultures, the criteria minimization of the flux distribution or maximization of biomass yield was determined as the most significant. On the other hand, the predictions obtained by flux balance analysis using additional combined standard constraints are not better than those obtained using the single constraint or even none of them.

Although some optimal criteria give reasonable prediction under certain conditions, there is no universal criteria that performs well under all conditions. Therefore, systems biologists should perform a careful evaluation and analysis of the objective functions case-by-case for each particular condition and application.

Matlab code for investigation the effect of cellular objective function and constrains on metabolic models has been also developed. This implementation of all objective functions and constraints can be easily adapted to test new metabolic systems and can be evaluated by comparing its results with those reported here. Ongoing efforts are currently directed for investigating the challenge of multi-objective optimization formulations.

Supporting Information

Supporting material for this article is freely available online at GitHub: https://github.com/hsnguyen/ObjComparison.

i) Supporting file 1 *matching.xlsx*: Mapping the model reactions to the experimental data.

ii) Supporting file 2 *plots.docx*: All generated figures are presented.

Acknowledgements. This work was supported by national funds through Fundação para a Ciência e a Tecnologia (FCT, Portugal) under contract (Pest-OE/EEI/LA0021/2013), LAETA Pest-OE/EME/LA0022 and European Union Framework Program 7 "BacHBerry", Project number FP7-613793.

RC was supported by the grant SFRH/BPD/80784/2011 from the FCT-Fundação para a Ciência e Tecnologia (http://www.fct.pt), SN thanks the master fellowship from euSYSBIO ERASMUS MUNDUS programme and AH receives PhD fellowship from FCT Portugal under the reference SFRH/BD/69336/2010. SV acknowledges support by Program Investigador FCT (IF/00653/2012) from FCT, co-funded by the European Social Fund (ESF) through the Operational Program Human Potential (POPH).

References

1. Becker, S.A., Feist, A.M., Mo, M.L., Hannum, G., Palsson, B.O., Herrgard, M.J.: Quantitative prediction of cellular metabolism with constraint-based models: the COBRA Toolbox. Nature Protocols 2, 727–738 (2007)
2. Bordbar, A., Monk, J.M., King, Z.A., Palsson, B.O.: Constraint-based models predict metabolic and associated cellular functions. Nature Reviews Genetics 15(2), 107–120 (2014)
3. Bonarius, H.P., Hatzimanikatis, V., Meesters, K.P., De Gooijer, C.D., Schmid, G., Tramper, J.: Metabolic flux analysis of hybridoma cells in different culture media using mass balances. Biotechnology and Bioengineering 50, 299–318 (1996)
4. Bornstein, B.J., Keating, S.M., Jouraku, A., Hucka, M.: LibSBML: an API library for SBML. Bioinformatics 24, 880–881 (2008)
5. Burgard, A.P., Maranas, C.D.: Optimization-based framework for inferring and testing hypothesized metabolic objective functions. Biotechnology and Bioengineering 82, 670–677 (2003)
6. Costa, R.S., Machado, D., Rocha, I., Ferreira, E.C.: Critical perspective on the consequences of the limited availability of kinetic data in metabolic dynamic modelling. IET Systems Biology 5(3), 157–163 (2011)
7. Emmerling, M., Dauner, M., Ponti, A., Fiaux, J., Hochuli, M., Szyperski, T., et al.: Metabolic flux responses to pyruvate kinase knockout in *Escherichia coli*. Journal of Bacteriology 184, 152–164 (2002)
8. Feist, A.M., Herrgard, M.J., Thiele, I., Reed, J.L., Palsson, B.O.: Reconstruction of biochemical networks in microorganisms. Nature Reviews Microbiology 7, 129–143 (2009)
9. Feist, A.M., Palsson, B.O.: The biomass objective function. Current Opinion in Microbiology 13, 344–349 (2010)
10. Feist, A.M., Henry, C.S., Reed, J.L., Krummenacker, M., Joyce, A.R., Karp, P.D., et al.: A genome-scale metabolic reconstruction for *Escherichia coli* K-12 MG1655 that accounts for 1260 ORFs and thermodynamic information. Molecular Systems Biology 3, 121 (2007)
11. Gianchandani, E.P., Oberhardt, M.A., Burgard, A.P., Maranas, D.C., Papin, J.A.: Predicting biologicsl system objectives *de novo* from internal state measurements. BMC Bioinformatics 9, 43–55 (2008)
12. Harcombe, W.R., Delaney, N.F., Leiby, N., Klitgord, N., Marx, C.J.: The Ability of Flux Balance Analysis to Predict Evolution of Central Metabolism Scales with the Initial Distance to the Optimum. Plos Computational Biology 9 (2013)
13. Holm, A.K., Blank, L.M., Oldiges, M., Schmid, A., Solem, C., Jensen, P.R., et al.: Metabolic and Transcriptional Response to Cofactor Perturbations in *Escherichia coli*. Journal of Biological Chemistry 285, 17498–17506 (2010)
14. Ibarra, R.U., Edwards, J.S., Palsson, B.O.: *Escherichia coli* K-12 undergoes adaptive evolution to achieve *in silico* predicted optimal growth. Nature 420, 186–189 (2002)
15. Ishii, N., Nakahigashi, K., Baba, T., Robert, M., Soga, T., Kanai, A., et al.: Multiple high-throughput analyses monitor the response of *E-coli* to perturbations. Science 316, 593–597 (2007)
16. Kauffman, K.J., Prakash, P., Edwards, J.S.: Advances in flux balance analysis. Current Opinion in Biotechnology 14, 491–496 (2003)
17. Knorr, A.L., Jain, R., Srivastava, R.: Bayesian-based selection of metabolic objective functions. Bioinformatics 23, 351–357 (2007)

18. Lewis, N.E., Nagarajan, H., Palsson, B.O.: Constraining the metabolic genotype-phenotype relationship using a phylogeny of *in silico* methods. Nature Reviews Microbiology 10, 291–305 (2012)
19. Lewis, N.E., Hixson, K.K., Conrad, T.M., Lerman, J.A., Charusanti, P., Polpitiya, A.D., Palsson, B.O., et al.: Omic data from evolved *E. coli* are consistent with computed optimal growth from genome scale models. Molecular Systems Biology 6(1) (2010)
20. Machado, D., Costa, R.S., Ferreira, E.C., Rocha, I., Tidor, B.: Exploring the gap between dynamic and constraint-based models of metabolism. Metabolic Engineering 14(2), 112–119 (2012)
21. Makhorin, A.: GLPK (GNU linear programming kit) (2008)
22. Marler, R.T., Arora, J.S.: Survey of multi-objective optimization methods for engineering. Structural and Multidisciplinary Optimization 26, 369–395 (2004)
23. Mahadevan, R., Schilling, C.H.: The effects of alternate optimal solutions in constraint-based genome-scale metabolic models. Metabolic Engineering 5(4), 264–276 (2003)
24. Molenaar, D., Van Berlo, R., Ve Ridder, D., Teusink, B.: Shifts in growth strategies reflect tradeoffs in cellular economics. Molecular Systems Biology 5 (2009)
25. Oberhardt, M.A., Palsson, B.O., Papin, J.A.: Applications of genome-scale metabolic reconstructions. Molecular Systems Biology 5, 320 (2009)
26. Orth, J.D., Thiele, I., Palsson, B.O.: What is flux balance analysis? Nature Biotechnology 28, 245–248 (2010)
27. Orth, J.D., Fleming, R.M.T., Palsson, B.O.: Reconstruction and use of microbial metabolic networks: the core *Escherichia coli* metabolic model as an Educational Guide. In: Escherichia Coli and Salmonella: Cellular and Molecular Biology, ASM Press (2010)
28. Ow, D.S.W., Lee, D.Y., Yap, M., Oh, S.K.W.: Identification of cellular objective for elucidating the physiological state of plasmid-bearing *E. coli* using genome-scale *in silico* analysis. AIChE 25, 61–67 (2009)
29. Perrenoud, A., Sauer, U.: Impact of global transcriptional regulation by ArcA, ArcB, Cra, Crp, Cya, Fnr, and Mlc o glucose catabolism in *Escherichia coli*. J. Bacteriol. 187, 3171–3179 (2005)
30. Price, N.D., Reed, J.L., Palsson, B.O.: Genome-scale models of microbial cells: evaluating the consequences of constraints. Nature Reviews Microbiology 2, 886–897 (2004)
31. Price, N.D., Papin, J.A., Schilling, C.H., Palsson, B.O.: Genome-scale microbial *in silico* models: the constraints-based approach. Trends in Biotechnology 21, 162–169 (2003)
32. Ramakrishna, R., Edwards, J.S., McCulloch, A., Palsson, B.O.: Flux-balance analysis of mitochondrial energy metabolism: consequences of systemic stoichiometric constraints. American Journal of Physiology-Regulatory, Integrative and Comparative Physiology 280(3), R695–R704 (2001)
33. Schuetz, R., Kuepfer, L., Sauer, U.: Systematic evaluation of objective functions for predicting intracellular fluxes in *Escherichia coli*. Molecular Systems Biology 3, 119 (2007)
34. Schuetz, R., Zamboni, N., Zampieri, M., Heinemann, M., Sauer, U.: Multidimensional Optimality of Microbial Metabolism. Science 336, 601–604 (2012)
35. Varma, A., Palsson, B.O.: Stoichiometric Flux Balance Models Quantitatively Predict Growth and Metabolic By-Product Secretion in Wild-Type *Escherichia-Coli* W3110. Applied and Environmental Microbiology 60, 3724–3731 (1994)

36. Van Gulik, W.M., Heijnen, J.J.: A Metabolic Network Stoichiometry Analysis of Microbial-Growth and Product Formation. Biotechnology and Bioengineering 48, 681–698 (1995)
37. Zhao, J., Shimizu, K.: Metabolic flux analysis of *Escherichia coli* K12 grown on C^{13}-labeled acetate and glucose using GG-MS and powerful flux calculation method. Journal of Biotechnology 101, 101–117 (2003)

Optimization Based Design of Synthetic Oscillators from Standard Biological Parts

Irene Otero-Muras and Julio R. Banga

BioProcess Engineering Group, IIM-CSIC,
Spanish Council for Scientific Research, Vigo, Spain
{ireneotero,julio}@iim.csic.es
http://www.iim.csic.es/~gingproc

Abstract. We consider the problem of optimal design of synthetic biological oscillators. Our aim is, given a set of standard biological parts and some pre-specified performance requirements, to automatically find the circuit configuration and its tuning so that self-sustained oscillations meeting the requirements are produced. To solve this design problem, we present a methodology based on mixed-integer nonlinear optimization. This method also takes into account the possibility of including more than one design objective and of handling both deterministic and stochastic descriptions of the dynamics. Further, it is capable of handling significant levels of circuit complexity. We illustrate the performance of this method with several challenging case studies.

Keywords: gene regulatory network, synthetic biology, multiobjective optimization, synthetic oscillator, optimization based design.

1 Introduction

Genetic oscillators play important regulatory roles in living organisms and, together with switches, are primary design targets in Synthetic Biology. From the first oscillator to be successfully implemented *in vivo* [1], known as *Repressilator*, a number of different working oscillators have been achieved in prokariotic [2] and eukariotic cells [3] with improved robustness and tunability. These designs are supported by mathematical models that predict the circuit behaviour *in silico* and help to understand the mechanistic principles leading to oscillations. Seminal theoretical studies of biochemical oscillators based on mathematical models go back decades [4,5] when phenomena like yeast glycolysis and periodic enzyme synthesis revealed the importance of clocks in molecular cell biology [6]. Although helpful for understanding, traditional modeling approaches have limitations from the design perspective since the system description provided is not directly or easily translatable into a circuit that can be implemented *in vivo*, i.e., into DNA sequences.

Synthetic circuits implemented to date, including oscillatory modules, are relatively simple, attending to the number of regulatory regions and also to the design engineering principles behind [7], and one of the challenges in synthetic biology is advancing towards higher order networks with programmable

P. Mendes et al. (Eds.): CMSB 2014, LNBI 8859, pp. 225–238, 2014.

functionalities and real world applicability [8]. In this regard, advanced mathematical and computational tools for modeling and optimization are needed to automatize the design and to explore complex circuit topologies beyond intuitive principles. Recently, great efforts have been made that contribute to the advance towards the automatic design of biocircuits. An increasing number of standard biological parts or DNA components is being characterized and made available at emerging catalogs like the library supported by the BioBricks Foundation [9] or the open source registry platform JBEI-ICEs [10]. On the other hand, modular modeling tools and formal programming languages like GenoCAD [11], GEC [12] or Eugene [13] are suitable for modeling biocircuits by combining standard parts from a library.

In this work we exploit these recent advances to address the automatic design of synthetic oscillators from standard biological parts. Modular programming languages for synthetic biology allow to translate systems described at the logical level of interactions (for example in a rule-based grammar) into combinations of devices compatible with the abstract circuit semantic. In [12] for example, the Repressilator circuit configuration is programmed in GEC language and the compilation of the program results in a series of systems complying with the Repressilator topology. Here we approach a different problem which consists of finding circuits with a particular performance starting from a given list of standard parts, without knowing *a priori* the abstract circuit configuration. The goal is to find the combination or combinations of components which optimally perform a given function or show a specific pattern of behaviour, and in particular, to find biocircuits which can work as self-sustained oscillators. The aim of this paper is developing a methodology for the design of oscillators with the following features:

- capability to handle high levels of circuit complexity,
- validity for both deterministic and stochastic descriptions of circuit dynamics,
- modularity and easy translation of the oscillators obtained into implementable circuits,
- possibility to incorporate more than one objective to the design.

To this aim the circuit design problem is formulated as a Mixed Integer Nonlinear Optimization Problem (MINLP) [14,15] and efficient global solvers are used to obtain the target oscillators. Finally an extension for multiobjective design is presented such that multiple criteria can be included into the oscillator design.

2   Methods

2.1   Modeling Framework

A gene regulatory network consists of a collection of DNA segments and their interactions which together control the expression levels of specific mRNA and proteins in a cell [16]. By controlling the amounts and temporal patterns in which gene products appear, gene regulatory networks regulate biological functions. A number of approaches are used to model gene regulatory networks including logic-based methodologies, continuous models of ordinary differential equations (ODEs) and stochastic models at the single cell level [16]. In this work we consider both the continuous description based on ODEs and the discrete stochastic description. For the automatic design of synthetic networks starting from a set of basic parts, the modeling framework needs to be suitable for modular model composition [17], such that individual components can be described as self-contained units and composed in different combinations [18].

Standard parts in synthetic biology are defined as DNA sequences encoding a function that can be assembled with other standard parts. Following the formalism from the Registry of Standard Biological Parts [9], we consider the following basic constitutive components of genetic circuits: *promoters* recruiting the transcriptional machinery which transcribes the downstream DNA sequence, *ribosome binding sites* controlling the accuracy and efficiency with which the translation of mRNA begins, *protein coding regions* containing the sequence information needed to create a functional protein chain and *terminators* signaling the end of transcription. The abstraction hierarchy proposed by Endy [19] classifies standard parts in three different layers: *parts*, defined as sequences with basic biological functions (like for example DNA binding proteins), *devices* (combinations of parts with a particular function) and *systems* (combinations of devices). In Fig. 1, we illustrate this hierarchy through the Repressilator regulatory system [1], where the different devices and parts are indicated. The system consists of three genes connected in a feedback loop. The first gene in the circuit expresses some protein A which represses the second gene, the second gene expresses a protein B which represses the third gene, and protein C expressed by the third gene closes the feedback loop by repressing the first gene. In order

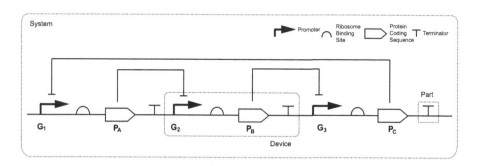

Fig. 1. Scheme of the Repressilator circuit (SBOL visual graphical notation [20])

to build a model capturing the dynamics of a synthetic gene network we need to know the full set of reactions of the system obtained by the combination of parts, where parts are defined by user specified properties, either from explicit declaration or from information available at standardized repositories [13].

A gene regulatory circuit is characterized by a reaction network consisting of a set of species $\mathcal{S} = \{X_1, \ldots X_s, \ldots, X_{n_s}\}$ interacting through a set of reactions $\mathcal{R} = \{R_1, \ldots R_r, \ldots, R_{n_r}\}$ of the form:

$$\sum_{s \in \mathcal{S}} \alpha_{sr} X_s \longrightarrow \sum_{s \in \mathcal{S}} \beta_{sr} X_s , \tag{1}$$

where α_{sr} and β_{sr} denote the molecularity as a reactant or product, respectively, of the species s in the reaction r. The state of the system at a given time is determined by the number of molecules $\zeta_1, \ldots, \zeta_{n_s}$ of the species $X_1, \ldots, X_{n_s}$, collected in the vector ζ. If a reaction r fires, the number of molecules of the system evolves according to the *vector of state change* [21]:

$$\nu_r = \beta_r - \alpha_r , \tag{2}$$

where α_r and β_r are the vectors containing the molecularities of educts and products of the reaction r, such that the state of the system goes from the state ζ to $\zeta + \nu_r$. Starting from the state change vectors of the reactions, the stochiometric matrix can be defined as:

$$N = [\nu_1, \ldots, \nu_{n_r}] . \tag{3}$$

In the discrete-stochastic approach, the numbers of molecules of every species $\zeta_1, \ldots, \zeta_{n_s}$ are considered as random variables and the evolution of the probability distribution on the state space over time is given by the Chemical Master Equation:

$$P(\dot\zeta, t) = \sum_{r \in \mathcal{R}} a_r(\zeta - \nu_r) P(\zeta - \nu_r, t) - \sum_{r \in \mathcal{R}} a_r(\zeta) P(\zeta, t) , \tag{4}$$

where a_r denotes the *propensity* [21] associated to the reaction r. Solving the Chemical Master Equation is often computationally intractable. Instead, single sample trajectories of the chemical process, i.e. single realizations of Eq. (4), can be computed with the exact stochastic simulation algorithm (SSA) by Gillespie [21].

In the continuous-deterministic approach, the dynamics of the network (1) are described through the evolution of the vector of concentrations z over time, ruled by a set of ODEs of the form:

$$\dot z(t) = N v , \quad z(0) = z_0 , \tag{5}$$

where N is the stoichiometric matrix (3) and v is a vector containing the reaction rates. At the thermodynamic limit, when the number of molecules is sufficiently large, both stochastic and deterministic descriptions become equivalent. Although the continuous approximation is widely used in systems biology, in the design of synthetic oscillators essential parts of the system (like bound repressors) appear always in low copy numbers and the effect of fluctuations may be critical for the dynamics. For this reason, our design method is conceived to handle both stochastic and deterministic descriptions.

Regarding the properties associated to each standard part type, we adopt the formalism proposed by Pedersen and Phillips [12]. The reactions and parameters corresponding to every standard part are included in the Appendix. We assume all reactions to be elementary and endowed with mass action kinetics. Within this formalism, further extensions can be easily considered to include cooperativity in the binding, degradation of intermediate complexes, etc. The optimization based design presented next is valid for any database complying with the formalism described. In this work, databases of standard parts are coded in Matlab, where, once a given set of parts is selected from the database, the model of the corresponding gene network is automatically generated (more details are given next), and the dynamics of the network are obtained numerically.

2.2 Optimization Based Design of Oscillatory Circuits

In order to formulate the automatic design of biological circuits as an optimization problem [14], the design criteria is encoded into an objective function (or functions) whose minimization under certain constraints (including the dynamics of the system) yields the desired circuit behavior. Both the circuit structure and the manipulable kinetic parameters need to be translated into decision variables of the optimization problem.

Decision Variables and Dynamic Constraints. We start by defining the decision variables and the dynamic constraints of the optimization problem according to the modeling framework described in the previous section. Let us assume that we start from a library containing g promoters, b ribosome binding sites, p protein coding regions and t terminators, the number of possible device configurations (in what follows we refer specifically to protein generator devices) is $n = g \times b \times p \times t$. Each standard part in the library is endowed with the corresponding properties, sets of reactions and values of associated parameters. In order to characterize a particular circuit configuration we can label every possible device with an integer index $i = 1 \ldots, n$ and build a vector $y \in \mathbb{Z}^n$ of binary variables such that:

$$\begin{cases} y_i = 1 \text{ , if the device } i \text{ is part of the circuit structure,} \\ y_i = 0 \text{ , otherwise.} \end{cases}$$

The vector y thus contains n binary decision variables for circuit optimization based design. The structure of a gene regulatory network will be completely defined by giving values to all the entries of the corresponding vector y. The kinetic parameters of each standard part need to be specified together with the set of reactions in the database. In case that some parameters can be tuned, they will constitute the real decision variables for the optimization. In what follows we denote by $x \in \mathbb{R}^m$ the vector of real tunable parameters and by $k \in \mathbb{R}^{n_r - m}$ the vector of parameters that remain fixed.

The optimization based design of genetic circuits is subject to the constraints imposed by the circuit's dynamics. For the continuous deterministic description, the dynamics are given by Eq. (5). With a slight abuse of notation let us rewrite the equations making explicit the dependencies on the decision variables and kinetic parameters as follows:

$$\dot{z}(t) = N(y)v(y,x,k) , \quad z(0) = z_0 , \tag{6}$$

The structure of the vector $v(y,x,k)$ depends on the active reactions in the circuit defined by the vector y while the mass action monomial coefficients vary according to the manipulable parameters x.

For a single cell stochastic description of a gene regulatory network the exact dynamics are obtained by the SSA algorithm [21]. In our implementation the inputs are the stoichiometric matrix $N(y)$, the matrix of molecularities of the reactants in all the reactions $\alpha(y)$ and the vector of stochastic kinetic rate constants $c(x,k)$ where the conversion of deterministic rate constants k to stochastic rate constants c is done according to the definitions by Gillespie [21]. In both cases, stochastic and deterministic, the output is encoded into a matrix containing the solutions (concentrations z in the deterministic description and numbers of molecules ζ in the stochastic description) obtained at discrete time points.

Objective Function. Now we need to define an objective function that, once minimized, will provide us with the desired circuit response consisting of a sustained oscillation. To this aim we introduce first the notion of autocorrelation which is used to evaluate the periodicity of time series data in different contexts, from signal processing to biochemical clocks [22] or neuronal responses [23].

Let $\{s_t; t = 1, \ldots, M\}$ be a time series corresponding to a process which is ergodic and stationary. The unbiased estimate of the autocorrelation function of s_t at a lag j can be defined as:

$$\Gamma(j \, ; M) = \frac{1}{M-j} \sum_{t=1}^{M-j} s_t s_{t+j} , \tag{7}$$

This function can be normalized $\Gamma_{norm}(j; M) = \Gamma(j; M)/\Gamma(0; M)$ such that the maximum value is $\Gamma_{norm}(0; M) = 1$. For s_t being the output of a deterministic simulation showing sustained oscillatory behaviour, the autocorrelation function will oscillate also in a sustained manner. For perfect (not damped) oscillations, the first peak in the normalized autocorrelation function, in what follows denoted by $P_{norm\Gamma}$, will take the maximum value 1.

Provided that s_t represents a realization of a stochastic process with a sustained oscillation, the autocorrelation estimate $\Gamma(j; M)$ will show a damped oscillation. This is due to the fact that stochastic fluctuations induce the phase diffusion of the oscillator and affect its periodicity [22]. The precision of the oscillators is often quantified through the so called quality factor, defined as $Q = 2\pi\gamma/T$ where γ is the inverse of the damping rate or characteristic time of the decay of the autocorrelation function [24] and T is the period of the oscillation [22]. In this way, the quality factor Q gives an estimation of the number of oscillations over which the periodicity is maintained [25]. We consider then that the lower the damping rate (and consequently the higher the first peak of the autocorrelation $P_{norm\Gamma}$) the better the oscillator is and, in order to establish an objective criterion to evaluate the oscillators, we measure the first peak of the autocorrelation $P_{norm\Gamma}$ [26].

MINLP Formulation. We are now in the position to formulate the design of a synthetic oscillator as an optimization problem. Let us start by the deterministic ODE framework, where the design can be formulated as the Mixed Integer Nonlinear Programming problem of finding a vector $x \in \mathbb{R}^m$ of continuous variables and a vector $y \in \mathbb{Z}^n$ of integer variables which minimize the objective function:

$$\min_{x,y} \quad - P_{norm\Gamma}(\dot{z}, z, y, x, k) \tag{8a}$$

subject to:

i) the circuit dynamics in the form of ODEs with the state variables z and additional parameters k:

$$\dot{z}(t) = N(y)v(y, x, k) , \quad z(t_0) = z_0 , \tag{8b}$$

ii) additional requirements in the form of equality and inequality constraints:

$$h(z, y, x, k) = 0 , \quad g(z, y, x, k) \leq 0 , \tag{8c}$$

iii) upper and lower bounds for the real and integer decision variables:

$$x_L \leq x \leq x_U , \quad y_L \leq y \leq y_U . \tag{8d}$$

For the discrete single cell stochastic description of the dynamics we use an analogue MINLP formulation, but in this case the minimization of the objective function (8a) is subject to the (noisy) circuit dynamics obtained by simulation with the SSA algorithm. As in the deterministic case, the minimization of the objective function is subject also to upper and lower bounds for the real and integer decision variables and it can be subject to additional requirements in form of equality and inequality constraints. This formulation of the design problem allows to impose a constraint on the maximum number of devices (D_{max}) allowed in the circuit:

$$\sum_i y_i \leq D_{max} \tag{9}$$

such that we can design oscillators with predefined complexity.

The solution of the resultant MINLP problem (both for the deterministic and stochastic descriptions of the dynamics) is challenging from the computational point of view. On the one hand, the design of gene circuits combines a high number of integer variables with real variables (tunable parameters), and on the other hand the dynamics of the systems under study are highly nonlinear resulting in an optimization problem which is non convex and multi-modal. Here we make use of a number of global MINLP solvers which have been tested in the context of biocircuit design with good results [28].

Multiple Design Criteria. The MINLP formulation of the design presented allows to achieve the primary goal of finding oscillators starting from a library of components. However, in a previous work [28] we have shown how the single objective formulation of the design of biocircuits might lead to arbitrariness when it comes to select the best solution and suggested the convenience of introducing additional competing criteria in order to provide more realistic design settings. In the presence of more than one competing objective the solution is not unique and every solution represents trade-off between different criteria. One design option is to consider the protein production cost as an additional criterion [28] competing with the oscillator performance. Another interesting multiobjective problem in oscillator design arises when it is needed to maximize the tunability of the oscillator's frequency without compromising its amplitude [29].

Following the ε-constraint proposed in [28], we can reduce the multiobjective optimization problem into a number of MINLP where each MINLP is obtained by minimizing one of the objectives and converting the rest of criteria into inequality constraints. Global MINLP solvers can then be used to find the Pareto optimal set of solutions.

3   Results and Discussion

As a proof of concept for the methodology presented we have set a number of design problems aiming to build genetic oscillators from a prototype library of standard parts in different scenarios. We have adapted the database from [12] to build a Matlab library containing 4 promoters: $P_1 = P_\lambda$, $P_2 = P_{tet}$, $P_3 = P_{bad}$, $P_4 = P_{lac}$, 1 ribosome binding site, 1 terminator and 11 protein coding regions for the proteins cIR, $tetR$, $araC$, $lacI$, $luxI$, $luxR$, $lasR$, $lasI$, $ccdB$, $ccdA$, $ccdA2$. We have used the kinetic parameters from Pedersen and Phillips [12].

3.1   Single Objective Design of Synthetic Oscillators in Deterministic Regime

First we search for sustained oscillators in the deterministic regime, using the MINLP formulation (8) in the Methods section. Starting from the prototype library, the number of integer decision variables (possible different devices) is $n = 44$. We set a maximum number of devices to $D_{max} = 3$. In order to solve the optimization problem we use the enhanced scatter search global solver eSS by Egea et al [30] with a multistart strategy, consisting of 20 runs of 600 seconds from different random initial guesses. No solution is found even when we increase the number of runs and the computation time per run. The same result is obtained with the mixed-integer tabu search solver (MITS) by Exler et al [31] and with the mixed-integer ant colony optimization solver (ACOmi) by Schlueter et al [32].

This result is coherent with the observations in [27], where they studied the effects of different model features in the appearance of oscillations for the Repressilator scheme. According to their results, in absence of cooperativity, the

degradation of the bound repressor is needed for oscillations in the deterministic regime. In view of this observation, we have extended the library to include
degradation of bound repressor. To this aim, the properties associated with the
negatively repressed promoter are modified by adding a degradation reaction as
indicated in the Appendix. Using a multistart strategy consisting of 20 runs of
600 seconds from different random initial guesses, the solver eSS found six different circuits, all of them endowed with the Repressilator topology illustrated
in Fig. 1 where P_A represses G_2, P_B represses G_3, and P_C represses G_1. We
include the solutions in Table 1. The values of the objective function are in all
cases close to -1 (the estimation of the autocorrelation is subject to numerical
error).

Table 1. Different gene circuits with Repressilator configuration found in the deterministic regime with degradation of the bound repressor, with $D_{max} = 3$

	G_1	P_A	G_2	P_B	G_3	P_C	Γ_{norm}
circuit 1	P_2	$lacI$	P_4	$araC$	P_3	$tetR$	0.9912
circuit 2	P_1	$araC$	P_3	$LacI$	P_4	cIR	0.9898
circuit 3	P_1	$lacI$	P_4	$tetR$	P_2	cIR	0.9887
circuit 4	P_1	$tetR$	P_2	$araC$	P_3	cIR	0.9836
circuit 5	P_4	$araC$	P_3	cIR	P_1	$LacI$	0.9836
circuit 6	P_1	$tetR$	P_2	$lacI$	P_4	cIR	0.9746

3.2   Single Objective Design of Synthetic Oscillators in Stochastic Regime

Now we aim to search for sustained oscillators in the stochastic regime starting from the original prototype library (without bound repressor degradation).
We formulate the MINLP problem as indicated in the Methods section where
the constraints imposed by the dynamics are obtained by simulation with the
stochastic Gillespie algorithm. Again we have that $n = 44$ and we set $D_{max} = 3$.

Although no oscillators appeared in the deterministic regime, sustained oscillators are found by solving the optimization problem in the stochastic regime.
Previous studies found that fluctuations modify the range of conditions in which
oscillations appear compared with the deterministic equations, and also that
deterministic and stochastic methods might not agree about the existence of
oscillations [27].

The best oscillator found by eSS in the stochastic regime consists of the three
devices P_4-rbs-$araC$-ter, P_3-rbs-cIR-ter, P_1-rbs-$LacI$-ter following the Repressilator configuration. The dynamics obtained by the Gillespie algorithm (for a
single realization) are depicted in Fig. 2 together with the autocorrelation function for the cIR stochastic dynamics.

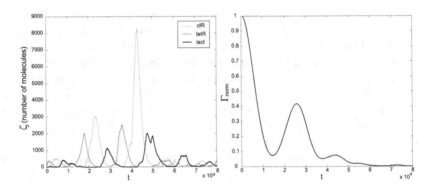

Fig. 2. Time course and autocorrelation for the optimal circuit in stochastic regime

3.3   Multiobjective Design of Synthetic Oscillators

One important property of genetic oscillators in some biological contexts is, as reported by [29], the ability to tune the oscillator's frequency without compromising its amplitude. We formulate here a multiobjective design problem aiming to maximize the period tunability (we will denote this magnitude by $|T|$) through the manipulation of a circuit parameter, in this case the $mRNA$ degradation constant, while minimizing the impact in the circuit output. Instead of amplitude differences, for simplicity we measure the differences between oscillation peaks, denoted by $|A|$. We start from the extended library (including bound repressor degradation) and set up a maximum number of devices of $D_{max} = 6$. The decision variables are $n = 44$ integers defining the circuit topology plus one real variable (we choose the degradation rate constant of the bound repressor k_{db} as the tunable parameter). We formulate the design problem as a multiobjective optimization one with two different objectives. The problem is solved by means of the ε-constraint strategy in combination with the global MINLP solver eSS, with $-|T|$ as the objective function to minimize at different intervals of $|A|$ defined as constraints. The Pareto front found is depicted in Fig. 3, and the corresponding circuits are included in Table 2.

Table 2. Circuits from the Pareto front in Fig. 3, $D_{max} = 6$

	device 1	device 2	device 3	device 4	device 5	device 6	k_{db}
circuit 1	P_1-araC	P_1-lacI	P_1-luxI	P_2-cIR	P_3-tetR	P_4-tetR	0.502
circuit 2	P_1-araC	P_1-lacI	P_2-cIR	P_3-tetR	P_3-luxI	P_4-tetR	0.515
circuit 3	P_1-araC	P_1-lacI	P_1-luxI	P_1-luxR	P_2-cIR	$P3$-tetR	0.455
circuit 4	P_1-araC	P_2-cIR	P_2-lacI	P_3-tetR	P_4-araC	P_4-ccdA	0.569
circuit 5	P_1-lacI	P_2-cIR	P_2-luxR	P_2-lasR	P_2-ccdB	P_4-tetR	0.587

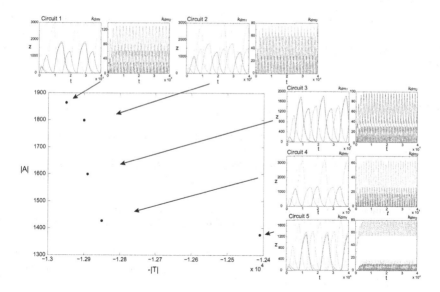

Fig. 3. Pareto Front of optimal solutions

4   Conclusions

In this work we develop a methodology for the automatic design of genetic oscillators based on mixed-integer nonlinear optimization. Starting from a library of standard parts our method allows us to find circuit configurations and tunings producing self-sustained oscillations with a number of predefined requirements.

The method is valid for any library or database complying with the formalism proposed by [12], where each standard part has associated a set of reactions and parameters and mass action kinetics are assumed. The advantages of this formalism are twofold: on the one hand, libraries can be easily extended to incorporate new elements and/or reactions, and on the other hand, the mass action assumption allows handling both deterministic and stochastic descriptions of the dynamics. This is of crucial importance in the design of genetic circuits where essential parts of the system appear always in low copy numbers and the effect of fluctuations may be critical for the dynamics.

Although optimization methods have been already used to find bifurcations in systems biology models [33,34], our approach, where the objective function is based on the autocorrelation function, results more practical in the design context (synthetic biology). For the solution of the resultant mixed-integer nonlinear optimization, we make use of a number of global MINLP solvers with proven efficiency in the context of biocircuit design, allowing to handle assemblies with significant levels of complexity.

Finally, the method allows including more than one design objective, providing a realistic design setting where every design solution represents a trade-off between different criteria.

Acknowledgments. This research received financial support from the Spanish Ministerio de Economía y Competitividad (and the FEDER) through the project MultiScales (DPI2011-28112-C04-03), and from the CSIC intramural project BioREDES (PIE-201170E018).

References

1. Elowitz, M.B., Leibler, S.: A synthetic oscillatory network of transcriptional regulators. Nature 403, 335–338 (2000)
2. Stricker, J., Cookson, S., Bennett, M.R., Mather, W.H., Tsimring, L.S., Hasty, J.: A fast, robust and tunable synthetic gene oscillator. Nature 456, 516–519 (2008)
3. Tigges, M., Dénervaud, N., Greber, D., Stelling, J., Fussenegger, M.: A synthetic low-frequency mammalian oscillator. Nucleic Acids Res. 38, 2702–2711 (2010)
4. Goodwin, B.C.: Oscillatory behaviour in enzymatic control processes. Adv. in Enzyme Regulation 3, 425–438 (1965)
5. Higgings, J.: The theory of oscillating reactions. Ind. Eng. Chem. 59, 18–62 (1967)
6. Tyson, J.J., Albert, R., Goldbeter, A., Ruoff, P., Sible, J.: Biological switches and clocks. J. R. Soc. Interface 6, S1–S8 (2008)
7. Purnick, P.E.M., Weiss, R.: The second wave of synthetic biology: from modules to systems. Nat. Rev. Mol. Cel. Biol. 10, 410–422 (2009)
8. Lu, T.K., Khalil, A.S., Collins, J.J.: Next-generation synthetic gene networks. Nat. Biotechnol. 27, 1139–1150 (2009)
9. Registry of Standard Biological Parts, http://partsregistry.org
10. Ham, T.S., Dmytriv, Z., Plahar, H., Chen, J., Hillson, N.J., Keasling, J.D.: Design, implementation and practice of JBEI-ICE: an open source biological part registry platform and tools. Nucleic Acids Res. 40, e141 (2012)
11. Cai, Y., Hartnett, B., Gustafsson, C., Peccoud, J.: A syntactic model to design and verify synthetic genetic constructs derived from standard biological parts. Bioinformatics 23, 2760–2767 (2007)
12. Pedersen, M., Phillips, A.: Towards programming languages for genetic engineering of living cells. J. R. Soc. Interface 6(suppl. 4), S437–S450 (2009)
13. Bilitchenko, L., Liu, A., Densmore, D.: The Eugene Language for Synthetic Biology. In: Voigt, C. (ed.) Methods in Enzymology, vol. 498, pp. 153–172 (2011)
14. Rodrigo, G., Carrera, J., Landrain, T.E., Jaramillo, A.: Perspectives on the automatic design of regulatory systems for synthetic biology. FEBS Lett. 586, 2037–2042 (2012)
15. Zomorrodi, A.R., Maranas, C.D.: Coarse-grained optimization-driven design and piecewise linear modeling of synthetic genetic circuits. Eur. J. Oper. Res. 237, 665–676 (2014)
16. Karlebach, G., Shamir, R.: Modelling and analysis of gene regulatory networks. Nat. Rev. Mol. Cell Biol. 9, 770–780 (2008)
17. Marchisio, M.A., Stelling, J.: Computational design of synthetic gene circuits with composable parts. Bioinformatics 24, 1903–1910 (2008)
18. Stewart, D.: Modular modelling in Synthetic Biology: Light-Based Communication in *E. coli*. Electron. Notes Theor. Comput. Sci. 277, 77–87 (2011)
19. Endy, D.: Foundations for engineering biology. Nature 438, 449–453 (2005)
20. Galdzicki, M., et al.: The Synthetic Biology Open Language (SBOL) provides a community standard for communicating designs in synthetic biology. Nature Biotech. 32, 545–550 (2014)

21. Gillespie, D.T.: Exact stochastic simulation of coupled chemical reactions. J. Phys. Chem. 81, 2340–2361 (1977)
22. Gaspard, P.: The correlation time of mesoscopic chemical clocks. J. Chem. Phys. 117, 8905–8916 (2002)
23. Gray, C.M., Konig, P., Engel, A.K., Singer, W.: Oscillatory responses in cat visual cortex exhibit inter columnar synchronization which reflects global stimulus properties. Nature 338, 334–337 (1989)
24. d'Eysmond, T., De Simone, A., Naef, F.: Analysis of precision in chemical oscillators: implications for circadian clocks. Phys. Biol. 10, 056005 (2013)
25. Gaspard, P.: Trace formula for noisy flows. J. Stat. Phys. 106, 57–96 (2002)
26. Min, B., Goh, K.I., Kim, I.M.: Noise characteristics of molecular oscillations in simple genetic oscillatory systems. J. Korean Phys. Soc. 56(3), 911–917 (2010)
27. Loinger, A., Biham, O.: Stochastic simulations of the repressilator circuit. Phys. Rev. E 76, 051917 (2007)
28. Otero-Muras, I., Banga, J.R.: Multicriteria global optimization for biocircuit design. arXiv:1402.7323 (2014)
29. Tsai, T.Y., Choi, Y.S., Ma, W., Pomerening, J.R., Tang, C., Ferrell, J.E.: Robust, tunable biological oscillations from interlinked positive and negative feedback loops. Science 321, 126–129 (2008)
30. Egea, J.A., Marti, R., Banga, J.R.: An evolutionary method for complex-process optimization. Comput. Oper. Res. 37, 315–324 (2010)
31. Exler, O., Antelo, L.T., Egea, J.A., Alonso, A.A., Banga, J.R.: A tabu search-based algorithm for mixed-integer nonlinear problems and its application to integrated process and control system design. Comput. Chem. Eng. 32, 1877–1891 (2008)
32. Schlueter, M., Egea, J.A., Banga, J.R.: Extended ant colony optimization for non-convex mixed integer nonlinear programming. Comput. Oper. Res. 36, 2217–2229 (2009)
33. Chickarmane, V., Paladugu, S.R., Bergmann, F., Sauro, H.M.: Bifurcation discovery tool. Bioinformatics 21, 3688–3690 (2005)
34. Levering, J., Kummer, U., Becker, K., Sahle, S.: Glycolytic oscillations in a model of a lactic acid bacterium metabolism. Biophys. Chem. 172, 53–60 (2013)

Appendix: Reactions Associated to Each Standard Part

Here we include the standard part properties taken from [12] together with the extension proposed to incorporate the degradation of bound repressor. Within this framework a promoter negatively regulated by a protein P has associated the reactions:

$$G + P \underset{k_u}{\overset{k_b}{\rightleftharpoons}} GP \xrightarrow{k_{tb}} GP + mP \tag{A1}$$

where G is the promoter, P is the protein, GP is the repressor-promoter complex and mP is the $mRNA$ of the transcribed protein. The parameters k_b, k_u and k_{tb} refer to the protein-promoter binding rate constant, protein-promoter unbinding rate constant and the rate of transcription in the bound state. The reactions corresponding to a promoter not regulated by any transcription factor are:

$$G \xrightarrow{k_t} G + mP \qquad mP \xrightarrow{k_{dm}} \emptyset \tag{A2}$$

where k_t is the constitutive rate of transcription in absence of transcription factors and k_{dm} is the degradation rate constant for the $mRNA$ degradation. Here it is important to note that a promoter may show also positive regulation, multi-regulation either positively or negatively by the levels of multiple transcription factors, and both constitutive and regulated transcription. The ribosome binding site part has one associated reaction:

$$mP \xrightarrow{k_r} mP + P \tag{A3}$$

where k_r is the rate constant corresponding to the translation of $mRNA$. Finally, the protein coding region part is endowed with:

$$P \xrightarrow{k_d} \emptyset \tag{A4}$$

where k_d is the degradation rate constant of the protein P. Starting from a library of genetic parts with their respective relevant properties, one can obtain the complete reaction network for composed devices and systems. For example, the set of reactions for a device consisting of a promoter G_1, repressed by a protein P_1, and a ribosome binding site for the expression of a downstream protein P_2 will read, in presence of the repressor protein P_1 as:

$$G_1 + P_1 \underset{k_{u1}}{\overset{k_{b1}}{\rightleftharpoons}} G_1 P_1 \xrightarrow{k_{tb1}} GP_1 + mP_2$$

$$mP_2 \xrightarrow{k_{r2}} mP_2 + P_2 \qquad P_2 \xrightarrow{k_{d2}} \emptyset \tag{A5}$$

In order to consider the degradation of the bound repressor, we add the reaction:

$$GP \xrightarrow{k_{db}} G \tag{A6}$$

to the previous scheme (A1), where k_{db} represents the rate constant for degradation of the bound repressor.

Modelling Polar Retention of Complexes in *Escherichia coli*

Abhishekh Gupta, Jason Lloyd-Price, and Andre S. Ribeiro

Laboratory of Biosystem Dynamics, Computational Systems Biology Research Group,
Department of Signal Processing, Tampere University of Technology,
P.O. Box 553, 33101 Tampere, Finland
abhishekh.gupta@tut.fi

Abstract. The cytoplasm of *Escherichia coli* is a crowded, heterogeneous environment. The spatial kinetics and heterogeneities of synthetic RNA-protein complexes have been recently studied using single-cell live imaging. A strong polar retention of these complexes due to the presence of the nucleoid has been suggested based on their history of positions and long-term spatial distribution. Here, using stochastic modelling, we examine likely sources, which can reproduce the reported long-term spatial distribution of the complexes. Based on the anisotropic displacement distribution observed at the border between the mid-cell and poles, we conclude that the original hypothesis that the observed long-term behavior is the result of macromolecular crowding holds.

Introduction. Even single-celled organisms, such as *Escherichia coli*, possess a far from random internal organization, as the cytoplasm is a crowded, heterogeneous environment. Some proteins preferentially locate at the cell poles (e.g. those involved in chemo-taxis), while others, e.g. those involved in gene expression, locate at mid-cell, within a structure known as the nucleoid.

Recent single-cell live microscopy measurements have studied the spatio-temporal distributions of a large complex, composed of a synthetic RNA tagged with multiple MS2-GFP proteins [1][2]. In one of these studies it was observed that, at short time scales, the motion of the complexes is sub-diffusive with an exponent that is robust to physiological changes and, at long time scales, the complexes tend to localize at the cell poles [1]. Further, it has been shown that these complexes are retained at the poles, as shown in Figure 1A, most likely due to the presence of the nucleoid at mid-cell [2]. This hypothesis arises from the observation of a strong anisotropy in the displacement distribution where the border of the nucleoid is expected to be (Figure 1B). However, the observed long-term spatial distribution of complexes could also, theoretically, arise from other sources, e.g. heterogeneities in the speed of the complexes along the major axis of the cell. Here, we use stochastic modelling to distinguish, from the observations, between the possible retention mechanisms taking place.

Methods. We model the cell as a compartmentalized 1-dimensional space which is divided into N homogeneous sub-volumes, indexed from $[1,N]$. The motion of

P. Mendes et al. (Eds.): CMSB 2014, LNBI 8859, pp. 239–243, 2014.

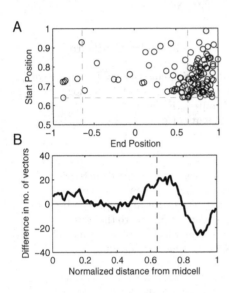

Fig. 1. (A) Relationship between the position along the major axis where each complex was last observed and the absolute position where it was first observed at a pole. Here, an end position of +1 indicates that the complex remained at the same pole as it was first observed in, while - 1 indicates that it traveled to the other pole. (B) Difference between the numbers of displacement vectors that are directed towards the poles and towards the mid-cell along the major cell axis. The differences in numbers were calculated from the displacement vectors originating within windows extending 0.05 normalized cell lengths around that point. All 160 cells were born during the measurement period and contained one complex in their lifetime. In both figures, the horizontal and vertical dashed lines represent the detected separation between the mid-cell and poles.

the complexes along the major cell axis is modeled with unimolecular reactions following the Reaction-Diffusion Master Equation [3]. We define $\overrightarrow{\alpha}(x)$ as the propensity of the forward reaction (modeling the motion of a complex from position x to position $x+1$) and $\overleftarrow{\alpha}(x)$ as the propensity of the backward reaction (from x to $x-1$). These propensity functions account for the combined effects of the rod shape of the cell and the nucleoid on the motions of the complexes.

Let $P(t)$ be the $N \times 1$ vector describing the probability of observing a complex in each sub-volume at time t, and A be the $N \times N$ transition rate matrix of propensities. $P(t)$ therefore evolves according to the following master equation, in matrix-vector form:

$$\frac{dP(t)}{dt} = AP(t). \tag{1}$$

Since a complex can travel from any sub-volume in the cell to any other sub-volume, given enough time, the system is ergodic and as $t \to \infty$, $P(t)$ will converge to a unique solution, P_{∞} . Solving the linear system of equations $0 = AP_{\infty}$, with the constraint that the total probability must sum to 1, we obtain the long-term spatial distribution of the complexes predicted by the model.

The propensities of the both diffusion reactions without accounting effects due to the rod shape and nucleoid are proportional to the diffusion constant of the complexes, D, given by:

$$\overrightarrow{\alpha}(x) = \overleftarrow{\alpha}(x) = \frac{N^2 D}{2}. \tag{2}$$

To account for the rod shape, i.e. a cylinder capped with two half-spheres, the length of the cell was parameterized by $B \in [0, 1]$, the normalized distance from midcell at which the cap begins.The forward propensities were attenuated by $\phi(x)$, the ratio between the areas of the cross sections of the cell (denoted $S(x)$) at adjacent positions. As such, $\overleftarrow{\alpha}(x)$ remains the same and $\overrightarrow{\alpha}(x)$ becomes:

$$\overrightarrow{\alpha}(x) = \frac{N^2 D}{2}\phi(x). \tag{3}$$

where,

$$\phi(x) = \frac{S(x+1)}{S(x)},$$

$$S(x) = \begin{cases} \pi & \text{if c(x) < B} \\ \pi\left[1 - \left(\frac{c(x)-B}{1-B}\right)^2\right] & \text{if c(x)} \geq \text{B} \end{cases}$$

Here, $c(x)$ translates the index of a sub-volume into the normalized distance from the midcell to the center of the sub-volume. In this case, $B = 1$ recovers the cylindrical cell from above, and $B = 0$ produces a spherical cell.

The effects of a nucleoid are introduced in the above model by adding a Gaussian function to $\overrightarrow{\alpha}(x)$ while subtracting it from $\overleftarrow{\alpha}(x)$. This anisotropy was parameterized with center $\mu \in [0, 1]$, standard deviation σ, and height h. Specifically:

$$\overrightarrow{\alpha}(x) = \frac{N^2 D}{2}\left[\phi(x) + h * exp\left\{\frac{-(c(x) - \mu)^2}{2\sigma^2}\right\}\right], \tag{4}$$

and

$$\overleftarrow{\alpha}(x) = \frac{N^2 D}{2}\left[1 - h * exp\left\{\frac{-(c(x) - \mu)^2}{2\sigma^2}\right\}\right]. \tag{5}$$

To fit the models to the measurements, we use the Earth-Mover's metric [4][5]:

$$W(F, G) = \int_{-\infty}^{\infty} |F(x) - G(x)| dx \tag{6}$$

where F and G are the cumulative distribution functions of the model and the measurements. This metric is a measure of the amount of work required to make two distributions identical.

Results. We constructed three 1-dimensional models to simulate the diffusion of the complexes within the cell. For all the models, we set N to 100, and D, the diffusion coefficient, to $1.43 * 10^{-2} \mu m^2/min$ based on previous measurements [1][2]. Two of the models contain spherical cell caps and their effects. We introduced a localized anisotropy in one of these models to test whether it, as observed in [2], can generate the observed long-term spatial distributions of the complexes (see Methods). In the last model, we set the forward and backwards propensities of diffusion events to be equal, and inversely proportional to the observed spatial distribution. Due to this, in the long term, the complexes tend to linger in the areas where they were observed with high probability.

Next, for each model, we varied all parameters and, for each set of values, obtained the distribution of complex positions that would be observed at infinite time. We then selected the set of parameters whose resulting distribution best fit the measured distribution of complex positions reported in [2].

The results from all three models, each using the best-fit parameter values, are shown in Figure 2. The model without the anisotropy fails to reproduce the displacement distribution (Figure 2A), and the consequent heterogeneity in the spatial distribution of complexes that favors their presence at the poles (Figure 2B). Meanwhile, the second model captures both of these properties of the dynamics of the complexes with significant accuracy. Interestingly, even though the third model reproduces the long-term spatial distribution exactly as observed (the lines are indistinguishable in Figure 2B), it produces a negligible anisotropy in the predicted displacement distribution (Figure 2A).

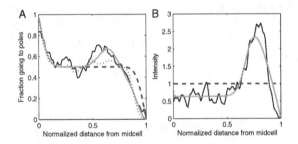

Fig. 2. (A)Measured fraction of displacement vectors originating within a window extending 0.05 normalized cell lengths around that point which are directed towards the pole (black line), model prediction with homogenous speed (without nucleoid (dashed line) and with nucleoid (gray line)), and with differing speed without nucleoid (dotted line). Note that the dashed line is superimposed by the gray line in the left side of the graph. (B) Measured spatial distribution of fluorescence intensities of complexes (black line) model prediction with homogenous speed (without nucleoid (dashed line) and with nucleoid (gray line)), and with differing speed without nucleoid (dotted line). Note that the dotted line is superimposed on the black line of the graph.

Using these models, with parameters tuned to match the measurements reported in [2], we show that both an anisotropy in the displacement vectors and a reduced speed at the poles produce good fits with the measurements. However, the model with varying speed along the major cell axis, at the time scale of the measurements, was unable to reproduce the observed anisotropic displacement distribution at the border between the mid-cell and poles. We conclude that polar retention most likely relies on these anisotropies in the displacement distribution rather than differences in speeds, consistent with the hypothesis that the observed long-term behavior is the result of macromolecular crowding, likely due to the nucleoid. Overall, the spatiotemporal kinetics of the complexes suggests that nucleoid occlusion is a source of dynamic heterogeneities of macromolecules in *E. coli* that ultimately generate phenotypic differences between sister cells.

References

1. Golding, I., Cox, E.: Physical Nature of Bacterial Cytoplasm. Phys. Rev. Lett. 96, 098102 (2006)
2. Gupta, A., Lloyd-Price, J., Neeli-Venkata, R., Oliveira, S.M.D., Ribeiro, A.S.: In Vivo Kinetics of Segregation and Polar Retention of MS2-GFP-RNA Complexes in *Escherichia coli*. Biophys. J. 106, 1928–1937 (2014)
3. Gardiner, C.W., McNeil, K.J., Walls, D.F., Matheson, I.S.: Correlations in Stochastic Theories of Chemical Reactions. J. Statiscal Phys. 14, 307 (1976)
4. Rubner, Y., Tomasi, C., Guibas, L.J.: The Earth Movers Distance as a Metric for Image Retrieval. Int. J. Comput. Vis. 40, 99–121 (2000)
5. Dobrushin, R.L.: Prescribing a system of random variables by conditional distributions. Theory Probab. Its Appl. 15, 458 (1970)

Extensible and Executable Stochastic Models of Fatty Acid and Lipid Metabolism

Argyris Zardilis[1,2,5], João Dias[3,4], Animesh Acharjee[5], and James Smith[1,4,5]

[1] Cambridge Computational Biology Institute, Department of Applied Mathematics and Theoretical Physics, University of Cambridge, UK
[2] SynthSys, C.H. Waddington Building, University of Edinburgh, UK
[3] Cambridge Systems Biology Centre, University of Cambridge, UK
[4] Department of Biochemistry, University of Cambridge, UK
[5] MRC Human Nutrition Research, Elsie Widdowson Laboratory, Cambridge, UK
`A.Zardilis@sms.ed.ac.uk`, `James.Smith@mrc-hnr.cam.ac.uk`

Abstract. Stochastic reaction-centric views are suitable for exploring hybrid minimal mechanism-statistical models of fatty acid and lipid metabolism, the basis of de novo lipogenesis. In this work, we demonstrate a reduced model for the core fatty acid synthesis and elongation process with a regulatory mechanism. This allows us to explore fatty acid profiles from lipid metabolomics data. This is part of a current study to assess the programming languages for capturing inherent probabilistic behaviour of the hierarchical chemical transformations of complex lipid species.

1 Introduction

Understanding lipid metabolism and its regulation is essential in metabolic disorders and diabetes-related diseases. Many of the metabolites and transformation reactions involved are still poorly characterised. Because of the low numbers of carbons flowing through the lipid pathways, the effect of the inherent probabilistic nature of chemical events is amplified. These two factors make lipid pathways difficult to analyse with current modelling approaches used for metabolic processes, such as constraint-based analyses (including FBA) [1], which is focused on deterministic average-case behaviour (of population growth). Here, we use an alternative stochastic and reaction-centric view to capture these pathways. In particular, we focused on Fatty Acid (FA) synthesis and elongation which is central in lipid metabolism. FAs are the core building blocks, modified for more complex lipids within the cell and tissues. FA synthesis and elongation pathways in particular display some characteristics that make models benefit from this alternative view: local iterative processes, probabilistic decisions at different levels and between pathway control mechanisms that affect decision making. We hope the eventual models will be useful for examining metabolomics data from model organisms, clinical trials or large-scale epidemiological cohort studies.

P. Mendes et al. (Eds.): CMSB 2014, LNBI 8859, pp. 244–247, 2014.
© Springer International Publishing Switzerland 2014

2   Methods

An important aspect of this work was to assess possible languages for capturing this stochastic behaviour in a reaction-centric projection. Recently, there has been a general trend towards constructing executable models, from the distributed systems world [2]. Here, we use Petri Nets [3] mainly and pi-calculus (SPiM [4] variant in particular) which is an example of a Process Algebra [5]. Petri Nets provide a vivid and intuitive graphical notation with a natural correspondence to chemical reactions and they have been used before for other metabolic pathways [6]. The main unit of definition for Petri Nets is the transition. The main unit of definition for Process Algebra is the species but the operational semantics are in terms of interactions, which again make them suitable for our reaction-centric view. Since these languages have been designed to handle distributed systems and therefore concurrency, non-determinism representing decisions is inherent in the structure of Petri Nets and in the syntax of pi-calculus. FA biosynthesis and elongation was taken from a reference annotation (KEGG hsa00062) with source and sink metabolites. Sinks were defined as the $C_n{:}0$, even-chain FAs. Elongation was combined with multiple steps in synthesis ignoring transport processes, under the assumption of non-reversable (net-forward) effects and constant reaction rates.

3   Results

We modelled the iterative FA elongation process as the combined effect of synthesis and elongation pathways, reduced down to a series of binary decisions or Bernoulli trials. An FA under elongation at each point makes a binary decision of whether to stay at its current length or continue to form longer FAs. This decision can be captured in the Petri Net language as a race condition between two enabled transitions. According to the operational semantics of the system the probabilities of the decision outcomes are controlled by the rates of the reactions corresponding to the transitions. The entire process can be seen as a series of binary decisions. An FA starts its "journey" in the net and moves along making the decisions along the way before getting trapped in one of sinks of the net that represents an FA reaching its final length (Figure 1).

We also modelled the Acetyl-CoA flow decision between the Krebs cycle and FA synthesis, controlled by the immediate energy requirements of the cell. Again this decision is captured very naturally as a race condition between two enabled transitions, the transition taking an Acetyl-CoA molecule to the first step of FA synthesis and the transition taking it towards the TCA cycle. The other pathways involved in this process were not modelled explicitly. Instead all the involved reactions were grouped into a single Petri Net transition respecting the stoichiometric constraints of the pathway. In this case, the transition rates are functions of ATP to display this control mechanism. The strength of the change in the likelihoods of the two outcomes can be captured by an exponent parameter on the two corresponding rates (Figure 2).

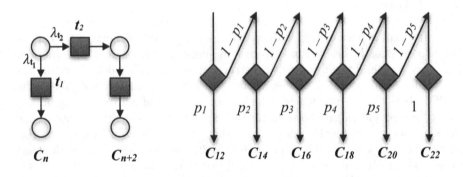

Fig. 1. On the left, the decision taken at a specific point by a metabolite during the elongation process. On the right, the entire process can then be seen as a series of binary decisions that the FA takes whether to stay at the current length or continue for further elongation.

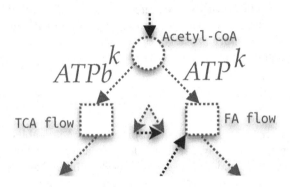

Fig. 2. Depending on the energy requirements of the cell, for which ATP levels act as a proxy, Acetyl-CoA can either go towards the TCA cycle to produce energy or towards FA synthesis. This decision is captured by a race condition. The likelihoods of the two outcomes are functions of ATP naturally. As ATP from the TCA feeds to the FA flow, this is described as a feed-forward motif (FFM).

The models were tuned with real experimental data, donated from a GC-FID mouse adipose metabolomics study. Relative metabolite abundances was used to parameterise the output states and therefore sample the relevant posteriors. Since the entire FA synthesis and elongation process can be seen as a series of Bernoulli trials, the only parameters of the model are the success probabilities for the decisions. A Maximum Likelihood estimation of these can be done from the data by recognising that the number of successes is the number of the corresponding species (for example C_4 for the first decision) and the number of trials the number of the corresponding species and the sum of the numbers of all longer species.

Non-parametric Dirichlet process mixture model clustering was used to partition the mouse adipose metabolite data, giving four clusters representing four distinct metabolic states and drug-dose treatment conditions. Ratios of even-chain FAs in each cluster were the posteriors per metabolic state for the models. Clustering of metabolic data adds an interesting dimension to the study as we can observe changes in the model parameters across different conditions and identify mechanistic detail or topological structures that show dynamical changes under perturbation.

4 Discussion and Conclusion

Reaction-centric network projections have advantages in model building. Pathways can be reduced to single composite transitions leaving interface metabolites (e.g. in the TCA cycle). To remain biochemically valid, stochastic statistical models are also hybrid, retaining mechanistic detail that can be parameterised to discriminate between metabolic states. Posteriors for model extensions are simply the ratios of additional metabolites taken from each metabolic state. In the next extensions to the current models, the additions could be parameterised from odd-chain, unsaturated or even combinations of these in the complex lipid species such as DAGs and phospholipids, sphingolipids and TAGs. Metabolic processes such as degradation of FAs and the anabolism and catabolism of more complex lipids can be added as extension modules to our core models. It is important that phenomenological-statistical models for metabolic processes retain some essential mechanistic detail that allow them to discriminate between metabolic states in health and disease. The impact to the lipid metabolomics community is that these models allow a re-examination of lipid metabolite profiles in large scale epidemiological data. Lipid metabolites profiled using extraction and separation of complex lipid species can be compared to model predictions of endogenous metabolic intermediates such as FAs in different metabolic states.

References

1. Orth, J.D., Thiele, I., Palsson, B.: What is flux balance analysis? Nature Biotechnology 28(3), 245–248 (2010)
2. Fisher, J., Henzinger, T.A.: Executable cell biology. Nature Biotechnology 25(11), 1239–1249 (2007)
3. Murata, T.: Petri nets: Properties, analysis and applications. Proceedings of the IEEE 77(4), 541–580 (1989)
4. Phillips, A., Cardelli, L., Castagna, G.: A graphical representation for biological processes in the stochastic pi-calculus. In: Priami, C., Ingólfsdóttir, A., Mishra, B., Riis Nielson, H. (eds.) Transactions on Computational Systems Biology VII. LNCS (LNBI), vol. 4230, pp. 123–152. Springer, Heidelberg (2006)
5. Fokkink, W.: Introduction to Process Algebra. Computer Science – Monograph (English), 2nd edn. Springer, Berlin (2007)
6. Baldan, P., Cocco, N., Marin, A., Simeoni, M.: Petri Nets for Modelling Metabolic Pathways: a Survey. J. Natural Computing (9), 955–989 (2010)

FM-Sim: Protocol Definition, Simulation and Rate Inference for Neuroscience Assays

Donal Stewart[1], Stephen Gilmore[2], and Michael A. Cousin[3]

[1] Doctoral Training Centre in Neuroinformatics and Computational Neuroscience,
School of Informatics, University of Edinburgh, Edinburgh, UK
[2] Laboratory for Foundations of Computer Science,
School of Informatics, University of Edinburgh, Edinburgh, UK
[3] Centre for Integrative Physiology,
School of Biomedical Sciences, University of Edinburgh, Edinburgh, UK

Abstract. Synaptic vesicle recycling at the presynaptic terminal of neurons is essential for the maintenance of neurotransmission at central synapses. Among the tools used to visualise the mechanics of this process is time-series fluorescence microscopy. Fluorescent dyes such as FM1-43, or engineered fluorescent versions of synaptic vesicle proteins such as pHluorins, have been employed to reveal different steps of this key process [3,7]. Predictive *in silico* modelling of potential experimental outcomes would be highly informative for these time consuming and expensive studies.

We present FM-Sim [9], user-friendly software for defining and simulating fluorescence microscopy experimental assays, with the following features: intuitive user definition of experimental protocols; automatic conversion of protocol definitions into time series rate value changes; domain-specific simulation model of a synaptic terminal; experimental data used for model parameter value inference; automatic Bayesian inference of parameter values [1,5] and reduction of inferred parameter set size for Bayesian inference.

1 The Synaptic Vesicle Cycle

Within chemical synapses of central nervous system (CNS) neurons, neurotransmitter is released from the presynaptic terminal to propagate the neural signal to the postsynaptic terminal of the following neuron. This neurotransmitter is stored in vesicles within the presynaptic terminal. These vesicles are exocytosed in response to an incoming action potential (Figure 1). To prevent vesicle depletion, compensatory endocytosis of plasma membrane allows regeneration of these vesicles. Two forms are studied within CNS nerve terminals:

- **Clathrin Mediated Endocytosis (CME)** [6]. Individual vesicles are reconstructed directly from the plasma membrane. Following reacidification of the vesicle contents and refilling with neurotransmitter, these vesicles rejoin the vesicle pools.

P. Mendes et al. (Eds.): CMSB 2014, LNBI 8859, pp. 248–251, 2014.

- **Activity Dependent Bulk Endocytosis (ADBE)** [4]. This is a second
  endocytosis mechanism triggered by periods of high stimulation. Here, large
  areas of plasma membrane are endocytosed as endosomes, which are later
  broken down into individual vesicles for reuse.

FM-Sim uses a hybrid stochastic model with delays of the vesicle cycle for
simulation and inference. The model supports the behaviour of different fluorescent probes. The kinetic rates and associated time delays of state transitions are
the parameters of the model.

2   Fluorescence Microscopy Imaging

Time-series fluorescent microscopy is one of the tools used to study the mechanisms of the synaptic vesicle cycle. Fluorescent probes added to nerve terminals
allow us to obtain time-series images of nerve terminal behaviour under stimulation. The two commonly used forms of fluorescent probes are FM dyes (such as
FM1-43 and FM2-10), and engineered pH-sensitive fluorescent synaptic vesicle
proteins (pHluorins).

The change in fluorescence of a nerve terminal as a whole over time and under
changing stimuli gives insight into internal behaviour. By using either FM dyes
or pHluorins in combination with chemical inhibitors, or on various knockdown
animal models, different aspects of the synaptic vesicle cycle can be isolated and
studied. It is this variety of potential experiments which makes FM-Sim useful at
the design phase. New experiments can be simulated based upon rate parameters
obtained from prior similar experiments.

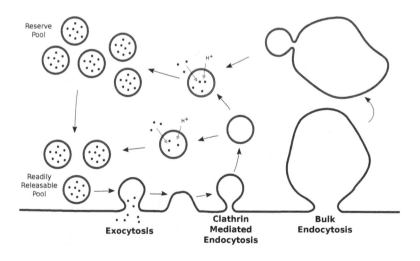

Fig. 1. The synaptic vesicle cycle, showing vesicle exocytosis and endocytosis

3   FM-Sim: Protocol Definition and Simulation

FM-Sim allows the definition of experimental protocols (Figure 2). These are timed sequences of events, including reagent addition, and electrical or chemical stimulation of neurons. Each protocol event can have rate parameter values set manually, inferred from observations, or inherited from protocol events already active.

Once defined, the set of protocol events are converted into a sequence of rate change events for simulation (Figure 3). At each rate change event, the set of rate values in effect are calculated, accounting for value inheritance. A single value is used for each inferred protocol event parameter when generating Bayesian inference proposals, ensuring consistency if that parameter value is used in multiple rate change events. This simplification of protocol definition entry and automatic rate event generation with inherited rate values is a feature not found in many of the general purpose simulators currently available, such as VCell [8].

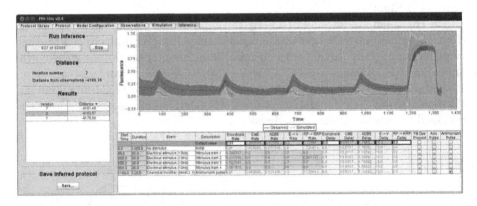

Fig. 2. Example parameter inference of a defined protocol, the parameter values in red are inferred from observed experimental data. The graph shows a sample simulation using these parameter values compared against the supplied experimental data.

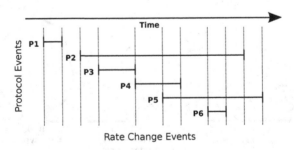

Fig. 3. Example rate change event generation. Protocol events P1,...,P6 are defined with start times and durations. These protocol events are then used to generate a sequence of events where rate values may change.

Protocols are simulated stochastically using the Delayed Stochastic Simulation Algorithm (DSSA) [2] with hybrid extensions, and the results of multiple simulation runs aggregated to provide mean and variance of the simulated model results. These results show both the expected fluorescence level, and the numbers of vesicles and endosomes at each stage of the synaptic vesicle cycle.

Rate parameters for a experimental protocol that have not been fixed by the user can be inferred from attempts to match a set of observed experimental data. A Bayesian approach to parameter inference is used, based on a Particle Marginal Metropolis-Hastings scheme using Sequential Monte Carlo estimates of marginal likelihoods [5,1].

Acknowledgements. Thanks to the members of the Cousin group for helpful discussions, and provision of experimental data. This work was supported by the DTC in Neuroinformatics and Computational Neuroscience, the EPSRC, the BBSRC, and the MRC.

FM-Sim is available at http://homepages.inf.ed.ac.uk/s9269200/software/.

References

1. Andrieu, C., Doucet, A., Holenstein, R.: Particle Markov chain Monte Carlo methods. Journal of the Royal Statistical Society: Series B (Statistical Methodology) 72(3), 269–342 (2010)
2. Barrio, M., Burrage, K., Leier, A., Tian, T.: Oscillatory regulation of Hes1: discrete stochastic delay modelling and simulation. PLoS Computational Biology 2(9), e117 (2006)
3. Cousin, M.: Use of FM1-43 and Other Derivatives to Investigate Neuronal Function. In: Current Protocols in Neuroscience, pp. 2–6 (2008)
4. Cousin, M.: Activity-dependent bulk synaptic vesicle endocytosis - a fast, high capacity membrane retrieval mechanism. Molecular Neurobiology 39(3), 185–189 (2009)
5. Golightly, A., Wilkinson, D.: Bayesian parameter inference for stochastic biochemical network models using particle Markov chain Monte Carlo. Interface Focus 1(6), 807–820 (2011)
6. McMahon, H., Boucrot, E.: Molecular mechanism and physiological functions of clathrin-mediated endocytosis. Nature Reviews Molecular Cell Biology 12(8), 517–533 (2011)
7. Miesenböck, G., De Angelis, D.A., Rothman, J.E.: Visualizing secretion and synaptic transmission with pH-sensitive green fluorescent proteins. Nature 394(6689), 192–195 (1998)
8. Moraru, I.I., Schaff, J.C., Slepchenko, B.M., Blinov, M., Morgan, F., Lakshminarayana, A., Gao, F., Li, Y., Loew, L.M.: Virtual cell modelling and simulation software environment. IET Systems Biology 2(5), 352–362 (2008)
9. Stewart, D., Gilmore, S., Cousin, M.A.: FM-Sim : A hybrid protocol simulator of fluorescence microscopy neuroscience assays with integrated Bayesian inference. To be published in Proceedings Third International Workshop on Hybrid Systems Biology, Vienna, Austria (July 2014)

Predictive Modelling of Mitochondrial Spatial Structure and Health

Arne T. Bittig[1], Florian Reinhardt[2], Simone Baltrusch[2],
and Adelinde M. Uhrmacher[1]

[1] Institute of Computer Science, University of Rostock
Albert-Einstein-Straße 22, 18059 Rostock, Germany
[2] Department of Medical Biochemistry and Molecular Biology, University of Rostock
Schillingallee 70, 18057 Rostock, Germany

Abstract. Mitochondria are mobile cellular organelles that form networks by fusion and fission. These events lead to an exchange of components responsible for maintaining membrane potential, i.e. mitochondrial health. Membrane potential can be disturbed by an imbalance of fission-triggering proteins. We expand an existing computational model of fusing and splitting mitochondria by representations of fission protein 1 (Fis1) and dynamin related protein 1 (Drp1) and perform parameter scans on simulations of it. Our relatively basic model already shows an effect of lower Fis1 and Drp1 recruitment rates, i.e. lower availability, on network structure and overall health. Various aspects of the real system can be incorporated into model, e.g. further regulatory proteins, a varying spatial distribution of Fis1 and Drp1, or consequences of changed mitochondrial network structure and health on their behaviour, e.g. under oxidative stress.

Keywords: spatial simulation, rule-based modeling, mitochondrial network, mitochondrial health, mitochondrial fission.

Background

Mitochondria are mobile organelles that exist in living cells as a tubular network. They continuously join the mitochondrial network by fusion and divide by fission events. Mitochondrial fission is mainly regulated by two nuclear-encoded proteins, fission protein 1 (Fis1) and dynamin related protein 1 (Drp1). Mitochondrial dynamics have been shown to be an essential quality control mechanism in order to maintain mitochondrial health. A proxy for mitochondrial health and integrity is the mitochondrial membrane potential [6]. Recent wet-lab studies have shown that the mitochondrial membrane potential is disturbed by an imbalance of the mitochondrial fission proteins. It is therefore the objective of this study to develop an in silico prediction model for the influence of Fis1 and Drp1 on mitochondrial spatial structure and health.

P. Mendes et al. (Eds.): CMSB 2014, LNBI 8859, pp. 252–255, 2014.
© Springer International Publishing Switzerland 2014

Our Approach

We here take an existing model of mitochondrial health maintenance [4], where mitochondria move in a random direction for random intervals of time (i.e. along not explicitly included microtubules) in an otherwise (for purposes of the model) empty 2D cell. Abstract health units mimic the functional state of mitochondria by representing the membrane potential. Mitochondrial fusion allows mitochondria to exchange components (here: health units) in order to maintain health.

Our model is implemented in ML-SPACE [1,2], which combines an attributed rule-based language for describing cell biological processes supporting binding and dynamic nestring of entities with a simulator for these in continuous or discretised (i.e. grid-based) space or a hybrid thereof. For the continuous part used here, spatial entities' positions are updated sequentially in fixed time steps, with collisions potentially triggering second-order reactions, while zeroth- and first-order reactions are executed as usual in stochastic simulation.

A simple way of describing a mitochondrial fusion with probability 1 on collision of moving mitochondria, including exchange of two health units (omitting, for simplicity, checks against exceeding the minimum, 0, and maximum, 10, for this attribute's value), and fission in our language of ML-SPACE would be

```
Mito()<bs:free> + Mito()<bs:free> -> Mito(velocity:0,health-=2)
    <bs:bind>.Mito(velocity:0,health+=2)<bs:bind> @ 1
Mito()<bs:Mito()> -> Mito()<bs:release> @ rFission
```

where the part in angle brackets indicates that the mitochondria bind to each other and the bond being released upon fission. Fused mitochondria become immobile here. Further rules include damage (a first-order reaction lowering health of a mitochondrion), autophagy (consumption of unhealthy mitochondrions) and replication (creation of a new, healthy mitochondrion, keeping the total number roughly constant.) We can reproduce basic findings of the original [4] despite some unresolved issues, e.g. whether fused mitochondria also get damaged.

We modified this model by representing the number of bound Fis1 and Drp1 molecules of mitochondria as attributes as follows.

```
Mito(f:=nFis<maxFis)<bs:Mito(nFis<8-f)> -> Mito(nFis+=1) @ rFisRecruit
Mito(f:=nFis,d:=nDrp)<bs:Mito(nFis>=4-f,nDrp<2-d)> -> Mito(nDrp+=1) @ ...
```

Only fused mitochondria facilitate Fis1 and Drp1 recruitment here and the number of Fis1 (Drp1) per mitochondria pair cannot exceed the fission threshold 8 (2). Additionally, we only allow Drp1 recruitment when a certain number of Fis1 is already bound (above: 4). The fission rule is more complex then:

```
Mito(f:=nFis,d:=nDrp)<bs:Mito(nFis>=8-f,nDrp>=2-d)>
    -> Mito(nFis:0,nDrp:0)<bs:release> @ Infinity
```

The nFis and nDrp attributes of the first mitochondrion are set to 0 in the process, indicating release of the previously bound Fis1 and Drp1. (Additional rules to "release" remaining Fis1 and Drp1 from the former partner omitted.)

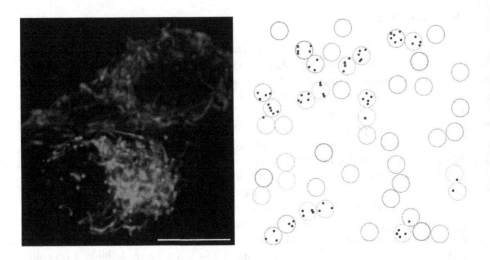

Fig. 1. Left: Microscopic image of the mitochondrial network in a glucose-responsive MIN6 beta cell. Scale bar 20 µm. Right: Simulation screenshot (cyan/green: healthy mitochondria, red: damaged; tiny circles/squares: recruited Fis1/Drp1 molecules of fused mitochondria).

With the above rules, Fis1 and Drp1 are essentially treated as ubiquitous and the recruitment rate constant choices are the only limitations to fusion. We also simulated a slightly more realistic model where Fis1 and Drp1 numbers are limited (but constant) and recruitment thus happens slower if there are already many fused mitochondria that have some Fis1 and/or Drp1 bound, but not enough for a fission event.

Results and Outlook

Related wet-lab experiments have shown that cells with reduced Fis1 or Drp1 expressions exhibited a significantly lower membrane potential and a heterogenic mitochondrial network [5].

In initial simulations of the simple model (Fig. 2 left), Fis1 and Drp1 recruitment were (predictably) negatively correlated with the ratio of fused against free mitochondria (the closest analogy to network structure in the simulation results) and positively correlated with mitochondrial health. This was when fused mitochondria did not loose health on their own like free mitochondria (round markers in Fig. reffig:results), so fewer fission events meant more mitochondria being safe from damage, which is not realistic. The positive correlation disappeared when damage to fused mitochondria was allowed (triangular markers), and became clearly negative when damaged parts of fused mitochondria could also undergo autophagy (which allowed a new, healthier one to be generated; squares).

In the simple model, average health varied only slightly overall. In the model with explicit Fis1 and Drp1 amounts, changes in the Fis1 amount had roughly

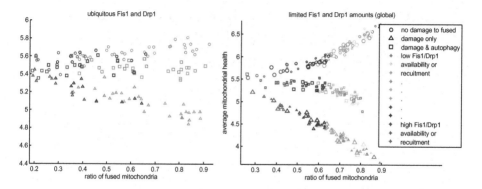

Fig. 2. Scatter plots of key results for the ubiquitous and limited Fis1/Drp1 scenario

the same effect as changes to the recruitment rate. By varying the amount and recruitment parameters, a much wider range of average mitochondrial health values was covered, and a more pronounced correlation of fused mitochondria ratio (and thus fission frequency) and average health could be observed (Fig. 2 right).

Recent studies indicate that adaptor proteins, namely Mff, MID49 and MID51 are important for Drp1 regulated mitochondrial fission. Thus, future research in this direction will include not only expanding the model by explicit fission protein entities whose spatial distribution may not be homogeneous (to be simulated also with our hybrid approach of continuous and discrete space), but also incorporating new wet-lab findings regarding mitochondrial fission. Once this model has been validated, ROS-related mechanisms for oxidative stress response [3] and adaptive processes to mitochondrial damage shall be simulated.

References

1. Bittig, A.T., Haack, F., Maus, C., Uhrmacher, A.M.: Adapting rule-based model descriptions for simulating in continuous and hybrid space. In: Proc 9th CMSB, CMSB 2011, pp. 161–170. ACM Press, New York (2011)
2. Bittig, A.T., Matschegewski, C., Nebe, J.B., Stählke, S., Uhrmacher, A.M.: Membrane related dynamics and the formation of actin in cells growing on microtopographies: a spatial computational model. BMC Systems Biology 8, 106 (2014)
3. Park, J., Lee, J., Choi, C.: Mitochondrial network determines intracellular ROS dynamics and sensitivity to oxidative stress through switching Inter-Mitochondrial messengers. PLoS One 6(8), e23211 (2011)
4. Patel, P.K., Shirihai, O., Huang, K.C.: Optimal dynamics for quality control in spatially distributed mitochondrial networks. PLoS Comput. Biol. 9(7), e1003108 (2013)
5. Reinhardt, F., Schultz, J., Waterstradt, R., Baltrusch, S.: The mitochondrial fission proteins Fis1 and Drp1 are important for glucose-induced insulin secretion in glucose-responsive INS1 832/13 cells. Exp. Clin. Endocrinol Diabetes 121(10), P17 (2013)
6. Schmitt, H., Lenzen, S., Baltrusch, S.: Glucokinase mediates coupling of glycolysis to mitochondrial metabolism but not to beta cell damage at high glucose exposure levels. Diabetologia 54(7), 1744–1755 (2011)

XTMS in Action: Retrosynthetic Design in the Extended Metabolic Space of Heterologous Pathways for High-Value Compounds

Baudoin Delépine, Pablo Carbonell, and Jean-Loup Faulon

Institute of Systems & Synthetic Biology
5 rue Henri Desbruères, Genopole Campus 1, Bât. 6,
91030 Évry Cedex, France
http://www.issb.genopole.fr/

Abstract. Despite the increase in recent years in the portfolio of added-value chemicals that can be microbially produced, the design process still remains a complex system, costly and rather slow. To overcome such limitations, the development of Computer-Aided-Design (CAD) tools is necessary to design production pathways that systematically screen metabolic databases to select best genes to import into chassis organisms. Here, we showcase the XTMS CAD tool for pathway design, which exploits the ability for pathway ranking in our RetroPath retrosynthetic algorithm within an extended metabolic space that considers putative routes through enzyme promiscuity. The validity of the ranking function for the production of malonyl-CoA, an important precursor for added-value compounds, is shown.

Keywords: synthetic biology, metabolic engineering, promiscuity, computer-aided-design.

1 Introduction

With recent advances in synthetic biology and metabolic engineering, synthetic production of high-value compounds in industrial hosts such as *Escherichia coli* or *Saccharomyces cerevisiae* is becoming more and more promising. Computer-Aided-Design pathway tools have been proposed to ease the metabolic engineering process [1–3]. Nevertheless, finding the best pathways achieving high-yield production is still challenging. In particular, the efficiency of such CAD tools has been often hindered so far due to the lack of high quality and exhaustiveness of reactome annotations. Missing annotations cause ineluctably missing potentially interesting pathways. Moreover, it is well-known that enzymatic reactions often display the ability of accepting several similar substrates (even un-natural ones), although this promiscuous capacity has been underexploited so far.

To overcome such limitations, we hereby present and demonstrate in operation the eXTended Metabolic Space server (XTMS, figure 1) [4], a novel CAD pathway tool that integrates our expertise on retrosynthesis of high-value compounds

P. Mendes et al. (Eds.): CMSB 2014, LNBI 8859, pp. 256–259, 2014.
© Springer International Publishing Switzerland 2014

(Retropath algorithm [5]) with an *in silico* metabolic space representation extended from endogenous compounds of the host organism. XTMS is designed to be user-friendly and pragmatic as it provides the user with critical information in order to assess generated pathways' quality, notably in the form of a ranking function. The validity of this ranking function has been recently highlighted by our group by the construction of several pathways producing malonyl-CoA [6].

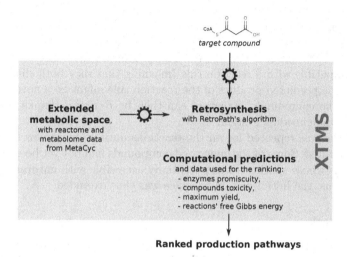

Fig. 1. XTMS' overview. An input compound is submitted to the server as a target, then XTMS retrieve candidates pathways from a pre-generated extended metabolic space, thanks to the molecular signature encoding representation (black cog). Computational predictions are used to rank the pathways witch are finally displayed to the user with relevant information about pathways confidence and putative efficiency.

2   Material and Methods

2.1   Molecular and Reaction Signature

The method that XTMS employs is based on the molecular signature. The molecular signature is a graph-based representation of a chemical compound by the topological neighbourhood of each atom (atomic signature). It is closely related to Morgan and ECFP fingerprints but has the originality to enable reverse engineering, meaning one can retrieve a compound's structure from its molecular signature [8].

A reaction signature is defined as the net difference between atomic signatures from products and substrates, which represents the changes occurring within the reactants in the reaction.

2.2 Generation of an Extended Metabolic Space

We used reactome and metabolome data from MetaCyc and EcoCyc to build our initial set of compounds and reactions covering the most part of current metabolic knowledge.

Those elements were encoded by their corresponding molecular and reaction signature.

This representation allowed us to infer promiscuous enzymatic activity, therefore expanding the number of reactions and compounds. For a given encoded reaction rule, we iterated though all encoded compounds in order to test the reaction rules with different substrates. Our assumption was that when a compound is compatible with a reaction rule (meaning that they both share atomic signatures), the predicted product of the reaction rule might be a novel encoded compound. The compounds' structure can then be retrieved thanks to the reversibility of the encoding system.

This process was repeated for all the reactions until no more new compound could be generated. Some of the generated compounds had not yet been reported as compounds whose synthesis was potentially accessible with natural enzymes. By these means, the initial metabolic space was thus extended.

2.3 Retropath

Retropath is a CAD software for embedded metabolic circuits. Retropath's retrosynthesis algorithm is at the core of XTMS, using the extended metabolic space generated through the molecular signature as a base to enumerating pathways.

RetroPath uses a two steps algorithm. First, a *forward* step generates all the reachable compounds from a list of chassis's endogenous compounds and reactions; this is the metabolic space expansion we described earlier. The ordered list of reactions producing each reachable compound is saved. Second, a *backward* step generates the production pathways for each reachable heterologous compound. The list of reactions generated at the forward step is scanned to retrieve all needed reactions in order to start the pathway only from endogenous compounds.

In that way, several pathways are usually retrieved to produce a target (heterologous) compound.

2.4 Pathway Ranking

Pathway ranking is carried out using a ranking function which compiles information inferred from several modules and data sources.

In XTMS, we estimate enzyme efficiency by predicting their promiscuity thanks to the tensor product technique; pathway's toxicity is computed by the EcoliTox server; maximum allowed yield is estimated through a flux balance analysis; and finally the free Gibbs energy are mined from MetaCyc. Those descriptors give insights about pathway's performance.

The number of enzymatic steps (and putative steps) and unfavourable reactions are also taken into account in order to select the best pathways.

3   Results and Discussion

XTMS is a user-friendly pathway CAD tool now available to the community (http://xtms.issb.genopole.fr/). Its strengths are to work on an extended metabolic space and to provide critical information to the user, as for example the predicted toxicity of generated pathways.

In order to evaluate XTMS's ranking score, we predicted and implemented several malonyl-CoA pathways in *E. coli* [6]. The lack of availability of malonyl-CoA precursor is often a bottleneck in the production of added-value compounds, since producing pathways must compete with those for fatty acids synthesis. Therefore, an embeded synthetic malonyl-CoA pathway with high efficiency will serve to boost availability of this precursor. A circuit consisting on a malonyl-CoA biosensor was then used to access pathways' yield and test their efficiency [7]. As we expected, we succeeded to predict the best pathways. Moreover, the order of the more productive pathways parallelled the one in our pathway ranking [6].

Even if the evaluation of the ranking function needs further validation, those results are encouraging. We hope that XTMS will help metabolic engineers to design efficient circuits able to produce high-value compounds and that their comments will help us to improve XTMS.

Acknowledgments. PC is supported by UPFellows program with the support of the Marie Curie COFUND program.

References

1. Cho, A., Yun, H., Park, J.H., Lee, S.Y., Park, S.: Prediction of novel synthetic pathways for the production of desired chemicals. BMC Systems Biology (2010)
2. McClymont, K., Orkun, S.: Metabolic tinker: an online tool for guiding the design of synthetic metabolic pathways. Nucl. Acids Res. (2013)
3. Campodonico, M.A., Andrews, B.A., Asenjo, J.A., Palsson, B.O., Feist, A.M.: Generation of an atlas for commodity chemical production in Escherichia coli and a novel pathway prediction algorithm, GEM-Path. Metabolic Engineering (2014)
4. Carbonell, P., Parutto, P., Herisson, J., Pandit, S.B., Faulon, J.L.: XTMS: pathway design in an eXTended metabolic space. Nucleic Acids Res. (2014)
5. Carbonell, P., Planson, A.G., Fichera, D., Faulon, J.L.: A retrosynthetic biology approach to metabolic pathway design for therapeutic production. BMC Syst. Biol. (2011)
6. Feher, T., Planson, A.G., Carbonell, P., Fernandez-Castane, A., Grigoras, I., Dariy, E., Perret, A., Faulon, J.L.: Validation of RetroPath, a computer aided design tool for metabolic pathway engineering. Biotech. J. (2014)
7. Liu, D., Xiao, Y., Evans, B., Zhang, F.: Negative feedback regulation of fatty acid production based on a Malonyl-CoA Sensor Actuator. ACS Synth. Biol. (2013)
8. Faulon, J.L., Visco, D., Pophale, R.: The Signature Molecular Descriptor. 1. Using Extended Valence Sequences in QSAR and QSPR Studies. Journal of Chemical Information and Computer Sciences (2003)

THiMED: Time in Hierarchical Model Extraction and Design[*]

Natasa Miskov-Zivanov[1,2], Peter Wei[1], and Chang Sheng Clement Loh[1]

[1] Electrical and Computer Engineering Department, Carnegie Mellon University
[2] Computer Science Department, Carnegie Mellon University,
Pittsburgh, USA
nmiskov@andrew.cmu.edu

Abstract. We describe our approach to modeling timing of cell signaling systems in which existing information about the system spans from detailed mechanistic knowledge to much coarser observations about cause and effect. The results for several models emphasize the fact that the selection of timing implementation can have both qualitative and quantitative effects on the model's transient behavior and its steady state.

Keywords: timing, cell signaling, stochastic model, delay.

1 Introduction

Time of occurrence and duration of events often play an important role in decision making in cell signaling networks [1]. Although timing of events can be modeled using reaction rates, exact element regulations are not always well understood, and even more, rates of reactions are not known. Still, to better understand how the overall system works, it is important to capture in the model much of the available knowledge about the system. When experimental observations provide insights into indirect cause-effect relationships only, and do not explain many of the detailed interaction mechanisms [2], our modeling approach accounts for (*i*) thresholds in element activity, thus discretizing model variables [3], (*ii*) relative delays between events and in element responses to regulation changes, thus capturing critical event timing.

2 Approach

We model system elements using multi-valued variables, and by using this approach we are able to capture multiple layers of cell signaling: interactions between receptors and external stimuli, intracellular signaling, gene regulation, cell's response to stimuli, and feedback to cell receptors [1][2]. Such an approach has been shown valuable in providing critical insights into system's transient behavior, when models are coarse-grained in parts or in whole due to available knowledge. To increase accuracy of the model, in our approach we allow for implementation of timing details that capture

---

[*] This work is supported in part by DARPA award W911NF-14-1-0422.

P. Mendes et al. (Eds.): CMSB 2014, LNBI 8859, pp. 260–263, 2014.

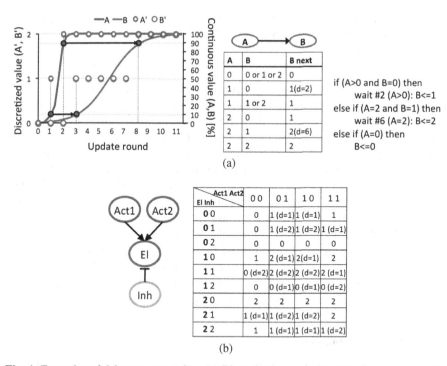

(a)

(b)

Fig. 1. Examples of delay representation. (a) Discretization and corresponding delay description. (b) Different delays defined for different regulator values.

relative delays between events. Once the delays are described formally (e.g., using delay truth tables [3]), our tool translates them into variable update rules. We identified three different methods to model delays that occur between a change in given element regulation (i.e., change in combination of regulator values and current element value), and a corresponding change in the element's value. We describe our approaches to delay modeling using the following two examples.

Example 1. Assuming that there are two elements, A and B, and that A positively regulates B, the time needed for B to respond to different changes in A may be different, depending on current values of A and B. Figure 1(a) (left) shows one scenario in which A increases from very low level (around 0% its maximum value) to high level (100% its maximum value). While B can relatively quickly follow the initial change in A, it takes longer for B to come close to 100% of its highest value. The two lines representing A and B can be discretized, assuming thresholds for values 1 and 2 (e.g., reaching 10% of highest activity or concentration can be a threshold for value 1 and reaching 90% of highest activity or concentration can be a threshold for value 2). Figure 1(a)(left) outlines discretization example for A and B, and the table in Figure 1(a)(middle) shows how current values of A and B can determine next value for B. Delays in changing B value are indicated by "d=2" and "d=6" which represent 2 time-unit and 6 time-unit delays, respectively. Our tools translate these tables into executable rules. In addition, these relationships can be described in code and translated into an executable model from the code. The description in Figure 1(a)(right) is very suitable for implementing in Hardware Description Languages (HDLs) that can be translated in an automated way into executable circuit models. Similar work on emulation of biological networks in Field Programmable Gate Arrays has been described in [4,5].

Example 2. Given a small regulatory network with four components (Figure 1(a), left), element (El), its two positive regulators (Act1 and Act2), and negative regulator (Inh), we draw a table (Figure 1(a), right) listing all combinations of regulator values including previous value of element El, and show the resulting new value for El. Note that El and Inh have three different levels of activity, 0 (not active), 1 (low activity), and 2 (high activity), while both activators are modeled only with two levels, 0 (no activity) and 1 (active). Table entries of the form "d=1" or "d=2" indicate that the transition from one value to another occurs after 1 time-unit delay or after 2 time-unit delays, respectively. For example, when El has value 2, Inh has value 1 and both activators, Act1 and Act2, have value 0, El will change value from 2 to 1 with some, short delay. For the same values of El and Inh, when one of the activators has value 0 and the other one value 1, then El will change from 2 to 1 with a longer delay.

The delay assumptions can be implemented in executable model generation and in simulation in several different ways, as shown in the following.

2.1   Forward Propagation

In the first delay implementation, all regulator value combinations that satisfy the same transition requirement in terms of previous and next element value and delay interval (i.e., all delay truth table entries with same output value, for example, "1(d=1)") are lumped into a single function. Such implementation assumes that measuring delay (lapsed time) is not reset even when the actual conditions change, as long as the outcome is same (e.g., when El=2, Inh=1, Act1=0, Act2=1, and then Act1 changes value to 1 and Act2 changes value to 0, the effect on El remains the same, and thus counting of steps to satisfy 2 time-unit delay, "1(d=2)", remains the same). This approach allows for minimizing element update functions, since multiple table entries can be lumped into a single function. Besides minimizing the function, this also requires smaller number of variables to be propagated from one simulation round to the next (thus the name for the method).

2.2   Backward Propagation

In contrast to the first approach, if the conditions change before the required delay interval has lapsed, even when the new output is same for the new conditions, measuring of delay interval is reset. This delay modeling approach requires different "memory" implementation compared to the first approach. In other words, this approach requires that, depending on how many delay steps are defined, the simulator checks variable values in the corresponding number of previous rounds. In this case, functions that are to be computed are simpler compared to the previous approach (forward propagation), but the number of variables increases.

2.3   Buffer Insertion

The third approach implements delays as "buffers" that add steps to the pathway, thus delaying propagation of any value of a regulator (for any combination with other regulators) to some or all of its downstream elements. In other words, the table created for this case will not have delay entries (e.g., "1(d=2)") but instead only discrete numbers without indication of delays. This approach can be used when modeling pathway sections without crosstalk or in the case where only indirect causal relationships are known while the overall timing of the pathway still needs to match the timing of other pathways in the network. This delay modeling approach was applied previously in [1] and it resulted in a good match with experimental results for situations where there are multiple competing pathways without significant crosstalk.

2.4   Simulation

We have also worked on simulation approaches to accurately account for these different delay modeling methods. Depending on the simulator setup, delay values in cell signaling models can be assumed exactly as defined, or can represent upper bounds or mean delay values.

3   Results

We applied the described timing modeling approaches in development and analysis of two models, T cell differentiation model [1] and immune crosstalk in malaria infection in mosquitoes [2]. We have shown that, depending on the delay implementation method, different delay values can affect results both qualitatively and quantitatively, and can change both transient behavior and steady state of individual elements, as well as of the system as a whole.

References

1. Miskov-Zivanov, N., et al.: The duration of T cell stimulation is a critical determinant of cell fate and plasticity. In Science Signaling 6, ra97 (2013)
2. Vodovotz, Y., et al.: Modeling Host-Vector-Pathogen Immuno-inflammatory Interactions in Malaria. In: Complex Systems and Computational Biology Approaches to Acute Inflammation, pp. 265–279. Springer, New York (2013)
3. Miskov-Zivanov, N., et al.: Dynamic behavior of cell signaling networks: model design and analysis automation. In: Proc. of Design Automation Conference (DAC), Article 8, 6 p. (June 2013)
4. Miskov-Zivanov, N., et al.: Regulatory Network Analysis Acceleration with Reconfigurable Hardware. In: Proc. of International Conference of the IEEE Engineering in Medicine and Biology Society (EMBC), pp. 149–152 (September 2011)
5. Miskov-Zivanov, N., et al.: Emulation of Biological Networks in Reconfigurable Hardware. In: Proc. of ACM Conference on Bioinformatics, Computational Biology and Biomedicine (ACM-BCB), pp. 536–540 (August 2011)

Author Index